Industrial Ecology and Environmental Management

Volume 7

Industrial Ecology and Environmental Management have been recognized as two powerful disciplines for solving the global challenges such as the depletion of non-renewable resources, global warming and energy/water shortage.

This book series includes informative books with comprehensive texts focusing on both theory and practice of industrial ecology and environmental management, it aims to present the multi-disciplinary frontiers of methods, models and technologies in industrial ecology and environmental management for scientists, engineers and researcher from academia, industry, government and institutions. It aims to build a platform for the worldwide scientists, engineers and researcher to share the frontiers of the methods, models and technologies in industrial ecology and environmental management for achieving a more sustainable future.

The books series will include both the theoretical tools and the practical applications of industrial ecology and environmental management for sustainable energy, industry and society.

Zhenfeng Wang • Zhanwu Wang • Heng Wang
Editors

Agricultural Waste Power Generation Supply Chain

Research on the Sustainable Development

Springer

Editors
Zhenfeng Wang
College of Mechanical and Electrical Engineering
Henan Agricultural University
Zhengzhou, Henan Province, China

Zhanwu Wang
College of Mechanical and Electrical Engineering
Henan Agricultural University
Zhengzhou, Henan Province, China

Heng Wang
College of Mechanical and Electrical Engineering
Henan Agricultural University
ZhengZhou, Henan Province, China

ISSN 2730-6070 ISSN 2730-6089 (electronic)
Industrial Ecology and Environmental Management
ISBN 978-3-032-20086-0 ISBN 978-3-032-20087-7 (eBook)
https://doi.org/10.1007/978-3-032-20087-7

This Springer imprint is published by the registered company Springer Nature Switzerland AG
The registered company address is: Gewerbestrasse 11, 6330 Cham, Switzerland

Foreword

Agricultural Waste Power Generation Supply Chain: Research on the Sustainable Development makes for accessible, important reading as we face into the climate emergency. The authors, based on the issues of global climate warming, the intensification of energy supply-demand contradiction, and the deterioration of people's living environment, believe that the utilization of agricultural waste can alleviate these problems. The sustainable development of the agricultural waste power generation supply chain is the key to resource utilization. Therefore, the authors deem this topic a necessary and timely area of research.

With the intensification of global energy supply-demand imbalances, environmental pollution and energy security challenges have become increasingly pressing. Establishing a sustainable society has thus become an urgent necessity. As a renewable and clean energy source, agricultural waste is recognized as a strategic option for addressing energy security concerns and reducing carbon emissions. The recycling and utilization of agricultural waste can help mitigate practices such as open-field burning and unregulated disposal, thereby lowering the risk of fires and supporting the development of green ecosystems and beautiful rural communities. Being a plentiful biomass resource, agricultural waste can substitute for coal in power generation, be converted into solid fuel, and serve as a feedstock for producing combustible gases, biodiesel, and fuel ethanol. Its fuel-oriented utilization contributes to reducing greenhouse gas emissions, decreasing fossil energy consumption, and expanding renewable energy supply, thereby promoting the realization of a green ecological environment, a circular economy, and an ecological civilization.

The agricultural waste supply chain functions as a critical bridge between raw material suppliers and end-users, and its sustainable development is essential for the effective resource utilization of straw. The agricultural waste power generation supply chain demonstrates complex characteristics, including dynamic evolution, nonlinear interdependencies, self-organization, openness, and system autonomy. Within this framework, each subsystem is interdependent, interacts with others, and exerts mutual constraints. However, when external conditions shift or competition intensifies among subsystems, the supply chain faces heightened operational risks, potentially leading to systemic disruptions. Moreover, the bulky and spatially dispersed nature of agricultural waste further compromises supply chain efficiency and amplifies operational risks. These risk factors collectively contribute to elevated operational costs, diminished efficiency, and intensified competition among stakeholders, ultimately posing a major barrier to the large-scale utilization of straw-based agricultural waste and the realization of dual carbon goals.

Against this backdrop, in-depth research on sustainable development models for the agricultural waste power generation supply chain carries substantial theoretical and practical significance. This book centers on holistic optimization of the entire supply chain, adopting the "resource–technology–economy–environment" four-dimensional synergy framework as its core analytical structure. It systematically examines the technological evolution of agricultural waste-based power generation, investigates the operational mechanisms and critical bottlenecks across supply chain stages, establishes a theoretical model encompassing stakeholder interactions, cost-sharing mechanisms, and risk mitigation strategies, and integrates representative case studies to explore regionally adaptive development pathways.

The authors anticipate that the research outcomes presented in this book will serve as a valuable reference for governmental agencies in formulating relevant industrial policies, provide actionable guidance for decision-making processes within power generation enterprises and other stakeholders along the supply chain, and offer fresh perspectives to advance academic research on the resource utilization of agricultural waste. Given the complexity and evolving nature of the research domain, certain limitations or omissions may exist. The authors sincerely welcome readers' constructive feedback and suggestions to collectively enhance the development of the agricultural waste power generation industry toward a more efficient, environmentally friendly, and sustainable trajectory.

College of Mechanical and Electrical Engineering
Henan Agricultural University
Zhengzhou, Henan Province, China
October 2025

Zhenfeng Wang

Competing Interests The authors have no competing interests to declare that are relevant to the content of this manuscript.

Contents

Chapter 1
Overview of the Agricultural Waste Power Generation Supply Chain

Zhanwu Wang and Zhenfeng Wang

Contents

Z. Wang (✉) · Z. Wang
College of Mechanical and Electrical Engineering, Henan Agricultural University, Zhengzhou, Henan Province, China
e-mail: wzw6997@henau.edu.cn

Z. Wang et al. (eds.), *Agricultural Waste Power Generation Supply Chain*, Industrial Ecology and Environmental Management 7,
https://doi.org/10.1007/978-3-032-20087-7_1

Abstract The agricultural waste power generation supply chain is a comprehensive system that converts agricultural waste into electricity. This supply chain mainly includes the collection, processing, storage, transportation, and biomass power generation of agricultural waste. Agricultural waste is the main raw material for biomass power generation. The construction and operation of the agricultural waste power generation supply chain require the participation of various parties, including farmers, research institutions, enterprises, and the government. The efficient operation of the agricultural waste power generation supply chain system requires the joint efforts of all stakeholders. Due to the different objectives and activities of the participants, the supply chain has many characteristics. Here, the agricultural waste power generation supply chain system will be studied by analyzing the supply chain model, supply chain characteristics, and influencing factors.

Keywords Agricultural waste · Biomass power generation · Agricultural waste supply chain · Supply chain model

1.1 Analysis of the Supply Chain Structure

1.1.1 Key Participants in the Supply Chain

The supply chain was based on the "economic chain" in its early days. Later, the "value chain" theory proposed by scholars such as Michael Porter promoted the evolution of the concept, and it eventually evolved into the modern network chain system covering the production network structure (Tan et al. 2017a, b). The supply chain refers to the process from raw materials to the final product being provided to consumers, involving a network chain structure of all activities. If the nature of the members at the same stage of the supply chain is the same, viewing them separately and as a whole will have the same effect (Wen 2024). This chapter divides the main participants in the raw material supply chain of agricultural waste power generation into three entities: farmers, brokers, and agricultural waste power plants.

1.1.1.1 Farmers

Farmers are the starting point of the agricultural waste power generation raw material supply chain and are mainly responsible for the collection of agricultural waste in the fields. As the main producers and suppliers of agricultural waste raw materials, the behaviors and decisions of farmers play a key role in the operation status of the entire agricultural waste raw material supply chain. Farmers can choose to sell the agricultural waste to brokers or agricultural waste power plants. If farmers choose to sell it directly to power plants, they need to undertake the transportation task. Unlike the large-scale planting forms in developed countries in Europe and

America, farmland planting in China is generally carried out on a household basis, with small scales, low yields, and relatively scattered farmlands. The degree of mechanization in crop planting and harvesting is relatively low. Agricultural waste is usually harvested at the same time as crops. Farmers have been accustomed to picking up agricultural waste in their spare time to feed livestock for many years, but most of the agricultural waste is scattered in the fields. During the busy farming season, when there is a shortage of labor and the intervals between crop planting are short, farmers will burn the agricultural waste all at once to save time. The reasons for this phenomenon are, on the one hand, that although farmers have a lot of agricultural waste in their hands, they lack the awareness and motivation to sell it. On the other hand, a large number of rural people have flocked to cities, and there is not enough labor force. Therefore, when dealing with agricultural waste, farmers tend to adopt the traditional way of returning it to the fields for utilization (Haiping 2015).

1.1.1.2 Broker (Third-Party Logistics)

Agricultural waste brokers play a crucial role in the raw material supply chain of agricultural waste power generation and act as intermediaries connecting farmers and agricultural waste power plants. Their main responsibility is to purchase agricultural waste from farmers and then resell it to power plants to earn the price difference. In the raw material supply chain of agricultural waste power generation, the cooperation between brokers and any participant is based on the premise of making profits. The main factors influencing their decisions include the purchase price of agricultural waste by power plants and the subsidies given by the government. Their costs mainly include the expenses for the storage, processing, and transportation of agricultural waste, as well as the purchase of agricultural waste from farmers. They provide strong support for the development of the agricultural waste power generation industry by integrating scattered resources, improving transaction efficiency, transmitting market information, etc. Therefore, if agricultural waste power plants want to establish a long-term and stable cooperative relationship with brokers, they need to effectively organize and coordinate them to enhance the willingness of brokers to participate in agricultural waste purchasing activities (Sun et al. 2021).

1.1.1.3 Agricultural Waste Power Plant

Agricultural waste power plants play a core terminal role in the agricultural waste power generation raw material supply chain. They are the ultimate demander of the agricultural waste power generation raw material supply chain and also a key link in realizing the value of the entire supply chain. An agricultural waste power plant is an enterprise that converts biomass resources such as agricultural waste into electricity. By purchasing biomass fuels like agricultural waste and using advanced power generation technologies and equipment, it converts biomass energy into electricity and supplies the electricity to users. The establishment of power plants requires a large amount of capital investment, and the cost of raw material acquisition is high. Therefore, they are relatively dependent on government subsidies to ensure operation. Although power plants are the core enterprises in the agricultural waste power generation raw material supply chain, due to their reliance on raw material suppliers, they do not have an advantage in the entire supply chain.

1.1.1.4 Government-Related Units

In the complex system of agricultural waste power generation supply chain, the government plays an extremely crucial and diverse role. The government is first and foremost the regulator of the market. It keeps a close eye on the operation of the agricultural waste power generation market and strictly controls every link. From whether the purchase price of agricultural waste raw materials is reasonable to whether the production process of power generation enterprises complies with environmental protection and safety standards, all fall within the scope of government supervision to ensure the orderly and healthy development of the market. The government is also a regulator. When there is an imbalance between supply and demand in the agricultural waste power generation market, such as insufficient agricultural waste supply affecting the power generation progress, or excessive power output causing price fluctuations, the government will regulate through administrative measures, economic levers, and other means. For instance, it may coordinate the supply channels of agricultural waste, encourage farmers to increase the amount of agricultural waste collected, or guide power generation enterprises to reasonably adjust production capacity (Xiaolu 2022). As policymakers, the government has laid the foundation for the long-term development of the agricultural waste power generation industry. It formulates subsidy policies to provide financial support to enterprises actively engaged in agricultural waste power generation and reduce their operating costs. It introduces preferential tax policies to encourage more enterprises to participate in the agricultural waste power generation industry, promote the continuous growth of this industry, and thereby achieve the resource utilization of waste while contributing to the optimization of the energy structure and the improvement of the ecological environment.

1.1.2 The Operation Process of the Agricultural Waste Power Generation Supply Chain

The operation of the agricultural waste power generation raw material supply chain can be briefly summarized as the one-way flow of agricultural waste raw materials. It mainly includes functional links such as field collection, logistics transportation, centralized storage, and power generation utilization. The introduction of each process is as follows:

1.1.2.1 Collection Stage

There are mainly two modes of agricultural waste collection: manual collection and mechanical collection.

Manual collection: In mountainous, hilly areas or small irregular farmlands, large-scale mechanical equipment is restricted by terrain and technology and cannot carry out collection work, so manual collection is required. After collecting the crops, farmers harvest and bundle the remaining agricultural waste in the fields and then use wheelbarrows and other tools to transport the agricultural waste to the collection points. Manual collection is highly flexible and can collect agricultural waste to the maximum extent in a targeted manner according to different terrains. However, manual collection is inefficient and requires a large amount of labor. Especially in large-scale planting areas with flat terrain, the manual collection mode cannot meet the collection demands. Therefore, the mechanical collection mode needs to be adopted.

Mechanical collection: Combine harvesters are used to collect agricultural waste, and balers are employed to compress the agricultural waste into bundles. Mechanical collection not only has high efficiency and saves labor costs but also has a high degree of standardization. The size and density of the bundles are more uniform, which is convenient for subsequent transportation and storage. However, mechanical collection requires expensive initial investment and maintenance costs. Therefore, an appropriate collection method can be selected based on the terrain and actual conditions.

1.1.2.2 Warehousing Link

The warehousing process consists of three parts: storage, distribution processing, and loading and unloading. Agricultural waste has the characteristics of large volume, low density, and flammability. Storage facilities should be scientifically planned based on the storage volume, storage period, and temperature and humidity conditions of agricultural waste. Common storage methods include open-air stacking, simple shed storage, and closed intelligent warehouse storage. Open-air stacking is cost-effective, but agricultural waste is prone to erosion by wind and rain, leading to spoilage, deterioration, and a decline in quality, and there is also a significant fire hazard. Simple shed storage has, to some extent, improved the drawbacks of open-air stacking, but agricultural waste still cannot avoid mold. Closed intelligent warehouse storage can control the temperature and humidity during storage, ensure the quality of agricultural waste, and use information technology to monitor the storage volume in real time, avoiding the occurrence of agricultural waste supply interruption and accumulation. However, the construction and operation costs are high. Therefore, closed intelligent warehouses are preferred for storing agricultural waste. After the agricultural waste is transported from the farmland to the warehouse, it needs to undergo pretreatment, crushing, drying, transmission, compression, quality inspection, and other treatments to improve transportation efficiency and extend storage time. The storage link is an important link to ensure the stable supply of agricultural waste.

1.1.2.3 Transportation Stage

Agricultural waste transportation refers to the process of agricultural waste moving from the fields to power plants, which includes two stages: primary transportation from the fields to the collection and storage stations and secondary transportation from the collection and storage stations to the power plants. The primary transportation methods mainly rely on rural roads and field roads. Agricultural tractors are flexible and convenient, suitable for short-distance transportation on field roads and rural roads, and can centrally transport scattered agricultural waste to the collection and storage stations. The secondary transportation mode mainly relies on roads. Large trucks have advantages such as good sealing and large loading capacity, making them suitable for long-distance transportation. They can transport agricultural waste from the collection and storage stations to the power plants, effectively reducing the loss of agricultural waste during transportation. Due to the large volume of agricultural waste, long-distance transportation leads to increased costs, with transportation costs accounting for approximately 60% of the total cost. Therefore, by integrating Geographic Information System (GIS) and Beidou Navigation Satellite System (BDS) technologies and comprehensively considering factors such as the geographical locations and traffic conditions of collection points, storage stations, and power plants, route planning is carried out to reduce transportation costs. Agricultural waste is flammable. Safety issues should always be noted during transportation (Zang et al. 2020).

1.1.2.4 Power Generation and Utilization Stage

The utilization of agricultural waste is the core link in the agricultural waste power generation supply chain, converting agricultural waste into electricity to achieve the effective utilization of biomass waste. Direct combustion of agricultural waste for power generation is one of the main ways of agricultural waste utilization. After arriving at the power plant, the agricultural waste undergoes quality inspection. The agricultural waste that meets the combustion standards is sent to a fixed site for crushing treatment. The processed agricultural waste is sent to the feeding device at the front end of the boiler and then enters the boiler combustion system. Under high-temperature conditions, it is fully combusted, and the superheated steam with certain temperature and pressure produced is sent to the steam turbine to drive the generator to produce electricity. The converted electricity is then transmitted to the power grid (Wang et al. 2022). Some of the ash and residue produced by burning agricultural waste can be reused, reducing environmental pollution and achieving comprehensive utilization of resources.

1.2 Analysis of the Supply Chain Model for Agricultural Waste Power Generation

An efficient supply chain can ensure the sustainable development of agricultural waste power generation. This chapter will conduct a detailed analysis of the three main operation models of the agricultural waste power generation supply chain: the traditional model, the cooperative contract model, and the integrated model. The advantages and disadvantages of each model will be discussed in detail to help better understand its performance in practical applications.

1.2.1 Traditional Operation Model

In the traditional agricultural waste power generation supply chain, the main participants include farmers, agricultural waste brokers, and power plants. The operation process of this model is relatively simple, but there are many problems. The operation process includes farmers collecting agricultural waste, agricultural waste brokers purchasing and transporting it, and power plants receiving it and generating electricity. First, after the crops are harvested, farmers collect the agricultural waste by themselves. Due to the lack of unified guidance and incentive mechanisms, farmers' enthusiasm for collecting agricultural waste is relatively low, resulting in unstable agricultural waste collection volume. Then, agricultural waste brokers purchase agricultural waste from farmers and usually lower the price according to market conditions to obtain higher profits. The purchased agricultural waste needs to undergo simple processing and packaging before being transported to the power plant. Due to high transportation costs, agricultural waste brokers often choose power plants that are closer for sales. Finally, after the power plant receives the agricultural waste sent by the agricultural waste broker, it conducts quality inspection and pretreatment. Then, the agricultural waste is sent into the boiler for combustion to generate electricity. The electricity generated is connected to the national power grid for sale.

1.2.2 Cooperative Contract Model

The cooperative contract model binds the interests of farmers, agricultural waste brokers, and power plants together by signing contracts, thereby improving the overall efficiency of the supply chain (Zhang et al. 2024). The operation process includes contract signing, collection and storage by farmers, purchase and transportation by agricultural waste brokers, and reception and power generation by power plants. Power plants sign long-term supply contracts with agricultural waste brokers, clearly stipulating the supply volume, quality standards, and prices of agricultural waste.

The agricultural waste brokers then sign agricultural waste collection contracts with farmers to ensure a stable supply of agricultural waste. Farmers collect agricultural waste as required by the contract and carry out preliminary treatment and storage. Power plants and agricultural waste brokers provide technical support and subsidies to enhance the enthusiasm of farmers. According to the contract requirements, agricultural waste brokers regularly purchase agricultural waste from farmers and are responsible for transporting it to power plants. Due to the signing of long-term contracts, agricultural waste brokers can optimize transportation routes and warehouse management to reduce costs. After receiving the agricultural waste, the power plant conducts quality inspection and pretreatment and then burns it to generate electricity. The power plant pays the agricultural waste broker the agricultural waste fee according to the contract and connects the electricity to the national power grid.

1.2.3 Integrated Model

The integrated model integrates all the links of the agricultural waste power generation supply chain to form an organic whole. This model integrates resources to achieve collaborative operation of the supply chain and enhance overall efficiency. The operation process includes supply chain integration, collection and storage by farmers, procurement and transportation by power plants, and reception and power generation by power plants. First of all, power plants directly sign agricultural waste collection contracts with farmers and establish their own collection networks. Meanwhile, the power plant invests in the construction of storage and transportation facilities to achieve centralized management and transportation of agricultural waste. Then, farmers collect the agricultural waste as required by the contract and carry out preliminary treatment and storage. The power plant provides technical support and subsidies to ensure the quality and quantity of agricultural waste. Then, the power plant directly purchases agricultural waste from farmers and transports it to the power plant through its own transportation fleet. Due to the reduction of intermediate links, transportation costs are significantly reduced. Finally, after receiving the agricultural waste, the power plant conducts quality inspection and pretreatment and then burns it to generate electricity. The power plant pays farmers the cost of agricultural waste according to the contract and connects the electricity to the national power grid.

1.2.4 Third-Party Logistics Model

The third-party logistics model of the agricultural waste power generation supply chain is a typical practice of enhancing industrial efficiency through specialized division of labor. The power plant outsources the agricultural waste collection, storage, and transportation links to qualified logistics enterprises, forming an efficient collaboration system of "focusing on core business + outsourcing of supporting services." The logistics companies undertaking the business first build a standardized collection and storage network, set up regional collection centers in the main agricultural waste production areas, and equip them with intelligent balers, dedicated transport vehicles, and other professional equipment. They standardize links such as farmers' sales, cooperative consolidation, transportation, and enterprise warehousing through unified operating procedures. In the transportation process, logistics companies use the Global Positioning System (GPS) and route optimization algorithms to conduct full-process monitoring of customized agricultural waste transport vehicles with heavy loads, ensuring that the one-way transportation cost is reduced by more than 20% compared with the traditional mode. The storage stage adopts standardized storage facilities, equipped with moisture-proof ventilation systems and intelligent monitoring equipment. Through the Warehouse Management System (WMS), the dynamic allocation of agricultural waste inventory is achieved, and the storage loss rate is controlled within 5%. This specialized operation model not only enables power plants to save 30–40% of the initial investment in building their own collection and storage systems (Wang et al. 2016) but also reduces the loss rate of agricultural waste storage.

1.2.5 Government-Led Model

Local governments play a key leading role in building an efficient and stable supply chain system. On the one hand, by delving into rural household groups, through promotional meetings and door-to-door exchanges, they explain the significance of participating in the system construction for enhancing the sales efficiency of agricultural products and ensuring income and stimulate the enthusiasm of farmers. For cooperatives, they assist in sorting out the management mechanism, optimizing the operation process, and enhancing their coordination ability. Meanwhile, they maintain close communication with enterprises such as power plants to explore cooperation opportunities. After integrating resources from all parties, a unified supply chain system is established. Farmers scientifically arrange planting and breeding based on market demand and orders from cooperatives. The cooperative centrally purchases farmers' products, completes sorting and initial processing, and, after ensuring that the quality meets the standards, distributes them to downstream enterprises such as power plants. Under this model, the government uses policy regulation and control (Hsu 2021). In terms of transportation subsidies, differentiated standards should be set based on mileage, vehicle specifications, and frequency to encourage logistics enterprises to participate. In terms of tax incentives, value-added tax on the sale of agricultural products is exempted for farmers and cooperatives. Enterprises that purchase local agricultural products are allowed to have their purchase costs deducted proportionally when calculating income tax, thereby reducing the burden on enterprises and enhancing their enthusiasm for cooperation.

1.2.6 Power Plant Self-Built Supply Chain Model

The power plant invests independently in the construction of agricultural waste collection, transportation, and storage equipment and establishes a complete supply chain. Power plants directly hire farmers or professional teams to manage agricultural waste, ensuring a stable supply and quality control of agricultural waste. In the field of agricultural waste resource utilization, power plants have established a standardized supply chain system integrating collection, transportation, and storage through full-chain independent investment and construction. The power plant first scientifically plans the layout of agricultural waste collection stations based on the distribution of surrounding agricultural planting areas and is equipped with intelligent agricultural waste harvesters, balers, crusher, and other professional collection equipment. At the same time, it purchases closed transportation vehicles, GPS, and on-board weighing equipment to ensure an efficient and controllable transportation process. For the storage stage, standardized storage facilities with moisture-proof, fireproof, and ventilation functions are constructed, along with equipment for detecting the moisture content of agricultural waste and pest control. Some power plants have also introduced automated three-dimensional storage systems to achieve digital management of agricultural waste storage.

1.2.7 Analysis of the Advantages and Disadvantages of the Agricultural Waste Power Generation Supply Chain Model

Under the global background of sustainable energy development, agricultural waste power generation, as an important way of biomass energy utilization, has attracted much attention for its supply chain model. Different agricultural waste power generation supply chains contain different operation processes, and the participating entities also vary (Cao et al. 2016; Kumar et al. 2023). The flowcharts of different agricultural waste power generation supply chain models are shown in Fig. 1.1. According to the different goals of the stakeholders in the agricultural waste power generation supply chain, the advantages and disadvantages of these agricultural waste power generation supply chain models are summarized in Table 1.1.

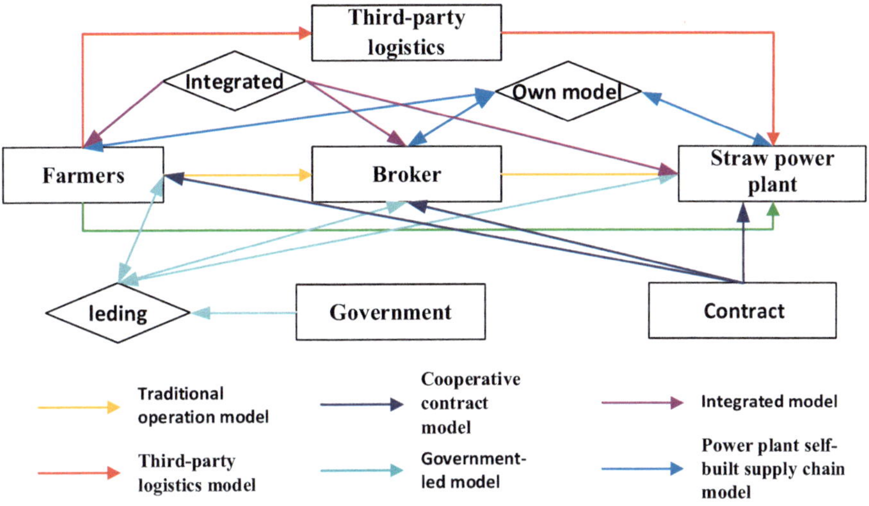

Fig. 1.1 Flowchart of the agricultural waste power generation supply chain model

Table 1.1 Advantages and disadvantages of the agricultural waste power generation supply chain model

Supply chain model	Advantages	Disadvantages
Traditional operation model	1. Flexible operation 2. Strong market adaptability	1. Low enthusiasm of farmers 2. Serious price pressure from agricultural waste brokers 3. High transportation costs 4. Insufficient supply of raw materials for the plant 5. Low overall efficiency
Cooperative contract model	1. Improved enthusiasm of farmers 2. Reduced costs for agricultural waste brokers 3. Stable supply of raw materials for power plants 4. Improved overall benefits of the supply chain	1. Complicated contract management 2. Farmers' default risk 3. Default risk of agricultural waste brokers 4. Dependence of power plants on agricultural waste brokers
Integrated model	1. Elimination of intermediaries 2. Improved supply chain stability 3. Improved overall efficiency 4. Enhanced farmers' willingness to cooperate	1. High investment costs 2. Complex management 3. Concentration of risk 4. Reduced flexibility
Third-party logistics model	1. Professional operation 2. Reduced operating pressure on power plants 3. Suitable for large-scale projects	1. High logistics costs 2. Cooperation risks

(continued)

Table 1.1 (continued)

Supply chain model	Advantages	Disadvantages
Government-led model	1. Strong policy support 2. High supply chain stability 3. Reduced operating costs	1. Low degree of marketization 2. Risk of administrative intervention
Power plant self-built supply chain model	1. Strong controllability of the supply chain 2. Low long-term operating costs	1. Large initial investment 2. Long recovery period 3. Complicated management

1.3 Analysis of the Characteristics of the Agricultural Waste Power Generation Supply Chain System

1.3.1 Development Characteristics of the Agricultural Waste Power Generation Supply Chain

The efficient operation of the agricultural waste power generation supply chain system requires the joint efforts of all participants. Due to the different purposes and activities of the participants, the supply chain has the following characteristics.

1. Dynamic development characteristics. During the operation of the agricultural waste power generation supply chain, the exchange of logistics, information flow, and cash flow between systems enables the supply chain system to function, transporting agricultural waste from farmland to power plants. The dynamic characteristics of the agricultural waste power generation supply chain enable each participating entity to have the right of independent choice (Koul et al. 2022).
2. Independent characteristics. Each subsystem of the agricultural waste power generation supply chain has independent characteristics. During the actual operation of the supply chain, all participants pursue the maximization of their own benefits. For example, farmers are the owners of agricultural waste resources and are only willing to sell agricultural waste when the price is high.
3. Nonlinear correlation. Each subsystem is interdependent, interacts with each other, and restricts each other. The nonlinear correlation among subsystems makes the system have complex characteristics such as diversity and unpredictability.
4. Self-organizing characteristics. The self-organizing characteristic of the system prompts the three subsystems to rely on and cooperate with each other, forming an agricultural waste power generation supply chain system with stable functions.
5. Open feature. The agricultural waste power generation supply chain system is constantly interrelated and interacting with the external environment and is influenced by it. If the supply chain system can develop in a more coordinated manner under the influence of policies, the utilization approaches of agricultural waste will change when there is competition for market resources.

The complex characteristics of the supply chain system hinder the sustainable development of the agricultural waste power generation supply chain. The function of the system is not simply the addition of the functions of subsystems. Therefore, when studying the agricultural waste power generation supply chain system, it is necessary to analyze the influencing factors of the system. The development of the agricultural waste power generation supply chain system is dominated by certain factors. After clarifying these dominant factors, the changing patterns of the system can be identified, and thus effective countermeasures can be proposed to promote the sustainable development of the agricultural waste power generation supply chain.

1.3.2 Uncertainty of the Agricultural Waste Power Generation Supply Chain System

The main function of the supply chain is the collection, storage, and transportation of agricultural waste. The realization of these functions requires the joint collaboration of the participants (Moncada et al. 2018). Due to problems such as market competition, difficulties in collection, storage and transportation, and asymmetry of supply and demand information in agricultural waste resources, the agricultural waste power generation supply chain has multiple uncertainties in its development process.

1.3.2.1 Uncertainty of Agricultural Waste Resource Supply

The energy utilization volume of agricultural waste resources is only 100 million tons, accounting for 11.4% of the collectible volume. It can be seen that the energy utilization level of agricultural waste is relatively low (Wen and Zhang 2015). Waste-to-energy power generation requires a stable supply of resources. However, agricultural waste resources have many uses, such as the production of organic fertilizers, biomass briquette fuels, biomass power generation, and biogas. This leads to a competitive relationship in the utilization methods of agricultural waste. If non-power generation utilization technologies in the market have greater advantages, the supply of agricultural waste for power generation will be reduced. Meanwhile, the collection of agricultural waste resources is seasonal and is only supplied in large quantities during the harvest season of crops. However, agricultural waste power plants operate throughout the year, and seasonal supply leads to uncertainty in raw material supply. In addition, the planting scale of farmers also affects the supply of agricultural waste resources. Some small-scale farmers directly use it for returning to the fields or discarding it, which directly affects the supply of agricultural waste resources. The supply of agricultural waste is uncertain, which seriously threatens the sustainable development of the agricultural waste power generation supply chain.

1.3.2.2 Uncertainty in the Selection of Supply Chain Models

Agricultural waste has multiple logistics supply models. Differences in pretreatment activities and processing procedures lead to diversity in supply chain logistics models. The supply of agricultural waste also directly affects the choice of supply chain models. The uncertainty of agricultural waste resource supply causes transportation difficulties. The supply situation of waste affects the construction of transportation systems and the choice of logistics transportation modes. Whether to build collection points and transfer stations is a problem that transportation enterprises must consider. In areas where agricultural waste resources are scattered, it is suitable to adopt a supply chain model based on constructing collection points and transfer stations. In areas rich in agricultural waste resources, mechanical collection can be carried out directly in the fields, then transported to transfer stations for pretreatment, and finally supplied as raw materials to power generation enterprises. This helps to improve transportation efficiency and reduce transportation costs. The setting of collection stations should be matched with the amount of agricultural waste resources. Too many stations result in high production costs, while too few stations tend to increase collection costs. The selection of transfer station locations needs to take transportation costs into account. Reasonable stations are conducive to the development of the waste supply chain. The selection of supply chain logistics models should be combined with regional conditions and the scale of power plants (Davtalab and Alesheikh 2018). Therefore, uncertainty in waste resource transportation is crucial to the selection of supply chain models. Building a sustainable supply chain requires choosing an appropriate logistics model.

1.3.2.3 Uncertainty of the Operating Load of Agricultural Waste Power Plants

Agricultural waste power plants are at the end of the power generation supply chain. The adequate supply of raw materials can ensure the year-round operation of power plants. However, in real life, the supply of raw materials is uncertain, so the operation of power plants is also uncertain. On the other hand, the damage to power plant equipment also affects the operation of the power plant. Although the power plant is equipped with a backup device that is alternately used and maintained, the tar produced by the burning of agricultural waste can easily cause corrosion to the equipment. Therefore, when selecting equipment, it is necessary to choose equipment suitable for agricultural waste burning. Agricultural waste supply and power plant equipment are the main factors affecting the operational load of power plants. During the site selection process of power plants, it is necessary to assess the agricultural waste resource status and power generation facilities of the region.

1.3.3 Sustainability of the Agricultural Waste Power Generation Supply Chain

The theory of sustainable development is widely applied in different fields. After the concept of green and sustainable development was proposed, the development of renewable energy has received increasing attention. Biomass power generation has been gradually promoted as an alternative to coal-fired power generation. The sustainability of the agricultural waste power generation supply chain has positive effects on regional economic development, social welfare, and ecological environment. The development of agricultural waste power generation technology can reduce the use of fossil fuels and contribute to the realization of the "carbon peak and carbon neutrality" goals.

The sustainable development of the agricultural waste power generation supply chain refers to the orderly, stable, and dynamic coordinated development of the system. The adequate supply of agricultural waste is a prerequisite for promoting the sustainable development of the power generation supply chain. The coordination of the agricultural waste power generation supply chain system, the efficient operation of power generation enterprises, and the higher energy conversion can enhance the sustainable performance of the agricultural waste power generation supply chain (Tan et al. 2017a, b). Economy, environment, technology, and society are the four important dimensions for evaluating the sustainable performance of a system:

1. Economic sustainability refers to the function of agricultural waste power generation projects to generate continuous income. The nature of agricultural waste being small in size and large in volume increases the costs of collection, storage, and transportation. The low energy conversion efficiency also leads to high production costs for enterprises, which results in relatively low profitability for power generation enterprises. The economic benefits prompt power generation enterprises to utilize agricultural waste. The higher profit-making ability can stimulate the utilization of agricultural waste resources and at the same time drive the growth of the regional economy. Therefore, the agricultural waste power generation supply chain system needs to optimize the supply mode and enhance the coordination ability among stakeholders to reduce the power generation cost. This is conducive to the improvement of economic performance and promotes the sustainable development of the agricultural waste power generation supply chain.
2. Environmental sustainability refers to the fact that the agricultural waste power generation supply chain has a relatively small negative impact on the environment during its operation. Agricultural waste has a higher emission reduction capacity than coal-fired power generation. Replacing coal with agricultural waste can reduce the emissions of harmful gases such as greenhouse gases and sulfur dioxide. Meanwhile, the recycling of agricultural waste lessens the harm caused by open-air burning. Agricultural waste power generation has good envi-

ronmental benefits, but it emits a large amount of pollutants during the collection, storage, and transportation stages. Therefore, in the research, the collection, storage, and transportation system should be optimized to further improve the environmental sustainability of the power generation system.

3. Social sustainability refers to the continuous provision of good social welfare by agricultural waste power generation activities to society. The utilization of agricultural waste resources can bring economic benefits to farmers. Meanwhile, the collection, storage, transportation, and power generation systems of the waste provide job opportunities for the people. It can be seen that the utilization of agricultural waste power generation has a good social impact. At the same time, the resource utilization of agricultural waste is in line with the goals of rural revitalization and building beautiful villages. Therefore, agricultural waste power generation is socially sustainable. The development of the agricultural waste power generation industry should provide more social welfare to society to enhance its sustainable development capacity.
4. Technological sustainability refers to the fact that agricultural waste power generation technology has a high level of maturity and technological innovation capabilities. At present, the technology of agricultural waste power generation has matured, but the energy conversion efficiency is relatively low, which indicates that the technology of agricultural waste power generation needs to be updated. Energy conversion rate is an important technical evaluation index. There are significant differences between agricultural waste power generation and coal-fired power generation technologies. Coal-fired power generation has a higher energy conversion efficiency, with the energy efficiency of coal-fired power generation reaching over 45% in the supercritical state. However, the energy conversion rate of agricultural waste power generation is between 18% and 30%. A higher conversion rate can enable agricultural waste power generation enterprises to gain more economic benefits. Therefore, agricultural waste power generation enterprises should innovate power generation technologies, improve energy conversion efficiency, and promote sustainable development of the agricultural waste power generation supply chain.

1.4 Analysis of Internal Factors in the Agricultural Waste Power Generation Supply Chain

The sustainable development of the agricultural waste power generation supply chain conforms to the national energy strategy and also contributes to the improvement of the ecological environment. The agricultural waste power generation supply chain connects farmers and power plants. The sustainability of the supply chain affects the supply of agricultural waste resources and the efficiency of power plants. Therefore, the sustainable development of the agricultural waste power generation supply chain requires the joint efforts of these participants (Zang et al. 2020; Zhao et al. 2022). During the production process, farmers, as the owners and suppliers of agricultural waste, affect the supply of raw materials for agricultural waste power plants. Agricultural waste brokers, as transporters and information transmitters, influence the behaviors of farmers and power plants. Agricultural waste power plants, as producers of electric energy, affect the enthusiasm, participation, and willingness of brokers and farmers to supply agricultural waste. Considering these issues, this paper analyzes the behaviors of these three main participants and selects the elements that affect the sustainable development of the agricultural waste power generation supply chain. The influencing factors of the system are shown in Table 1.2 (Zang et al. 2020).

Table 1.2 Influencing factors of agricultural waste power generation supply chain

	No.	Factors	Influencing factor description
Farmers	1	Age	Farmers over 40 years old have a higher willingness to supply agricultural waste
	2	Whether village cadres	Village cadres have higher awareness than ordinary people to accept and participate in supply chain sustainability construction
	3	Source of household income	Farmers who depend on planting income are more supportive of agricultural waste utilization
	4	Planting scale	Farmers with large planting scale support the government's no-burning policy and support the use of agricultural waste
	5	Perception of waste value	Farmers who can recognize the use value of agricultural waste will have a higher willingness to supply
	6	Farmers' environmental awareness	Strong environmental awareness among farmers will reduce open burning of agricultural waste and support recycling
	7	Village collective participation	More village collective participation in agricultural waste supply can drive more farmers to join
	8	Understanding of the policy	Farmers who know the relevant subsidy policies of the government have a higher willingness to supply agricultural waste
	9	Cooperation consciousness	Strong cooperative consciousness of farmers can improve the sustainable development of the agricultural waste power supply chain

(continued)

Table 1.2 (continued)

	No.	Factors	Influencing factor description
Agricultural waste broker	10	Transportation efficiency	High transportation efficiency of agricultural waste can reduce transportation costs and ensure the supply of raw materials
	11	Transportation energy consumption	Low transportation energy consumption can reduce transportation costs and reduce pollutant emissions during the transportation phase
	12	Management capability	Higher management capabilities of brokers can optimize operational costs and improve the sustainability of the system
	13	Cooperative consciousness	High collaborative awareness contributes to the sustainable development of agricultural waste power supply chain
	14	Pretreatment	Agricultural waste is transported over a distance of 25 km, and pretreatment can effectively improve the economy
	15	Storage management	High-capacity storage management can guarantee the quality and quantity of agricultural waste
	16	Production safety	Spontaneous combustion occurs easily when agricultural waste is stored, and high safety can improve sustainability
	17	Information sharing	Information sharing by brokers helps farmers understand the development of the agricultural waste power generation industry
	18	Credit rating	A high degree of credit can promote cooperation among participants and enhance the sustainability of the system
Agricultural waste power plant	19	Investment costs	High investment cost, long recovery cycle, and high operation cost are not conducive to the sustainable development of the industry
	20	Production cost	Higher production costs compress the profits of enterprises and affect the earnings of other participants
	21	Employee motivation	Higher employee motivation can improve production efficiency and reduce the cost of power plants
	22	Social contribution	The more social contribution power plants provide, the more government policy support and farmers' participation will be attracted
	23	Power plant management	The management ability of agricultural waste power plants can enhance the competitiveness and promote the sustainability of the system
	24	Innovation input	Large investment in enterprise innovation is conducive to the renewal of power generation technology and the sustainable development of power generation enterprises
	25	Core competitiveness	The greater the core competitiveness is, the stronger the profitability of the enterprise will be, which is conducive to the sustainable development of the system

The internal factors of the agricultural waste power generation supply chain refer to the influencing factors generated by the interaction between the constituent elements and links of the supply chain system itself. The internal factors of the agricultural waste power generation supply chain are interrelated and mutually transmitted: the characteristics of raw materials are the fundamental constraints that determine the direction of the internal structure design and technological adaptation of the supply chain; structure and collaborative mechanisms are the core bottlenecks that affect internal resource allocation and process efficiency; technology and equipment are the supporting foundation, determining the efficiency and cost of each link's operation; and management and funding are guarantee conditions that affect the internal optimization and sustainable operation of the supply chain.

1.5 Analysis of Problems in the Agricultural Waste Power Generation Supply Chain

The business objects and goals of the agricultural waste power generation supply chain are different from those of other supply chains. Agricultural waste has the characteristics of strong seasonality and easy spoilage and deterioration. At the same time, it is affected by uncertain factors such as weather, climate, and informatization. Agricultural waste power generation still needs to face many difficulties. Through current situation analysis and literature summary, the main problems existing in the agricultural waste power generation supply chain are as follows:

1. It is difficult to collect and store agricultural waste

   Agricultural waste has the characteristics of seasonality and dispersion. The output of agricultural waste produced by agricultural planting is limited. To collect and stably supply the agricultural waste for power generation in power plants, large areas of farmland need to be covered, which increases the time and difficulty of agricultural waste collection. As a major agricultural province, Henan Province mainly practices two harvests a year in most areas. The time interval between harvesting crops and sowing the next crop is relatively short. Moreover, due to the policy of prohibiting the burning of agricultural waste, most farmers choose the traditional method of returning agricultural waste to the field to save labor and agricultural waste disposal costs. As a result, farmers have a weak willingness to sell, which increases the difficulty of agricultural waste collection. The maturation of crops has seasonal characteristics. As an agricultural by-product, agricultural waste is supplied in large quantities in a short period of time. This centralized supply model requires the supply chain to complete the utilization of agricultural waste in a short time, which brings huge pressure to the supply chain system. Agricultural waste is piled up in the fields and exposed to the open air, making it impossible to store properly. Moreover, the freshly collected agricultural waste has a high water content and is prone to spoilage and deterioration, resulting in a decline in both the quantity and quality of the agricultural waste. During storage, agricultural waste is prone to spoilage and deterioration due to environmental factors, which not only reduces the calorific value of combustion but also generates harmful substances. Ensuring a stable supply of agricultural waste and guaranteeing its quality and safety are the urgent problems to be solved in the agricultural waste power generation supply chain.

2. The degree of information sharing and the efficiency of collaboration among various links are low

The application of modern information technology can promote the development of various industries, but the informatization level of the agricultural waste power generation supply chain is relatively low. The lack of an effective information sharing platform among farmers, agricultural waste brokers, and power plants leads to information asymmetry in the supply chain. Changes in market demand and price fluctuations are difficult to resolve promptly and quickly. Coordination among various links in the supply chain is also difficult, making it hard for participants to respond to emergencies in a timely manner and reducing the flexible adaptability of the supply chain. When agricultural waste brokers collect agricultural waste, due to the lack of data on the demand for agricultural waste from power plants, it increases the difficulty of decision-making, leading to large fluctuations in market agricultural waste prices and imbalances between supply and demand, thereby affecting power plants. The cooperation mechanism among the participants in the supply chain is not sound, and there is a lack of long-term and stable cooperation, which leads to an unstable supply of agricultural waste and makes it difficult for agricultural waste power generation to provide a stable power supply. Due to the lack of information sharing and collaborative mechanisms, the efficiency of agricultural waste resource allocation is low. The collection, storage, and transportation of agricultural waste lack planning. For instance, transport vehicles are prone to empty loads and detours, and mechanical equipment is idle, resulting in resource waste and increased costs. The degree of information sharing and collaborative efficiency will lead to problems such as low supply chain efficiency, increased operating costs, and market instability, which are the issues that the agricultural waste power generation supply chain needs to solve promptly at present.

3. The infrastructure of the transportation network is poor and the transportation cost is high

Agricultural waste transportation is a key link in the supply chain. Due to the fact that agricultural planting is generally carried out in rural areas and the road infrastructure construction is in poor condition, small agricultural tractors are needed for primary transportation from the fields to the collection and storage stations. Moreover, agricultural waste has a low density and large volume, and the number and transportation capacity of agricultural tractors are limited, making it difficult to meet the large-scale agricultural waste transportation demands and resulting in low transportation efficiency. In rural areas, modern transportation technology is less applied. The transportation tools are relatively traditional and lack corresponding supporting facilities. For instance, the storage environment and temperature during transportation are not suitable, which affects the quality of agricultural waste.

During the secondary transportation process, due to the incomplete transportation network, not only will the empty load rate of trucks be high, but also the transportation time and frequency will increase. Moreover, the agricultural waste resources are widely distributed and scattered, resulting in longer transportation distances and increased transportation costs. The transportation cost of agricultural waste accounts for a large proportion of the total cost, approximately 60%, and there is a considerable amount of agricultural waste loss. Therefore, improving transportation conditions, rationally planning transportation routes, and building transportation networks are particularly important for the agricultural waste power generation supply chain.

4. Insufficient policy support and an imperfect market mechanism

   Policy support is an important driving force for the development of the agricultural waste power generation industry. However, at present, the policy support for the agricultural waste power generation industry is insufficient, lacking clear policies and incentive mechanisms. The subsidy policies for agricultural waste power generation mainly focus on the power generation link, and the national policies and subsidies will also gradually withdraw, resulting in increased cost pressure at each link of the supply chain. Tax preferential policies only target some large-scale enterprises. Small enterprises have high investment and operation costs, which leads to difficulties in capital turnover for enterprises, intensifies the situation of unequal competition within the industry, and restricts the overall development of the agricultural waste power generation industry. Due to insufficient enforcement by the government or inadequate supervision, policies cannot be effectively implemented. During the sale of agricultural waste, the lack of unified standards and norms can easily lead to a chaotic market trading environment and opaque agricultural waste prices, causing fluctuations in agricultural waste prices and posing significant operational risks to supply chain participants. The lack of effective market supervision is prone to trigger vicious competition, making it difficult to form a market scale and affecting the stability of the market. Insufficient policy support and an imperfect market mechanism have intensified the vicious competition within the industry, increased the operating costs of the agricultural waste power generation supply chain, and affected the economic feasibility of agricultural waste power generation. These are important issues restricting the development of the agricultural waste power generation supply chain.

5. The profit distribution mechanism is unreasonable

   The price of raw materials is the core incentive for farmers to participate, but the current price formation mechanism lacks transparency: Power plants suppress purchase prices to control costs, while collectors and transport companies must bear intermediate costs, resulting in low actual returns for farmers and insufficient motivation, leading to situations such as "holding back sales during peak seasons and shortages during off-peak seasons." The high proportion of small-scale suppliers and the lack of professional cooperatives or leading enterprises to integrate raw material resources result in insufficient economies of scale in collection, transportation, and storage.

6. Insufficient stability of policy support

Profit pressure brought by subsidy reduction: Early projects relied on renewable energy electricity price subsidies, but after 2020, subsidies gradually decreased, and some projects experienced cash flow disruptions due to subsidy arrears or cancellations. Inconsistent implementation of local policies: Some regions lack clear policies such as subsidies for raw material collection, green transportation channels (such as exemption of high-speed fees for straw transportation), and land security for storage, or policy implementation is inadequate (such as delayed subsidy distribution), increasing supply chain uncertainty. Compliance pressure from increased environmental policies: With the improvement of environmental requirements, standards for manure treatment, straw burning control, and power generation exhaust emissions have tightened. Power plants need to invest more funds in environmental renovation to further squeeze profits.

The agricultural waste power generation supply chain is a complex system involving "raw material collection, storage, transportation, pretreatment, power generation, and by-product utilization," and its main pain points are concentrated in four dimensions: raw material characteristics constraints, high costs, low collaborative efficiency, and insufficient policy and technology adaptation. The core contradiction of the agricultural waste power generation supply chain is the mismatch between the "dispersion and instability of raw materials" and the "scale and continuity of power generation," coupled with high costs, insufficient coordination, and poor policy and technology adaptability, resulting in low overall efficiency and weak profitability of the supply chain.

1.6 Conclusion

The agricultural waste power generation supply chain is an important bridge connecting the raw material production and utilization ends, and its sustainable development plays a significant role. In this chapter, the structure, operation mode, and influencing factors of the agricultural waste power generation supply chain are mainly analyzed. The main participants and operation processes of the agricultural waste power generation supply chain are introduced. The advantages and disadvantages of six agricultural waste power generation supply chain models were analyzed. Meanwhile, the development characteristics and system uncertainties in the development process of the agricultural waste power generation supply chain are summarized. The internal influencing factors of the supply chain system are analyzed from three main subjects: farmers, agricultural waste brokers, and power generation enterprises, and an in-depth analysis of the external environment of the agricultural waste power generation supply chain is conducted. Through an in-depth analysis of the agricultural waste power generation supply chain, the problems existing in its development were summarized.

References

Cao J, Pang B, Mo X, Xu F (2016) A new model that using transfer stations for straw collection and transportation in the rural areas of China: a case of Jinghai, Tianjin. Renew Energy 99:911–918. https://doi.org/10.1016/j.renene.2016.07.061

Davtalab M, Alesheikh AA (2018) Spatial optimization of biomass power plant site using fuzzy analytic network process. Clean Techn Environ Policy 20(5):1033–1046. https://doi.org/10.1007/s10098-018-1531-5

Haiping Y (2015) Research on multi-stakeholder group synergistic income distribution in biomass power generation supply chain. (Master), Retrieved from https://kns.cnki.net/kcms2/article/abstract?v=ywQ10qv17SGbNThBZGDeHfCLAyAsN1nwk9Z7eiwFQtujtslbkFg13Ofa6T_bHw0HQyMzQcvb7QTswVQbFcq7ZnvVNiwRwu9MipoQ1kaFWA5SMtSSiYXHOT_gPOVG50iWr403s3Wd8Ivu1mK_i7QpkgEcFoZFeUNBVh0tN2jgBBRDcXySqn8yg55MeOUxNqNQN9qizEPzlU8=&uniplatform=NZKPT&language=CHS. Available from Cnki

Hsu E (2021) Cost-benefit analysis for recycling of agricultural wastes in Taiwan. Waste Manag 120:424–432. https://doi.org/10.1016/j.wasman.2020.09.051

Koul B, Yakoob M, Shah MP (2022) Agricultural waste management strategies for environmental sustainability. Environ Res 206:112285. https://doi.org/10.1016/j.envres.2021.112285

Kumar JA et al (2023) Agricultural waste biomass for sustainable bioenergy production: feedstock, characterization and pre-treatment methodologies. Chemosphere 331:138680. https://doi.org/10.1016/j.chemosphere.2023.138680

Moncada JA, Verstegen JA, Posada JA, Junginger M, Lukszo Z, Faaij A, Weijnen M (2018) Exploring policy options to spur the expansion of ethanol production and consumption in Brazil: an agent-based modeling approach. Energy Policy 123:619–641. https://doi.org/10.1016/j.enpol.2018.09.015

Sun Y et al (2021) Emergy evaluation of straw collection, transportation and storage system for power generation in China. Energy 231:120792. https://doi.org/10.1016/j.energy.2021.120792

Tan Q, Wang T, Zhang Y, Miao X, Zhu J (2017a) Research on the supply chain model of biomass power generation fuel under fuzzy supply. China Soft Sci 02:123–131

Tan Q, Wang T, Zhang Y, Miao X, Zhu J (2017b) Fuel supply chain optimization for multi-phase biomass power generation. Indu Technol Econ 36(11):21–28

Wang H et al (2016) Policies and regulations of crop straw utilization of foreign countries and its experience and inspiration for China. Trans CSAE 32(16):216–222. https://doi.org/10.11975/j.issn.1002-6819.2016.16.030

Wang J, Xu M, Deng Y (2022) Closed-loop supply chain pricing and coordination contract under information asymmetry. J Syst Eng 37(05):617–631. https://doi.org/10.13383/j.cnki.jse.2022.05.004

Wen S (2024) Research on supply chain innovation based on new quality productivity drive. Special Zone Economy 08:68–71

Wen W, Zhang Q (2015) A design of straw acquisition mode for Chinas straw power plant based on supply chain coordination. Renew Energy 76:369–374. https://doi.org/10.1016/j.renene.2014.11.062

Xiaolu R (2022) Research on risk identification and assessment of agricultural product supply chain in shandong province. (Master), Retrieved from https://link.cnki.net/doi/10.27740/d.cnki.gqhdx.2022.000610. Available from Cnki

Zang G, Zhang J, Jia J, Lora ES, Ratner A (2020) Life cycle assessment of power-generation systems based on biomass integrated gasification combined cycles. Renew Energy 149:336–346. https://doi.org/10.1016/j.renene.2019.12.013

Zhang H, Gao X, Wang H, Wang Z, Qu Q (2024) Research on supply chain model of straw power generation based on life cycle assessment. J Henan Agric Univ 58(04):663–673. https://doi.org/10.16445/j.cnki.1000-2340.20240429.002

Zhao B, Wang H, Huang Z, Sun Q (2022) Location mapping for constructing biomass power plant using multi-criteria decision-making method. Sustain Energy Technol Assess 49:101707. https://doi.org/10.1016/j.seta.2021.101707

Chapter 2
Research on Node Site Selection of Agricultural Waste Power Generation Supply Chain

Jing Gao, Chao Wang, Zhanwu Wang, Jin Li, Runkai Zhang, Xin Wu, Guangyin Xu, and Zhenfeng Wang

Contents

J. Gao
College of Logistics and E-Commerce, Henan University of Animal Husbandry and Economy, Zhengzhou, China
e-mail: 251006@hnuahe.edu.cn

Z. Wang · J. Li · R. Zhang · X. Wu · G. Xu
College of Mechanical and Electrical Engineering, Henan Agricultural University, Zhengzhou, Henan Province, China
e-mail: wzw6997@henau.edu.cn

C. Wang
Mineral Resources Exploration Center of Henan Geological Bureau, Zhengzhou, China

Z. Wang (✉)
College of Mechanical and Electrical Engineering, Henan Agricultural University, Zhengzhou, Henan Province, China

Henan Provincial Cold Chain Information and Equipment Laboratory for Logistics of Agricultural Products, Henan Agricultural University, Zhengzhou, China
e-mail: zfking@henau.edu.cn

Z. Wang et al. (eds.), *Agricultural Waste Power Generation Supply Chain*, Industrial Ecology and Environmental Management 7,
https://doi.org/10.1007/978-3-032-20087-7_2

Abstract Biomass energy has made significant contributions to energy demand and environmental protection. The purpose of this study is to establish a new integrated method adapted to the site selection of biomass cogeneration projects. The highlight lies in the integration of Multi-criteria Decision-Making methods with Geographic Information System at different stages. This study takes Henan Province as an example to introduce the application of this model. Stage I preliminary screening stage: Use Geographic Information System and Fuzzy Best-Worst Method to classify the suitability level of project site selection within the research area. Identify alternative solutions in areas with high suitability and large patch area. Stage II selection stage: Calculate the indicator weights using Criteria Importance Through Intercriteria Correlation method first; then use the Multi-attributive Border Approximation Area Comparison method to select the optimal factory site. Using Geographic Information System technology, the factory is selected to be built in an area with a suitability value greater than 7 in Xiangcheng County, Zhoukou City. Finally, the obstacle model diagnoses the direction of optimizing the location environment. This study provides a theoretical reference for the new construction, renovation, and expansion of biomass cogeneration projects.

Keywords Biomass cogeneration · GIS-FBWM · CRITIC-MABAC · Obstacle degree · Sustainability

2.1 Introduction

With the increasing contradiction of global energy supply and demand, fossil fuels can lead to environmental problems such as air pollution and global warming. However, more than 80% of the world's energy production comes from traditional fossil fuels (Shao et al. 2020). The world is facing a severe energy crisis. Renewable energy is inexhaustible. The United Nations Sustainable Development Goal is to promote the use of clean energy globally and reduce dependence on fossil fuels by 2030. As a result, growing environmental problems, energy crises, and high energy prices are driving the exploration of renewable energy.

The output of biomass energy is large and widely distributed and is considered the fourth largest energy reserve in the world after oil, coal, and natural gas. The potential of biomass energy in achieving the United Nations Sustainable Development Goals has attracted worldwide attention (Edrisi and Abhilash 2016). As a key source of biomass energy, the energy utilization of agricultural waste can not only effectively alleviate environmental pollution and energy shortages but also greatly avoid the waste of agricultural waste resources and the various hazards caused by incineration. It is widely used in power generation, solid molding, fuel, ethanol, and other fields. For example, Kengpol et al. (2022) proposed a sustainability assessment method for the site selection of biomass power plants. Ji et al. (2023) studied an integrated decision support method for strategic planning and tactical management of regional biomass power plants under uncertain conditions. Zhang et al. (2022) conducted a systematic evaluation of the biomass briquette fuel supply system using economic costs and greenhouse gas emissions as evaluation indicators. Zheng et al. (2021) studied the site selection of an ethanol refinery in Alberta, Canada, based on Geographic Information System (GIS) technology.

The energy conversion rate of agricultural waste cogeneration can reach more than twice that of pure power generation. China has abundant agricultural waste resources and great potential for energy utilization. At present, the Chinese government advocates the transformation and upgrading of agricultural waste power generation to cogeneration and reasonably arranges agricultural waste energy projects (Gao et al. 2024).

Due to low density, dispersed distribution, and difficulty of preservation of agricultural waste, there are many problems in the process of agricultural waste utilization, such as unreasonable site selection of projects, large acquisition radius, insufficient supply of raw materials and a tendency to intensify, relatively high cost of collection, storage, and transportation, low economic efficiency, and other issues. However, site selection is one of the key steps in building renewable energy projects, not just biomass. Wu et al. (2019) proposed a framework for site selection of biomass cogeneration projects and successfully applied it in China. Adamski and Magnuszewski (2020) present issues related to the preparation of investments in renewable energy—specifically, the selection of a location. The reasonable layout of biomass projects is not only related to the investment amount but also related to later operating costs and other factors. Therefore, the rational layout of Agricultural Waste Cogeneration Projects (AWCPs) requires scientific site selection, and, at the same time, it is necessary to promote according to local conditions and local development status.

Many scholars have studied the site selection of the project. Site selection for agricultural waste cogeneration is influenced by many factors. Al Garni and Awasthi (2017) established an index system including environmental, economic, climatic, geographic, and location factors to apply to the site selection of photovoltaic (PV) power generation projects. Colak et al. (2020) consider factors such as solar energy potential, roads, slope, aspect, land cover, rivers, energy grids, and residential areas in photovoltaic power generation site selection. Shahdabadi et al. (2021) consider factors such as installation spots, weather conditions, risk factors, and raw material factors at the site of biomass power plants in Iran. The previous power generation site selection index system could not be well adapted to existing cogeneration projects. Wu et al. (2019) consider the impact of social demands, raw material supplies, and economic factors on the site selection of biomass cogeneration projects but do not consider the role of carbon emissions in the site selection process of agricultural waste cogeneration. Therefore, it is necessary to establish a systematic and complete index system suitable for the site selection of Agricultural Waste Cogeneration Projects.

GIS has great advantages in the operation, storage, management, and analysis of geospatial data (Chukwuma et al. 2021). Multi-criteria Decision-Making (MCDM) provides a series of algorithms for assigning weights to indicators, evaluating decision plans, and determining priority of plans (Khanlari and Nazari 2022). The combination of GIS and MCDM can achieve complementary advantages (Rodrigues et al. 2020). Therefore, this integrated method is becoming increasingly popular in the macro site selection decision-making of renewable energy projects (Shao et al. 2020). Al Garni and Awasthi (2017) adopted the Geographic Information System-Analytic Hierarchy Process (GIS-AHP) method to classify land suitability into five levels in the selection of solar PV power sites. Colak et al. (2020) used the GIS-AHP method in photovoltaic power generation site selection. A GIS-based offshore wind site selection model using fuzzy Multi-criteria Decision-Making was proposed by Sanchez-Lozano et al. (2022). Razeghi et al. (2023) applied GIS-MCDM to the location selection of wind farms. Waewsak et al. (2020) applied the GIS-AHP model to the site selection of biomass power plants using rubber trees as raw materials. Nuno-Villanueva et al. (2023) proposed a new method using geospatial tools to select feasible areas for geothermal heating. At present, GIS-MCDM is widely used in renewable energy sites such as wind energy, solar energy, and geothermal energy. However, the application of site selection for Agricultural Waste Cogeneration Projects using agricultural waste materials is relatively limited. Therefore, integrating GIS and MCDM technology into the field of agricultural waste cogeneration site selection is a good direction.

The MCDM method is a decision-making method that takes into account multiple conflicting goals and the preferences of decision makers. Common MCDM sequencing techniques include Technique for Order Preference by Similarity to an Ideal Solution (TOPSIS), VIseKriterijumska Optimizacija I Kompromisno Resenje (VIKOR), Multi-attributive Border Approximation Area Comparison (MABAC), and Multi-objective Optimization on the Basis of Simple Ratio Analysis (MOOSRA). Different sorting methods often result in different sorting results. Shahdabadi et al. (2021) used the TOPSIS model to rank the site selection plans for biomass power plants in India. Lee et al. (2017) used the VIKOR method to determine the ranking of photovoltaic solar power plant locations. Wu et al. (2021) proposed a Best-Worst Method-Criteria Importance Through Intercriteria Correlation-Multi-attributive Border Approximation Area Comparison (BWM-CRITIC-MABAC)-based decision-making framework for the location selection of photovoltaic hydrogen production projects. Pamucar and Cirovic (2015) compare Simple Additive Weighting (SAW), TOPSIS, MOOSRA, and VIKOR, and the results show that the MABAC method is a useful and reliable decision-making tool. Deveci et al. (2021) applied MABAC to the field of offshore wind farm site selection in the United States, and the weight sensitivity analysis results show that the MABAC method is not affected by weight. GIS and MABAC technology have been successfully applied in various fields. For example, Gigovic et al. (2017) applied the MABAC method to wind farm site selection in combination with Geographic Information System-Analytic Network Process (GIS-ANP). Tulun et al. (2022) combined GIS-CRITIC-MABAC to empirically study biogas facility site selection in Turkey.

In the process of determining the index weight, although the subjective empowerment method refers to expert opinion, there are problems of heavy workload or low consistency in pairwise comparisons, such as Analytic Hierarchy Process, Best-Worst Method, Fuzzy Analytic Hierarchy Process (FAHP), and Fuzzy Best-Worst Method (FBWM), etc. Al Garni and Awasthi (2017) used the AHP method to calculate the weight of geographic layers and applied it to the solar PV power plant site selection. Kheybari et al. (2021) used the Best-Worst Method (BWM) to evaluate the importance of indicators and applied it to the location selection of corn planting areas. At the same time, it is found that the application of fuzzy theory to MCDM can compensate for the shortcomings of a single method, which can better deal with ambiguity and uncertainty in information (Khanlari and Nazari 2022). Guo and Zhao (2017) verified that FBWM has higher consistency than BWM. FBWM has fewer comparisons, is easier to operate, and is more consistent than FAHP. Objective weighting method can make full use of the original data, but it is prone to weight distortion, including entropy method, correlation coefficient method, and Criteria Importance Through Intercriteria Correlation method. Comprehensive weighting takes into account the advantages of both subjective and objective weights. For example, Wu et al. (2020) successfully applied the combination of Decision-Making Trial and Evaluation Laboratory (DEMATEL) and entropy to calculate indicator weights in wind farm site selection. Wu et al. (2019) used BWM and entropy methods to calculate the comprehensive weight of indicators and applied it to the site selection of biomass cogeneration projects.

From the existing literature, foreign AWCPs-related research is more mature than domestic research. At present, the site selection of agricultural waste cogeneration is a hot topic. Research can be conducted from the following points of entry: establishing a sound site selection index system, constructing a site selection model suitable for cogeneration projects, fully utilizing data, and scientifically calculating indicator weights.

The main contribution of this study is to establish a new project site selection integration method for selecting the optimal installation location for Agricultural Waste Cogeneration Project. This method is easy for researchers to copy or modify. The highlight of this study is the combination of multi-criteria FBWM, CRITIC, and MABAC methods with GIS at different stages, achieving functional complementarity of different technologies. This study was successfully applied to a typical site selection case of Agricultural Waste Cogeneration Project in Henan Province. It is expected to enable people to better understand the potential of agricultural waste. Finally, sensitivity analysis of indicator weights verifies the stability of the decision.

The sorting results of MABAC method were compared with SAW and MOOSRA methods, confirming the effectiveness of using MABAC method. The obstacle model is used to analyze the weaknesses of the project's environment and find breakthroughs in optimizing the project site selection environment, in order to improve the sustainability of the project (Table 2.1).

2.2 Construction of a Two-Stage Site Selection Model for Agricultural Waste Power Generation Supply Chain Nodes

The process of selecting sites for Agricultural Waste Cogeneration Projects constitutes a highly intricate system engineering endeavor. This study considers the impact of multiple attributes such as economy, society, environment, and geography to reduce project investment risks. The main advantage of integrating GIS and MCDM methods is that their specific functions can complement each other. The site selection framework established in this chapter is shown in Fig. 2.1, and the established indicator system is shown in Table 2.2.

2.2.1 Two-Stage Site Selection Framework

Stage I: Preliminary Selection of Agricultural Waste Cogeneration Projects Based on GIS Spatial Analysis
This stage utilizes the capabilities of GIS in spatial data management. First, spatial data of the research area are collected, as shown in Sect. 2.2.4. Covering nine indicators in economic, geographic, and location environments, the indicator system is shown in Table 2.2. Second, this chapter utilizes the unique advantages of GIS in processing spatial data for data preprocessing and spatial analysis (Sarikaya et al. 2022), calculates the weights of each indicator using the FBWM method with high consistency, and then conducts weighted overlay analysis of GIS for each geographic layer. The spatial analysis results are visualized in the form of a (grid or vector) map. The scope of influence and unsuitable areas of completed projects are then erased. Finally, the potential areas are classified into levels, and alternative solutions are extracted for decision-making in the second stage. Here, it is necessary to provide two or more options for the second stage of decision-making; otherwise, it is necessary to return to adjusting the GIS spatial processing parameters until the requirements are met. The process is shown in Fig. 2.1.

Stage II: Location Selection
Based on the alternative solutions obtained in the previous stage, on-site research and data collection were completed within the study area. The data sources are shown in Sect. 2.2.4. Covering 4 primary indicators and 13 secondary indicators on

Table 2.1 Site selection methods

References	Region	Site selection areas	GIS	MCDM methods									
				AHP	FAHP	ANP	BWM	Entropy	CRITIC	TOPSIS	SAW	MOOSRA	MABAC
Waewsak et al. (2020)	Thailand	Biomass power plant	✓	✓									
Ali et al. (2019)	Thailand	Wind and solar plant	✓	✓									
Colak et al. (2020)	Turkey	PV power plant	✓	✓									
Razeghi et al. (2023)	Iran	Wind farm	✓	✓									
Waewsak et al. (2020)	Thailand	Power plant	✓		✓								
Wu et al. (2021)	China	Hydrogen					✓		✓				✓
Rahimi et al. (2020)	Iran	Solid waste	✓				✓					✓	
Hooshangi et al. (2023)	Iran	Solar farm	✓							✓			
Wu et al. (2019)	China	Biomass cogeneration					✓						✓
Shahdabadi et al. (2021)	Iran	Biomass power plant						✓		✓	✓		
Konstantinos et al. (2019)	Greece	Wind farm	✓	✓						✓			
Gigovic et al. (2017)	Serbia	Wind farm	✓			✓							✓

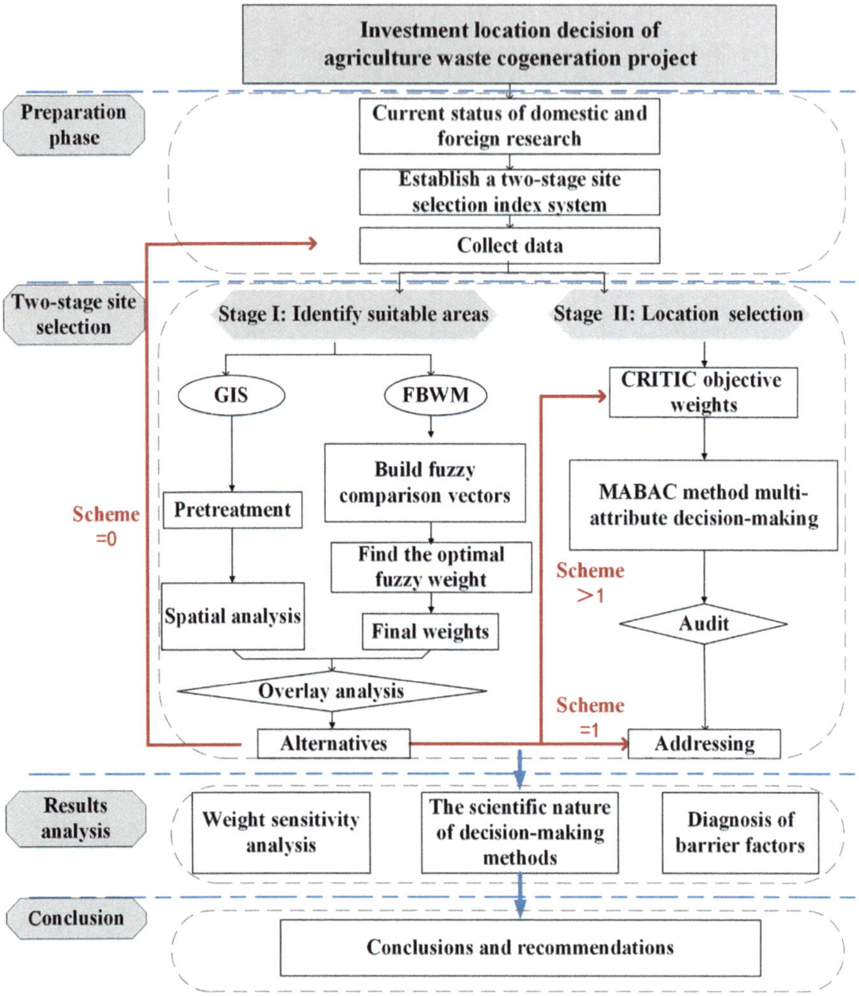

Fig. 2.1 Site selection of Agricultural Waste Cogeneration Projects

the supply side, demand side, social side, and environmental side of raw materials, the specific indicator system is shown in Table 2.2. In the location selection stage, the CRITIC method is used to calculate the weights of each indicator, and then the MABAC method is used to evaluate and rank each scheme. The point with the highest suitability level within the area is selected as the optimal project location. Then, the results are analyzed from three aspects: sensitivity analysis of weights, effectiveness of sorting methods, and diagnosis of obstacle factors, in order to verify the stability and effectiveness of the method and provide suggestions for optimizing the project location environment. Finally, conclusions are drawn. The process is shown in Fig. 2.1.

Table 2.2 Specific index and analysis methods for each stage

	Factors	Subfactors	Applicability analysis
Stage I: Identify suitable areas	Economic environment C_1	Road C_{11}	GIS spatial quantification
		Railway C_{12}	
		Water C_{13}	
		Power lines C_{14}	
	Geographic environment C_2	Slope C_{21}	
		Aspect C_{22}	
		Land use C_{24}	
	Location environment C_3	Residential C_{31}	
		Regional CO_2 emissions C_{32}	
Stage II: Location selection	Demand side C_4	Regional GDP C_{41}	Based on the results of the stage I evaluation, the MABAC comprehensive evaluation was used to determine the addressing scheme
		Power consumed C_{42}	
		Heating rate C_{43}	
	Supply side C_5	Agricultural waste resources C_{51}	
		Agricultural waste utilization rate C_{52}	
		Agricultural waste density C_{53}	
		Highway density C_{54}	
	Social side C_6	Government industrial policy C_{61}	
		Transportation cost C_{62}	
		Labor force C_{63}	
		Land price C_{64}	
	Environmental side C_7	Carbon emission pressure C_{71}	
		Carbon benefit C_{72}	

2.2.2 *Site Selection Model Construction*

This section first introduces the process of identifying suitable areas in order to screen out alternative solutions for the next stage. Next, the process of sorting alternative solutions is introduced. Finally, the best location for the project is selected.

2.2.2.1 Stage I: Identify Suitable Areas

In this section, GIS-FBWM technology is used to identify areas suitable for establishing Agricultural Waste Cogeneration Projects. The following introduces the process of GIS spatial analysis and FBWM weight calculation.

Steps of GIS Spatial Analysis

Step 1: Data collection. Raster or vector data are obtained from institutions such as the Geospatial Data Cloud, Henan Provincial Department of Transportation, Henan Electric Power Company, or other official websites. Refer to Sect. 2.2.4 (Data Source Section).

Step 2: Data preprocessing. Water systems are extracted from Digital Elevation Model (DEM) for slope and aspect analysis. Data such as highways, water, transmission lines, land use, residential areas, and regional CO_2 emissions are unified in coordinate system, and converting grids and cutting according to masks are preprocessed.

Step 3: GIS data spatial analysis. Data spatial analysis, such as Euclidean distance analysis, establishment of multilevel buffer analysis, and reclassification, is performed. Constraints and restrictions are considered, such as avoiding construction land and bad land. Agricultural waste power plants shall not be repeatedly built within a radius of 50 km, and other restrictions shall also be considered.

Step 4: FBWM technology is used to determine the weight of indicators. An expert team is established, including government officials in relevant fields, experts in the field of agricultural waste resource utilization, and leaders of agricultural waste energy resource utilization enterprises. Trigonometric fuzzy theory is introduced to calculate FBWM weights. For specific steps, see the FBWM method.

Step 5: AWCPs are suitable for regional initial screening models.

$$F = \sum_{i=1}^{n} W_i G_i \tag{2.1}$$

where F is the score of the comprehensive evaluation of regional suitability; W_i is the ith factor fuzzy weight; and G_i is the suitability scale score for the ith factor; n is the total number of factors.

Calculation Steps of FBWM Weights

FBWM has been successfully applied in many fields, such as prioritization of sludge-to-energy technologies (Liu et al. 2020), sustainable supplier selection (Fallahpour et al. 2021), green supplier evaluation and selection (Liou et al. 2021), and sustainability prioritization of energy systems (Xu 2020).

Step 1: The decision-making indicators are determined and denoted by $\{c_1, c_2, \cdots, c_n\}$, where n is the number of evaluation indicators.

Step 2: The best indicator C_B and the worst indicator C_W are determined by experts among the evaluation indicators.

Step 3: The preference of the best indicator over other indicators, as well as the preference of all indicators and the worst indicator, is determined. These preferences are converted into a trigonometric fuzzy function according to Table 2.1 to construct a fuzzy comparison vector $\tilde{A}_B = \left(\tilde{a}_{B1}, \tilde{a}_{B2}, \cdots \tilde{a}_{Bn}\right)$' $\tilde{A}_W = \left(\tilde{a}_{1W}, \tilde{a}_{2W}, \cdots \tilde{a}_{nW}\right)$. Here, $\tilde{a}_{Bj}$ represents the fuzzy preference of optimal indicator C_B compared to other indicators C_j. $\tilde{a}_{iW}$ indicates the degree of fuzzy preference of other indicator C_i compared to the worst indicator C_W. It should be noted in particular that $\tilde{a}_{BB} = (1,1,1), \tilde{a}_{WW} = (1,1,1)$.

Step 4: The optimal fuzzy weight $\left(\tilde{\omega}_1^*, \tilde{\omega}_2^*, \cdots \tilde{\omega}_n^*\right)$ is obtained using the objective programming model. The planning model is:

$$
\begin{aligned}
& \min \tilde{\xi} \\
& s.t. \\
& \left| \frac{\tilde{\omega}_B}{\tilde{\omega}_j} - \tilde{\omega}_{Bj} \right| \le \tilde{\xi} \\
& \left| \frac{\tilde{\omega}_j}{\tilde{\omega}_W} - \tilde{\omega}_{jW} \right| \le \tilde{\xi} \\
& \sum_{j=1}^{n} R\left(\tilde{\omega}_j\right) = 1 \\
& 0 \le l_j^W \le m_j^W \le u_j^W, j = 1, 2, \cdots, n
\end{aligned}
\tag{2.2}
$$

Here, $\tilde{\xi} = \left(l^{\xi}, m^{\xi}, u^{\xi}\right)$. Since $l^{\xi} \le m^{\xi} \le u^{\xi}$, let $\tilde{\xi}^* = \left(k^*, k^*, k^*\right), k^* \le l^{\xi}$ (Guo and Zhao 2017); Eq. (2.2) can be converted to (2.3).

$$
\begin{aligned}
& \min \tilde{\xi} \\
& s.t. \\
& \left| \frac{\left(l_B^W, m_B^W, u_B^W\right)}{\left(l_j^W, m_j^W, u_j^W\right)} - \left(l_{Bj}, m_{Bj}, u_{Bj}\right) \right| \le \left(k^*, k^*, k^*\right) \\
& \left| \frac{\left(l_j^W, m_j^W, u_j^W\right)}{\left(l_W^W, m_W^W, u_W^W\right)} - \left(l_{jw}, m_{jw}, u_{jw}\right) \right| \le \left(k^*, k^*, k^*\right) \\
& \sum_{j=1}^{n} R\left(\tilde{\omega}_j\right) = 1 \\
& 0 \le l_j^W \le m_j^W \le u_j^W, j = 1, 2, \cdots, n
\end{aligned}
\tag{2.3}
$$

The optimal fuzzy weight $\left(\tilde{\omega}_1^*, \tilde{\omega}_2^*, \cdots \tilde{\omega}_n^*\right)$ is obtained using solution formula (2.3).

Step 5: Eq. (2.4) is adopted to convert the weight represented by the triangular fuzzy into the exact value $R\left(\tilde{\omega}_j\right)$.

$$R\left(\tilde{\omega}_j\right) = \left(l_i + 4m_i + u_i\right) / 6 \quad i = 1, 2, \cdots, n \tag{2.4}$$

Step 6: A consistency check is conducted. The consistency ratio C_R of the FBWM method and the consistency index are shown in Table 2.3, Fig. 2.2

$$C_R = \xi^{l*} / C_I \tag{2.5}$$

2.2.2.2 Stage II: Location Selection

In the position selection stage, a weight calculation model based on the CRITIC method was first introduced, and then the MABAC technology was used for sorting.

Steps for Calculating the CRITIC Weight

CRITIC is an objective empowerment approach proposed by Diakoulaki. It comprehensively measures the importance of indicators based on the intensity of comparison and the conflict between indicators. The variability of indicators is taken into account, and meanwhile, the correlation between indicators is also considered. The objective attributes of the data itself are used for scientific evaluation. CRITIC has been successfully applied to wind farm site selection (Zhao et al. 2024), hospital site selection (Wang et al. 2022), and photovoltaic hydrogen production project site selection (Wu et al. 2021).

The usual steps for CRITIC are as follows:

Step 1: If the indicator dimensions are different, the evaluation index matrix needs to be standardized. Cost-based indicators are converted into benefit-based indi-

Table 2.3 Transformation rules and consistency indicators of fuzzy language variables

Fuzzy linguistic terms	Membership function	C_I
Equally important (EI)	(1, 1, 1)	3.00
Weakly important (WI)	(2/3, 1, 3/2)	3.8
Fairly important (FI)	(3/2, 2, 5/2)	5.29
Very important (VI)	(5/2, 3, 7/2)	6.69
Absolutely important (AI)	(7/2, 4, 9/2)	8.04

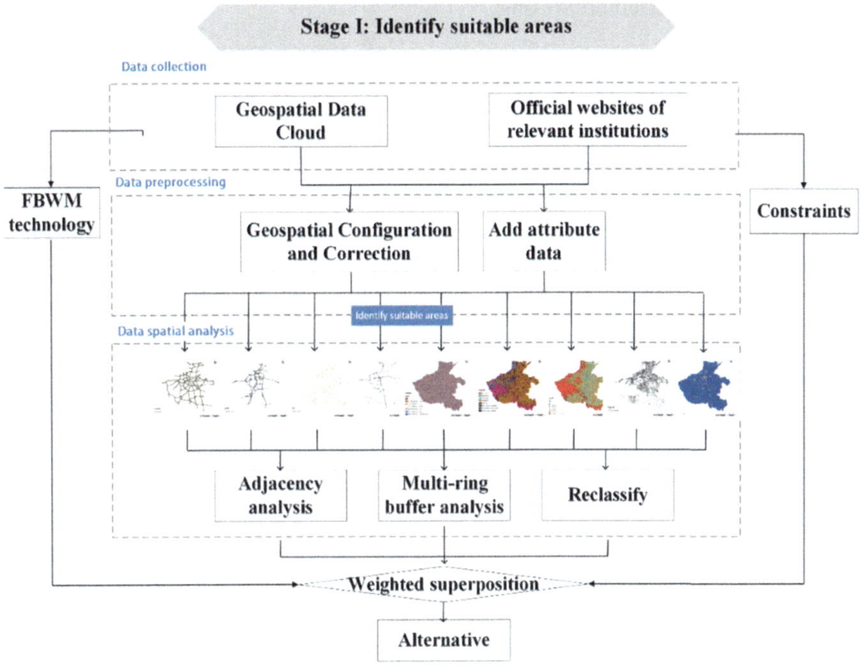

Fig. 2.2 Data analysis flowchart

cators by taking the reciprocal. The proportion of the ith scheme under the jth indicator is calculated as follows:

$$z_{ij} = \frac{y_{ij}}{\sqrt{\sum_{i=1}^{m} y_{ij}^2}} \quad i = 1, \cdots, m; j = 2, \cdots, n \tag{2.6}$$

Step 2: The amount of information C_i for the ith indicator is calculated.

$$C_i = \sigma_i \sum (1 - r) \tag{2.7}$$

Here, $\sum(1 - r)$ indicates indicator conflict, and r represents the correlation coefficient between indicators. Variability is expressed by standard deviation σ_i. The larger the C_i, the more important the ith indicator.

Step 3: The objective weight of the ith indicator is calculated.

$$\alpha_i = \frac{C_i}{\sum_{i=1}^{n} C_i} \tag{2.8}$$

Steps for MABAC Method Sorting

MABAC was first proposed in 2015. This method takes into account potential losses and returns, which is simple to calculate and has strong stability in the results. Jiang et al. (2022) used the MABAC method to select suppliers. Rahim et al. (2020) applied MABAC sorting technology and found that "biomass energy" is the best solution for sustainable energy. Wu et al. (2021) applied MABAC to determine the sequencing of biomass power plants. Wu et al. (2019) applied MABAC sorting technology to the field of site selection of biomass cogeneration projects.

The calculation steps of MABAC sorting technology are as follows:

Step 1: Standardization. The calculation method is the same as the first step of calculating weights of CRITIC.

Step 2: After the matrix is standardized, a weighted decision matrix $\hat{Z} = \left[\hat{z}_{ij}\right]_{mn}$ is constructed. Here, $\hat{z}_{ij}$ is obtained by formula (2.9). α_j is the weight of attribute j.

$$\hat{z}_{ij} = \alpha_j \left(1 + z_{ij}^*\right) \tag{2.9}$$

Step 3: Determine the approximate boundary region matrix G. For each criterion, the boundary approximation region matrix and its corresponding elements are defined by formula (2.10), which forms the boundary approximation region matrix G.

$$g_j = \left(\prod_{i=1}^{m} \hat{z}_{ij}\right)^{1/m} \tag{2.10}$$

$$\begin{matrix} c_1\, c_2\, c_3 \cdots c_n \\ G = \left[g_1\, g_2\, g_3 \cdots g_n\right] \end{matrix} \tag{2.11}$$

Step 4: Calculate the distance Q between the alternative solution and the approximate boundary area. The distance between the candidate point and the boundary approximation region is expressed by the distance between the elements in the weighted gauge matrix and the boundary approximation matrix G, as shown in formulas (2.12) and (2.13).

$$q_{ij} = \hat{z}_{ij} - g_j, i = 1, 2, \cdots, m; j = 1, 2, \cdots, n. \tag{2.12}$$

$$Q = \hat{Z} - G = \begin{bmatrix} \hat{z}_{11} - g_1\ \hat{z}_{12} - g_2 \cdots \hat{z}_{1n} - g_n \\ \hat{z}_{21} - g_1\ \hat{z}_{22} - g_2 \cdots \hat{z}_{2n} - g_n \\ \cdots \quad \cdots \quad \cdots \quad \cdots \\ \hat{z}_{m1} - g_1\ \hat{z}_{m2} - g_2 \cdots \hat{z}_{mn} - g_n \end{bmatrix} \tag{2.13}$$

Step 5: Calculate the total distance D_i from each scheme to the approximate boundary area, as shown in eq. (2.14). The solution with the highest final value is the optimal solution (Fig. 2.3).

$$D_i = \sum_{j=1}^{n} q_{ij}, i = 1, 2, \cdots, m. \tag{2.14}$$

2.2.3 Obstacle Degree Model

On the basis of two-stage site selection, this study examines the obstacle factors that affect the sustainability of AWCP site selection schemes. This chapter uses an obstacle degree model to diagnose the obstacle factors of each alternative plan (Gao et al. 2022). The diagnostic results provide direction for optimizing the site selection environment for each plan.

$$\sigma_j = \frac{\left(1 - \upsilon_{ij}\right) * \rho_j}{\sum_{j=1}^{n} \left(1 - \upsilon_{ij}\right) * \rho_j} \tag{2.15}$$

where υ_{ij} is the normalized standard value; $1 - \upsilon_{ij}$ is the deviation degree of the indicator; $\rho_i = \omega_j$, ρ_i is the factor contribution degree, and ω_j is the weight of the ith indicator; σ_j is the specific obstacle level of indicator j, and the sum of obstacle levels for all indicators in each region is 1.

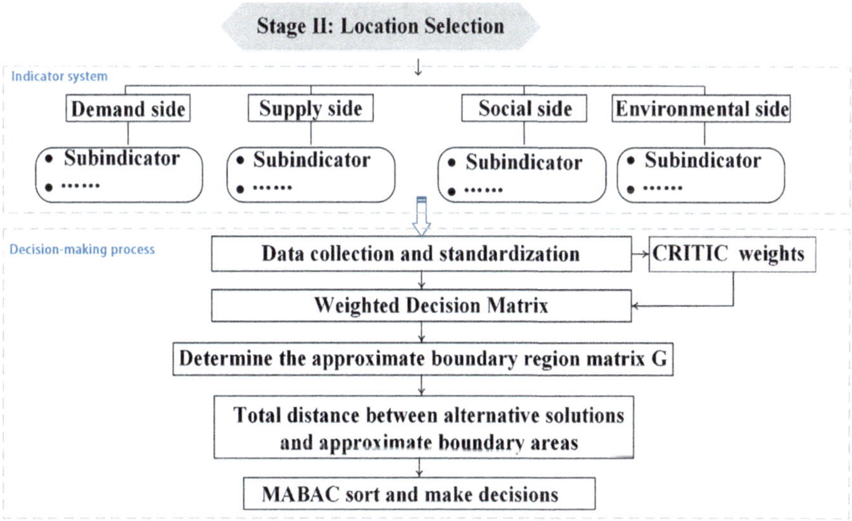

Fig. 2.3 Location selection flowchart

2.2.4 *Data Sources*

The data mainly come from field research, statistical yearbooks, the Geospatial Data Cloud, the national base geodatabase, and official agency websites involved.

2.2.4.1 Geographic Layer Data Source

The Digital Elevation Model (DEM) of Henan Province (30 M resolution) and the European Space Agency (ESA) total feature vector data of land use in 2020 (10 M resolution) come from the Geospatial Data Cloud. Data on road, railway, power lines, water, residential areas, and carbon emissions are mainly collected by relevant official institutions. The specific sources of data are shown in Table 2.4.

2.2.4.2 Data Sources for the Selected Stages of AWCPs

Quantitative data are mainly compiled by consulting relevant statistical data and are shown in Table 2.5. Qualitative indicators are obtained through expert scoring. In view of the specific situation of each city, the average expert scoring is adopted.

2.3 Case Study on Site Selection of Agricultural Waste Power Generation Supply Chain Nodes

This case study aims to verify the effectiveness and feasibility of the site selection model. This section first introduces the research area and then introduces spatial geographic data preprocessing.

2.3.1 *Study Area*

Henan Province is a major agricultural province in China, located between 31°23′ to 36°22′ north latitude and 110°21′ to 116°39′ east longitude. This region has a large population base, flat terrain, convenient transportation, fertile land, and abundant water resources. In 2021, the total crop planting area in the province was 14.71 million hectares, and the total grain planting area was 10.78 million hectares, of which wheat and corn accounted for approximately 89% of the total grain planting area in the province. Therefore, the Henan region is rich in agricultural waste and

Table 2.4 Geographic layer data source

Subcriteria	Data sources	Notes	Type	References
Road C_{11}	Henan provincial department of transportation	Distance range to road: 0.1–8 km	Vector data	Al Garni and Awasthi (2017), Chukwuma et al. (2021)
Railway C_{12}	Henan provincial department of transportation	Distance range to railway: 0.1–8 km	Vector data	Al Garni and Awasthi (2017)
Water C_{13}	DEM extraction	Suitable water intake distance: 0.5–6 km	Vector data	Waewsak et al. (2020)
Power lines C_{14}	Henan electric power company	Distance range to power lines: 0.1–25 km	Vector data	Al Garni and Awasthi (2017), Chukwuma et al. (2021)
Slope C_{21}	National base geodatabase	Slope: 0–10%	Raster data	Al Garni and Awasthi (2017), Chukwuma et al. (2021)
Aspect C_{22}	The geospatial data cloud	Slope direction range: East, west, southeast, southwest, and south	Raster data	Al Garni and Awasthi (2017)
Land use C_{24}	The geospatial data cloud	Tree cover, shrubland, grassland, cropland, and bare/sparse vegetation	Raster data	Al Garni and Awasthi (2017), Chukwuma et al. (2021)
Residential C_{31}	DEM extraction	Distance range to residential: 1–6 km	Vector data	Chukwuma et al. (2021)
Regional CO_2 emissions C_{32}	https://ngdc.noaa.gov/eog/download.html	The higher the light intensity, the greater the carbon dioxide emissions and energy consumption	Raster data	Waewsak et al. (2020)

has broad development space for agricultural waste utilization. The application of the AWCPs two-stage site selection framework proposed in this chapter in Henan Province is as follows (Fig. 2.4).

2.3.2 Spatial Geographic Data Preprocessing

For this work, nine different map layers were prepared for use in the GIS database. The data sources of the following geographic layers are shown in Table 2.4. ArcGIS software was used to perform spatial registration and correction of data with different coordinates. All spatial data were converted into the China Lambert Conformal Conic coordinate system and integrated into a geographic database for use in ArcGIS 10.8. The following content first introduces the evaluation criteria for nine spatial geographic layers, then clarifies the limitations for selecting agricultural

Table 2.5 Statistical data sources

Subcriteria	Data sources	Notes	Type
Regional GDP (C_{41})	2019 Statistical Yearbook of Henan Province	The higher the GDP, the larger the market for the project	Quantitative (benefit)
Power consumed (C_{42})	2019 Statistical Yearbook of Henan Province	Areas with high electricity demand also have a large market for projects	Quantitative (benefit)
Heating rate (C_{43})	Henan Province's 13th Five-Year Plan for Electric Power Development	The lower the heating rate, the larger the market for clean heating	Quantitative (cost)
Agricultural waste resource (C_{51})	2019 Statistical Yearbook of Henan Province; Henan Province's 13th Five-Year Plan for Energy Development	Indicates raw material supply potential	Quantitative (benefit)
Agricultural waste utilization rate (C_{52})	Henan Province's 12th Five-Year Plan for Comprehensive Utilization of Crop Agricultural Waste	Reflects the competition in agricultural waste utilization	Quantitative (cost)
Agricultural waste density (C_{53})	2019 Statistical Yearbook of Henan Province; Henan Province's 13th Five-Year Plan for Energy Development	The higher the density of agricultural waste, the more secure the supply of resources	Quantitative (benefit)
Highway density (C_{54})	2019 Statistical Yearbook of Henan Province	Convenient transportation can reduce transportation costs	Quantitative (benefit)
Government industrial policy (C_{61})	Development and Reform Committees of Henan Province and Various Cities	Measures the government's support for the project	Qualitative (benefit)
Transportation cost (C_{62})	2019 Statistical Yearbook of Henan Province	Reflects transportation costs	Quantitative (cost)
Labor force (C_{63})	2019 Statistical Yearbook of Henan Province	Measures labor costs during collection and handling	Quantitative (cost)
Land price (C_{64})	Official website	Reflects differences in acquisition costs across regions	Quantitative (cost)
Carbon emission pressure (C_{71})	2019 Statistical Yearbook of Henan Province	Reflects the intensity of regional carbon emissions	Quantitative (benefit)
Carbon benefit (C_{72})	2019 Statistical Yearbook of Henan Province	Reflects regional carbon emission reduction preferences	Quantitative (benefit)

waste cogeneration sites, as shown in Table 2.6, and finally presents the preprocessing results of spatial geographic data in Fig. 2.5.

Railway and road networks: The condition of the road network is an important criterion for project investment. Convenient transportation can reduce transportation costs and is economically feasible.

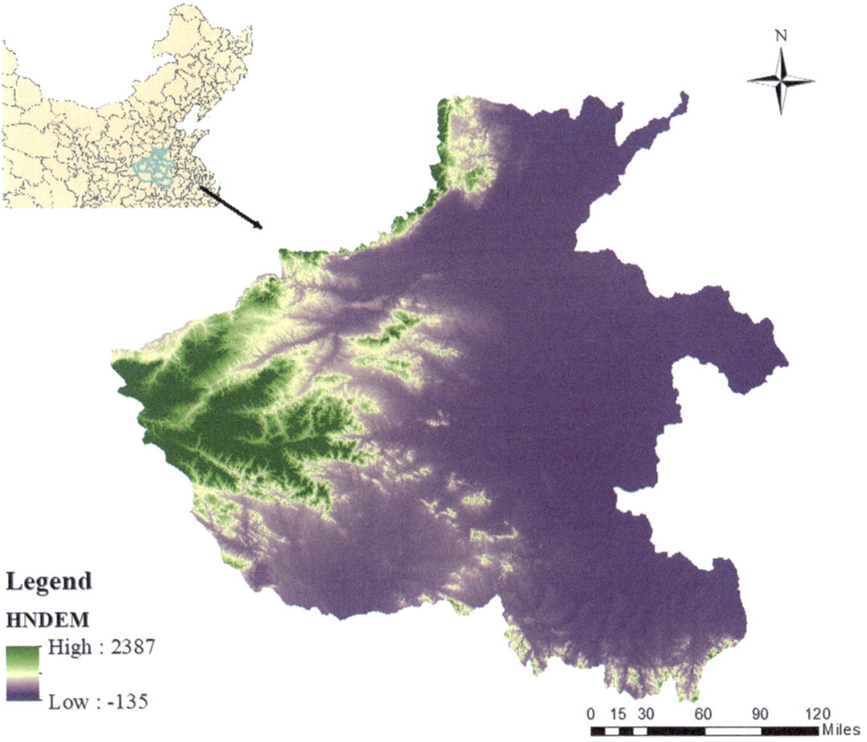

Fig. 2.4 Henan province DEM

Water system: The distance from the water source is 500–6000 m, which is a reasonable distance for project water use.

Power lines: A reasonable range of distance from the transmission line is considered. This helps reduce energy loss and reduce costs.

Slope and aspect: The DEM of Henan Province is used to generate slope and aspect. Flat areas can effectively reduce the construction cost of AWCPs. The slope of the site should not exceed 10%. A slope facing south and less than 10° is the ideal orientation for Agricultural Waste Cogeneration Projects.

Land use: Land use data are one of the most important criteria for project site selection. The site selection of agricultural waste cogeneration plants should avoid built-up areas and unfavorable land, and there should be no repeated construction within the influence range of AWCPs. AWCPs should occupy wasteland as much as possible and reduce or eliminate the use of arable land, forest land, and other areas that need to be protected. Long-term needs should be considered, and room should be appropriately left for future development.

Residential areas: The future development trend of residential areas should be considered, and necessary buffer zones should be reserved to prevent occupation of future residential areas. To ensure that energy demand and economic costs are met, agricultural waste cogeneration plants should be located near rural residential areas.

Table 2.6 Factors and restrictions for agricultural waste cogeneration site selection

Factors	Subfactors	Points	Factors	Subfactors	Points
Roads (km)	0–0.1	0	Railway (km)	0–0.1	0
	0.1–2	10		0.1–2	10
	2–4	8		2–4	8
	4–6	6		4–6	6
	6–8	4		6–8	4
	>8	0		>8	0
Land use	Tree cover	6	Aspect	North	0
	Shrubland	10		Northeast	0
	Grassland	10		Northwest	0
	Cropland	4		East	1
	Built-up	0		West	1
	Bare/sparse vegetation	10		Southeast	6
	Permanent water bodies	0		Southwest	6
	Herbaceous wetland	0		South	10
Slope (%)	0–2	8	Power lines (km)	0–0.1	0
	2–4	10		0.1–5	10
	4–6	8		5–10	8
	6–8	6		10–15	6
	8–10	2		15–20	4
	>10	0		20–25	2
				> 25	0
Water (km)	0–0.5	0	Residential (km)	0–1	0
	0.5–2	10		1–2	4
	2–4	8		2–3	10
	4–6	6		3–4	8
	>6	0		4–5	6
				>5	0
Restricted areas					
Slope > 10%					
Road distance: 0–0.1 km					
Residential area distance: 0–0.5 km					
Biomass power plants: No construction allowed within a radius of 50 kilometers					
Adverse land: Avoid wetlands, water bodies, and buffer zones					
Construction land: Avoid land and buffer zones covered by buildings, highways, and railways					

CO_2 emissions: These data are spatial grid data of carbon dioxide emissions from fossil fuel combustion. Carbon dioxide emissions are calculated based on the regional energy balance table in the China Energy Statistical Yearbook. Carbon emission grid data for Henan Province are further pruned, synthesized, and extracted. In order to alleviate the impact of global climate change, regions with higher regional carbon emission intensity need to improve the utilization rate of agricultural waste.

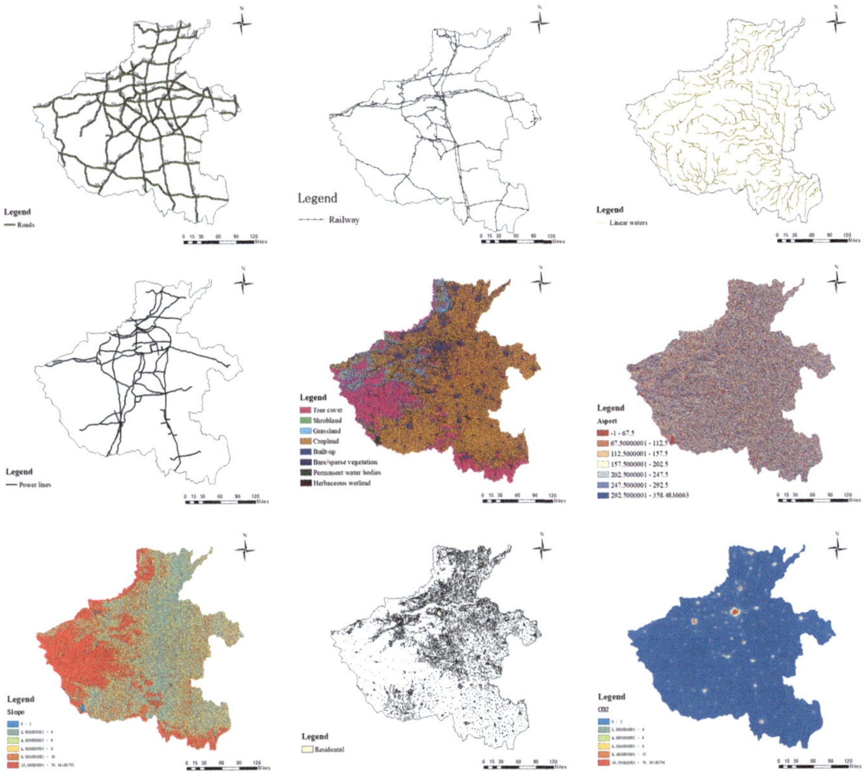

Fig. 2.5 Preprocessed geographic map layers

2.4 Results of Two-Stage Site Selection for Agricultural Waste Power Generation Supply Chain Nodes

First, GIS 10.8 is applied to identify areas suitable for the construction of Agriculture Waste Cogeneration Projects and extract alternative solutions. In the second stage, the solutions are sorted and the positioning results are obtained. The calculation results are as follows.

2.4.1 Stage I: Identify Suitable Areas

In order to determine the suitable area for AWCPs in Henan Province, the geographic layers are preprocessed and subjected to GIS spatial analysis, using the FBWM method to determine the weight of indicators. These grid buffer surfaces are weighted and overlaid for analysis in ArcGIS 12.8 to obtain an AWCPs regional suitability index map.

2.4.1.1 GIS Spatial Analysis

After map layer preprocessing, spatial analysis is performed such as multi-ring buffer analysis, neighbor analysis, and reclassification, as shown in Fig. 2.6. The different radiation ranges of each indicator are rated, and the values ranging from 0 to 10 are assigned. The radiation range rating of each index during spatial analysis is shown in Table 2.6.

2.4.1.2 Fuzzy Best-Worst Method Weights Calculation

Weights are calculated using the Fuzzy Best-Worst Method (FBWM) as detailed in Sect. 2.2.2.1. First, the importance membership function of each indicator is obtained based on expert scoring, as detailed in Appendix A–E. Expert language is converted into triangular fuzzy numbers. The nonlinear programming model is constructed by formulas (2.2) and (2.3), and the optimal fuzzy weight is obtained by solving it. The Graded Mean Integration Representation (GMIR) method in formula (2.4) is used to convert the fuzzy weights into accurate numerical values. Among them, the optimal weight for the first-level indicator is $\tilde{\omega}_1 = 0.5020, \tilde{\omega}_2 = 0.3564, \tilde{\omega}_3 = 0.1417$.

Here, $C_I = 6.69$, $C_R = \frac{0.5505}{6.69} = 0.0829 < 0.1$; consistency check is passed. The exact fuzzy weight values of the secondary indicators are calculated in the same way. The total weight is obtained as shown in Table 2.7.

2.4.1.3 Weighted Overlay Analysis and Extraction of Alternative Solutions

Because raster data can perform numerical and weighting operations on input and output data, vector buffer data are converted into raster format. Formula (2.1) is used to weight and overlay the map layers. Due to the inability to repeat the construction of the completed Agricultural Waste Cogeneration Project within a radius of 50 km, the affected area of the existing projects is erased. A suitability index map is generated, divided into high suitability areas (value above 7), general suitability areas (value between 4 and 7), and unsuitable areas (value between 0 and 4), as shown in Fig. 2.7. The initial screening plan includes nine regions: Luoyang (S1), Anyang (S2), Xinxiang (S3), Sanmenxia (S4), Nanyang (S5), Shangqiu (S6), Xinyang (S7), Zhoukou (S8), and Zhumadian (S9).

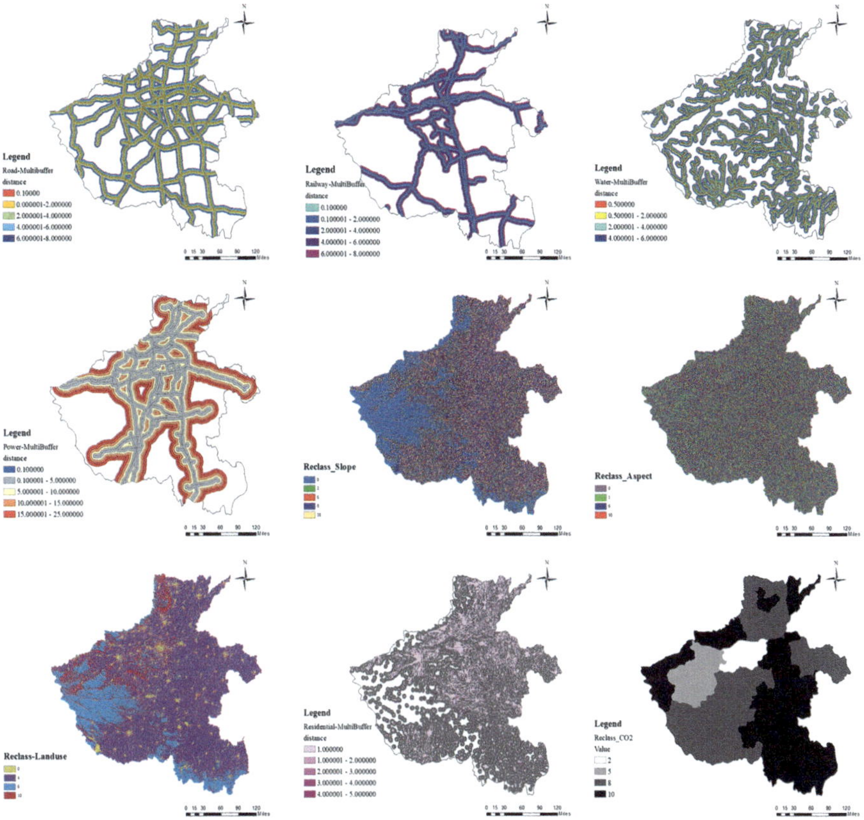

Fig. 2.6 GIS spatial analysis map

2.4.2 Stage II: Location Selection

In the location selection stage, according to the second-stage location selection model in Sect. 2.2.2.2, the CRITIC objective weights of the indicators are first calculated, then the solutions are sorted using MABAC technology, and finally addressed.

2.4.2.1 Criteria Importance Through Intercriteria Correlation Weights Calculation

Relevant data on the supply side, demand side, social side, and environmental side of alternative solutions are collected. The data sources are shown in Table 2.5. Among them, Government industrial policy C_{61} is a qualitative indicator. Domain experts, government officials, and business leaders with solid theoretical knowledge or practical experience are invited for scoring, as shown in Appendix A. The average

Table 2.7 Fuzzy weights and consistency testing

Criteria	$\tilde{\omega}^*$	$R(\tilde{\omega})$	Subcriteria	$\tilde{\omega}^*$	$R(\tilde{\omega})$	W_t
C1	(0.4349,0.5073,0.5477)	0.502	C11	(0.3588,0.3716,0.4324)	0.3796	0.190
			C12	(0.1347,0.1368,0.1623)	0.1407	0.070
			C13	(0.1550,0.1755,0.2006)	0.1763	0.088
			C14	(0.2873,0.2895,0.3755)	0.3035	0.152
C2	(0.2804,0.3500,0.4580)	0.356	C21	(0.2356,0.2400,0.2962)	0.2486	0.088
			C22	(0.2871,0.3126,0.4185)	0.3260	0.116
			C23	(0.4057,0.4073,0.5176)	0.4254	0.151
C3	(0.1356,0.1429,0.1429)	0.141	C31	(0.4932,0.4932,0.4932)	0.4932	0.069
			C32	(0.3288,0.4932,0.7397)	0.5068	0.071

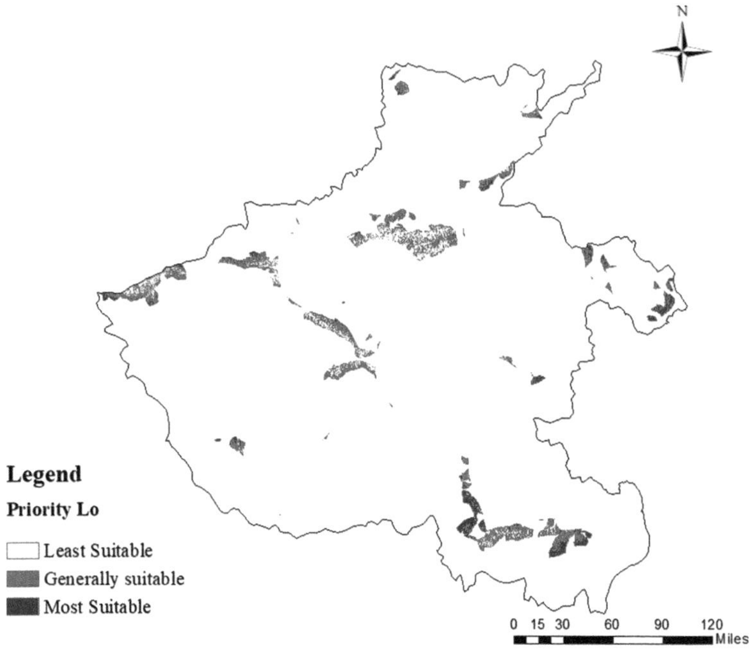

Fig. 2.7 Suitability index results using FBWM weights

score is taken, with a standard of 0–100; 0 indicates that the government has not adopted any supportive policies, and 100 represents strong government support. Qualitative data are converted into quantitative data, and the data are standardized using formula 2.6. The CRITIC weights of each indicator are calculated using formulas (2.7) and (2.8). The standardized matrix and the weight calculation results are shown in Table 2.8.

Table 2.8 Standardized matrix and CRITIC weights

Scheme	Criteria												
	C41	C42	C43	C51	C52	C53	C54	C61	C62	C63	C64	C71	C72
α^T	0.0612	0.1184	0.0563	0.0830	0.0412	0.1599	0.1310	0.0346	0.0208	0.0284	0.0496	0.1221	0.0933
S1	0.5328	0.6455	0.2464	0.1220	0.2982	0.0767	0.2513	0.3288	0.3517	0.2837	0.3036	0.1001	0.1605
S2	0.2359	0.3073	0.2156	0.2223	0.2319	0.1929	0.2281	0.3493	0.3203	0.2502	0.2753	0.0871	0.2467
S3	0.3088	0.3810	0.2464	0.2747	0.2982	0.3515	0.3505	0.3699	0.2584	0.3206	0.1655	0.1485	0.4280
S4	0.1528	0.1703	0.2875	0.0360	0.3479	0.0102	0.0582	0.2466	0.2927	0.2425	0.2838	0.1938	0.1040
S5	0.4037	0.3547	0.3450	0.3955	0.3479	0.1024	0.2103	0.3904	0.3775	0.4355	0.3771	0.1987	0.4280
S6	0.3081	0.2594	0.3450	0.4328	0.2982	0.1331	0.1549	0.3082	0.3152	0.3546	0.4671	0.1872	0.3715
S7	0.2919	0.1988	0.4928	0.1010	0.4175	0.0155	0.0832	0.2466	0.3527	0.3650	0.2926	0.3120	0.2734
S8	0.3385	0.1831	0.4312	0.5279	0.2982	0.8762	0.8072	0.3082	0.3368	0.3594	0.3515	0.8069	0.1456
S9	0.2902	0.2205	0.2875	0.4758	0.4175	0.1927	0.1783	0.4110	0.3762	0.3429	0.3931	0.3159	0.5498

2.4.2.2 Ranking

A weighted matrix $\hat{Z}$ is constructed from formula (2.9). Formulas (2.10) and (2.11) are used to determine the approximate boundary area of each indicator and obtain the approximate boundary area G. Then, formulas (2.12)–(2.14) are used to calculate the total distance between the alternative solution and the approximate boundary area. The D values and final ranking results of each alternative solution are shown in Table 2.9.

The results presented in Table 2.9 show that Zhoukou (S8), ranked first, is the most suitable location for AWCPs. The evaluation value of Zhoukou is 0.2536, which is the optimal plan. Zhumadian (S9) is a suboptimal solution, while Sanmenxia (S4) is the least suitable option.

2.4.3 *Addressing*

Based on the comprehensive consideration of regional suitability index and scheme ranking, this study further narrows the scope of factory construction. The results show that the rose-red marked area in Zhoukou City is the most suitable area. Therefore, the Agricultural Waste Cogeneration Project should be located in Xiangcheng County, Zhoukou City, as shown in Fig. 2.8.

Zhoukou City has the highest agricultural waste density in the province, reaching 29162.51 tons/square kilometer, which ensures the supply of raw materials for AWCPs in Zhoukou City. The highway network has the highest density and convenient transportation, which is conducive to reducing transportation costs. Labor costs during project operation are relatively low. At the same time, the heating rate in Zhoukou City is significantly lower than in other schemes, and the construction of AWCPs can effectively improve resource utilization and increase heating rate. Based on the above favorable conditions, constructing an Agricultural Waste Cogeneration Project in Zhoukou City is a good decision.

Table 2.9 MABAC method scheme sorting table

Scheme	D value	Sort
S1	−0.0130	6
S2	−0.0477	7
S3	0.0361	3
S4	−0.1313	9
S5	0.0213	4
S6	−0.0058	5
S7	−0.0601	8
S8	0.2536	1*
S9	0.0396	2

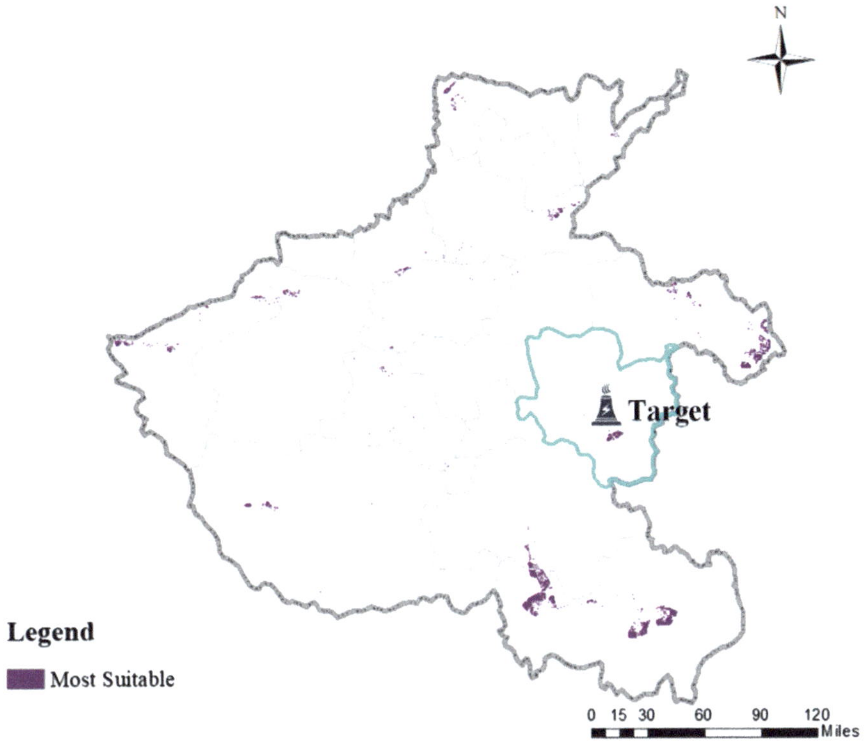

Fig. 2.8 Project final target

Additionally, the site selection results are consistent with the energy development plan of Henan Province during the 13th Five-Year Plan period. This also verifies the accuracy of the decision-making framework in this chapter.

2.5 Discussion on the Two-Stage Site Selection Results of Agricultural Waste Power Generation Supply Chain Nodes

The results were analyzed from three perspectives: sensitivity analysis of geographic layer weights on regional suitability, effectiveness of sorting methods, and diagnosis of obstacle factors affecting site selection.

2.5.1 Weight Sensitivity Analysis

Sensitivity analysis is mainly used to test the sensitivity and stability of conclusions. Indicator weights are used to measure the relative importance of indicators. Weight changes may alter the ranking results of alternative solutions. Therefore, it is necessary to study the sensitivity of indicator weights. If the sensitivity of alternative scheme extraction to indicator weights is low, it indicates that the decision result is stable.

This chapter conducts sensitivity analysis on the initial screening results by changing the weights of the primary indicators in the initial selection stage, as shown in Table 2.10. The results in Fig. 2.9 indicate that the suitability index of the region changes slightly with the weights of the indicators, which hardly changes the extracted alternative solutions. Therefore, the initial screening results are relatively stable.

2.5.2 Effectiveness of Sorting Methods

The results of comparative analysis of methods show that the optimal solutions selected by SAW method, MOOSRA method, and MABAC method are all S_8, as shown in Table 2.11.

At the same time, the decision-making η of MABAC method is up to 84.4%, which is significantly higher than that of SAW and MOOSRA. Among them, $S_{\max}$ is the evaluation value of the optimal solution, S_{sec} is the evaluation value of the sub-optimal solution, and η is the degree of differentiation of scheme selection, as shown in Table 2.12. Therefore, MABAC ranking is more stable, scientific, and reasonable.

2.5.3 Obstacle Factor Diagnosis and Suggestions

The obstacle level of criteria and subcriteria is calculated using formula 2.15, and the results are shown in Table 2.13. Obstacle factors are identified from the criteria, subcriteria, and scheme levels.

Table 2.10 Sensitivity analysis of weights

	Cluster weights		
Scenarios	Economy	Geographic factors	Location
A-uniform weights	0.3333	0.3333	0.3333
B-priority Ec	0.5020	0.3564	0.1417
C-priority Ge	0.1417	0.5020	0.3564
D-priority Lo	0.3564	0.1417	0.5020

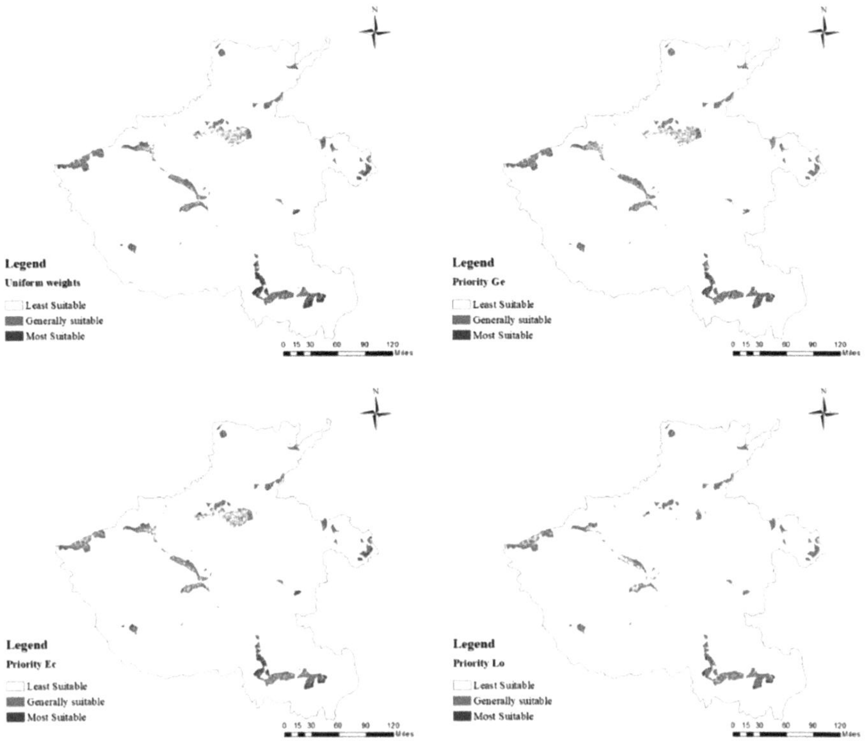

Fig. 2.9 FBWM weight sensitivity analysis

Table 2.11 Comparative analysis of methods

Scheme	MABAC	SAW	MOOSRA
S1	6	6	6
S2	7	7	8
S3	3	3	4
S4	9	9	9
S5	4	4	3
S6	5	5	5
S7	8	8	7
S8	1	1	1
S9	2	2	2

2.5.3.1 Criteria Layer Diagnostic Results

Overall, the obstacle level on the environmental side is the highest, at 10.84%. The supply side obstacle level is second, at 10.58%. The demand side obstacle level is third, at 7.74%. The social side obstacle level is the lowest, at 3.20%.

Table 2.12 Differentiation of scheme selection

Methods	S_{max}	S_{sec}	$S_{max} - S_{sec}$	η
MABAC	0.2536	0.0396	0.2140	84.4%
SAW	0.5261	0.3121	0.2140	40.7%
MOOSRA	5.6317	2.6762	2.9555	52.3%

Table 2.13 Criteria and subcriteria layer obstacle degree (%)

	Demand side			Supply side				Social side				Environmental side	
Scheme	C41	C42	C43	C51	C52	C53	C54	C61	C62	C63	C64	C71	C72
S1	3.86	5.67	5.73	9.84	3.91	19.94	13.25	3.14	1.82	2.75	4.67	14.84	10.58
S2	6.03	10.58	5.70	8.33	4.08	16.65	13.05	2.91	1.82	2.75	4.64	14.38	9.07
S3	6.12	10.60	6.14	8.71	4.18	15.00	12.31	3.15	2.23	2.79	5.99	15.04	7.72
S4	6.04	11.44	4.67	9.32	3.13	18.43	14.37	3.04	1.71	2.51	4.14	11.47	9.74
S5	5.17	10.82	5.22	7.11	3.81	20.33	14.65	2.99	1.83	2.27	4.38	13.86	7.56
S6	5.78	11.96	5.03	6.42	3.94	18.91	15.10	3.27	1.94	2.50	3.61	13.54	8.00
S7	5.50	12.05	3.63	9.48	3.05	19.99	15.25	3.31	1.71	2.29	4.46	10.67	8.61
S8	8.55	20.42	6.76	8.27	6.10	4.18	5.33	5.05	2.91	3.84	6.79	4.98	16.83
S9	6.32	13.42	5.83	6.33	3.49	18.77	15.65	2.96	1.89	2.71	4.38	12.14	6.11
$\bar{\sigma}$	5.93	11.89	5.41	8.20	3.97	16.91	13.22	3.31	1.99	2.71	4.78	12.32	9.36

2.5.3.2 Subcriteria Layer Diagnostic Results

The Subcriteria that affect the sustainable development of agricultural waste cogeneration include electricity consumption C_{42}, agricultural waste density C_{53}, road density C_{54}, and carbon emission reduction pressure C_{71}. Their average obstacle levels are all greater than 10%, which are 11.89%, 16.91%, 13.22%, and 12.32%, respectively, as shown in Table 2.13.

2.5.3.3 Scheme Layer Diagnostic Results

Specifically, Zhoukou City, as the optimal site selection plan, has the advantages of high agricultural waste density, dense road network, lower transportation costs, relatively low labor costs, and low heating rates. However, compared to other cities, Zhoukou City has a maximum obstacle level of 20.42% in electricity consumption C_{42}, followed by a barrier of 8.55% in GDP of C_{41}. Agricultural waste resource C_{51} is the main obstacle factor on the supply side in Zhoukou City. Land price C_{64} is a weakness on the social side, and reducing land price is a key indicator for improving the site selection environment on the social side. Carbon benefit C_{72} is the main obstacle factor on the environmental side of Zhoukou City. According to Table 2.13, other solutions can also be analyzed and are omitted here.

Based on the above obstacle factor analysis, the main paths for improving the environmental sustainability of AWCPs site selection in Henan Province in the future include: promoting clean production in accordance with the requirements of "output efficiency and ecological sustainability," accelerating road network construction and reducing transportation costs, increasing planting density, improving farmland utilization rate, and ensuring supply. The specific suggestions are as follows:

To further improve the sustainability of the site selection environment, the following suggestions are proposed:

1. The supply side is the primary factor affecting Agricultural Waste Cogeneration Projects. On the one hand, the supply of agricultural waste raw materials through land consolidation is increased (Edrisi et al. 2022). On the other hand, the convenience of agricultural waste cogeneration is relatively low. The low density and scattered distribution of agricultural waste can easily cause transportation difficulties. By optimizing the road network and improving the convenience of transportation, agricultural waste supply can be ensured, which in turn affects the operational efficiency of the project.
2. Electricity consumption is the "wind vane" of economic operation and a characterization of industrial structure optimization and upgrading. Affected by the epidemic, the growth rate of electricity consumption in various regions is relatively low. With the help of expanding domestic demand policies, the recovery of electricity consumption is driven and the ability of demand side to adjust resources is leveraged. A supply-demand friendly interactive system is built to promote the sustainability of AWCPs.
3. According to the requirements of sustainable development theory, the greater the pressure on regional carbon reduction, the higher the preference for clean energy projects. Therefore, the government can effectively regulate corporate behavior and improve the carbon trading market system by imposing carbon taxes. It can also encourage the development of renewable energy projects, promote industrial and energy structure adjustments, and reduce carbon dioxide emissions.

2.6 Conclusion and Limitations

Drawing on relevant research on site selection for other renewable energy projects, the main contribution of this study is to present an integrated method suitable for site selection of Agricultural Waste Cogeneration Projects. Preliminary screening stage: The regional suitability index is visualized using GIS weighted overlay analysis based on FBWM weights. This process fully utilizes spatial data to improve the accuracy of selecting alternative solutions. Addressing stage: Based on four aspects—supply side, demand side, social side, and environmental side—the CRITIC-MABAC model is used to sort each alternative area. The site selection framework has been verified to have strong operability and stable site selection results through examples. During the diagnosis process of obstacle factors, it was found that electricity consumption C_{42}, agricultural waste density C_{53}, highway density C_{54}, and carbon emission reduction pressure C_{71} are the main obstacle factors for site selection of various cities. Targeted suggestions are provided to optimize the site selection environment and enhance project sustainability. The research conclusion of this chapter theoretically compensates for the shortcomings of relevant research fields, and, in practice, it also has certain guiding significance for reasonable site selection.

The diagnosis of obstacle factors for each scheme belongs to static diagnosis, which is a limitation of this study. Therefore, in the following research, a dynamic management system for obstacle factors will be established to control and manage the obstacle factors that affect site selection and achieve continuous improvement of the site selection environment through inspection and management.

Acknowledgments This chapter has been reprinted from J. Gao, C. Wang, Z Wang, et al. *Site selection decision for biomass cogeneration projects from a sustainable perspective: A case study of China*. Energy, 2024, 286:129518. Copyright (2024), with permission from Elsevier.

Appendices

Appendix 1: Experts Group

Sample	Quantity	Source
Government officials	10	Provincial Development and Reform Commission & Organs
Domain experts	30	Member of Henan Agricultural Engineering Society
Business leaders	40	Leading members of biomass power plants

Appendix 2: Fuzzy Comparison of Primary Indicators

Criteria	Best	Worst
Economic environment	EI	VI
Geographic environment	FI	VI
Location	VI	EI

Appendix 3: Fuzzy Comparison of Economic and Environmental Indicators

Subcriteria	Best	Worst
Road C_{11}	EI	VI
Railway C_{12}	VI	EI
Water C_{13}	FI	WI
Power lines C_{14}	WI	FI

Appendix 4: Fuzzy Comparison of Geographic Environmental Indicators

Subcriteria	Best	Worst
Slope C_{21}	FI	WI
Aspect C_{22}	WI	EI
Land use C_{23}	EI	FI

Appendix 5: Fuzzy Comparison of Location Environmental Indicators

Subcriteria	Best	Worst
Residential area C_{31}	FI	EI
Regional CO_2 emissions C_{32}	EI	FI

Appendix 6: Nomenclature and Abbreviations

Nomenclature	Abbreviations
Geographic Information System	GIS
Multi-Criteria Decision-Making	MCDM
Agricultural Waste Cogeneration Projects	AWCPs
Analytic Hierarchy Process	AHP
Decision-Making Trial And Evaluation Laboratory	DEMATEL
Fuzzy Best-Worst Method	FBWM
Criteria Importance Through Intercriteria Correlation	CRITIC
Multi-Attributive Border Approximation Area Comparison	MABAC
Technique for Order Preference by Similarity to an Ideal Solution	TOPSIS

References

Adamski M, Magnuszewski J (2020) Investment processes in renewable energy (in polish). Ciepłownictwo, Ogrzewnictwo, Wentylacja:10–17. https://doi.org/10.15199/9.2020.12.3

Al Garni HZ, Awasthi A (2017) Solar PV power plant site selection using a GIS-AHP based approach with application in Saudi Arabia. Appl Energy 206:1225–1240. https://doi.org/10.1016/j.apenergy.2017.10.024

Ali S, Taweekun J, Techato K, Waewsak J, Gyawali S (2019) GIS based site suitability assessment for wind and solar farms in Songkhla, Thailand. Renew Energy 132:1360–1372. https://doi.org/10.1016/j.renene.2018.09.035

Chukwuma EC, Okey-Onyesolu FC, Ani KA, Nwanna EC (2021) GIS bio-waste assessment and suitability analysis for biogas power plant: a case study of Anambra state of Nigeria. Renew Energy 163:1182–1194. https://doi.org/10.1016/j.renene.2020.09.046

Colak HE, Memisoglu T, Gercek Y (2020) Optimal site selection for solar photovoltaic (PV) power plants using GIS and AHP: a case study of Malatya Province, Turkey. Renew Energy 149:565–576. https://doi.org/10.1016/j.renene.2019.12.078

Deveci M, Erdogan N, Cali U, Stekli J, Zhong SY (2021) Type-2 neutrosophic number based multi-attributive border approximation area comparison (MABAC) approach for offshore wind farm site selection in USA. Eng Appl Artif Intell 103. https://doi.org/10.1016/j.engappai.2021.104311

Edrisi SA, Abhilash PC (2016) Exploring marginal and degraded lands for biomass and bioenergy production: an Indian scenario. Renew Sustain Energy Rev 54:1537–1551. https://doi.org/10.1016/j.rser.2015.10.050

Edrisi SA, Dubey PK, Chaturvedi RK, Abhilash PC (2022) Bioenergy crop production potential and carbon mitigation from marginal and degraded lands of India. Renew Energy 192:300–312. https://doi.org/10.1016/j.renene.2022.04.109

Fallahpour A, Wong KY, Rajoo S, Fathollahi-Fard AM, Antucheviciene J, Nayeri S (2021) An integrated approach for a sustainable supplier selection based on industry 4.0 concept. Environ Sci Pollut Res. https://doi.org/10.1007/s11356-021-17445-y

Gao J, Wang ZF, Wang ZW, Wang C, Zhang RK, Xu GY, Wu X (2022) Macro-site selection and obstacle factor extraction of biomass cogeneration based on comprehensive weight method of game theory. Energy Rep 8:14416–14427. https://doi.org/10.1016/j.egyr.2022.10.409

Gao J et al (2024) Site selection decision for biomass cogeneration projects from a sustainable perspective: a case study of China. Energy 286. https://doi.org/10.1016/j.energy.2023.129518

Gigovic L, Pamucar D, Bozanic D, Ljubojevic S (2017) Application of the GIS-DANP-MABAC multi-criteria model for selecting the location of wind farms: a case study of Vojvodina, Serbia. Renew Energy 103:501–521. https://doi.org/10.1016/j.renene.2016.11.057

Guo S, Zhao HR (2017) Fuzzy best-worst multi-criteria decision-making method and its applications. Knowl-Based Syst 121:23–31. https://doi.org/10.1016/j.knosys.2017.01.010

Hooshangi N, Gharakhanlou NM, Razin SRG (2023) Evaluation of potential sites in Iran to localize solar farms using a GIS-based Fermatean fuzzy TOPSIS. J Clean Prod 384. https://doi.org/10.1016/j.jclepro.2022.135481

Ji L, Zheng ZX, Huang YN, Xie YL, Sun LJ, Huang GH (2023) An integrated decision support method for strategic planning and tactical management of regional biomass power plants under uncertainties. Journal of Cleaner Production 388. https://doi.org/10.1016/j.jclepro.2023.135968

Jiang ZW, Wei GW, Guo YF (2022) Picture fuzzy MABAC method based on prospect theory for multiple attribute group decision making and its application to suppliers selection. J Intell Fuzzy Syst 42(4):3405–3415. https://doi.org/10.3233/jifs-211359

Kengpol A, Rontlaong P, Elfvengren K (2022) Sustainable assessment for biomass power plant location during the COVID-19 pandemic. Ind Eng Manag Syst Int J 21(1):20–42

Khanlari A, Nazari MA (2022) A review on the applications of multi-criteria decision-making approaches for power plant site selection. J Therm Anal Calorim 147(7):4473–4489. https://doi.org/10.1007/s10973-021-10877-1

Kheybari S, Javdanmehr M, Rezaie FM, Rezaei J (2021) Corn cultivation location selection for bioethanol production: an application of BWM and extended PROMETHEE II. Energy 228. https://doi.org/10.1016/j.energy.2021.120593

Konstantinos I, Georgios T, Garyfalos A (2019) A decision support system methodology for selecting wind farm installation locations using AHP and TOPSIS: case study in eastern Macedonia and Thrace region, Greece. Energy Policy 132:232–246. https://doi.org/10.1016/j.enpol.2019.05.020

Lee AHI, Kang HY, Liou YJ (2017) A hybrid multiple-criteria decision-making approach for photovoltaic solar plant location selection. Sustainability 9(2). https://doi.org/10.3390/su9020184

Liou JJH, Chang MH, Lo HW, Hsu MH (2021) Application of an MCDM model with data mining techniques for green supplier evaluation and selection. Appl Soft Comput 109. https://doi.org/10.1016/j.asoc.2021.107534

Liu Y, Ren JZ, Man Y, Lin RJ, Lee CKM, Ji P (2020) Prioritization of sludge-to-energy technologies under multi-data condition based on multi-criteria decision-making analysis. J Clean Prod 273. https://doi.org/10.1016/j.jclepro.2020.123082

Nuno-Villanueva N, Mate-Gonzalez MA, Nieto IM, Blazquez CS, Martin AF, Gonzalez-Aguilera D (2023) GIS-based selection methodology for viable district heating areas in Castilla y Le & PRIME;on, Spain. Geothermics 113. https://doi.org/10.1016/j.geothermics.2023.102767

Pamucar D, Cirovic G (2015) The selection of transport and handling resources in logistics centers using multi-attributive border approximation area comparison (MABAC). Expert Syst Appl 42(6):3016–3028. https://doi.org/10.1016/j.eswa.2014.11.057

Rahim N, Abdullah L, Yusoff B (2020) A border approximation area approach considering bipolar Neutrosophic linguistic variable for sustainable energy selection. Sustainability 12(10). https://doi.org/10.3390/su12103971

Rahimi S, Hafezalkotob A, Monavari SM, Hafezalkotob A, Rahimi R (2020) Sustainable landfill site selection for municipal solid waste based on a hybrid decision-making approach: fuzzy group BWM-MULTIMOORA-GIS. J Clean Prod 248. https://doi.org/10.1016/j.jclepro.2019.119186

Razeghi M, Hajinezhad A, Naseri A, Noorollahi Y, Moosavian SF (2023) Multi-criteria decision for selecting a wind farm location to supply energy to reverse osmosis devices and produce freshwater using GIS in Iran. Energ Strat Rev 45. https://doi.org/10.1016/j.esr.2022.101018

Rodrigues A, Gonçalves AB, Casquilho M, Gomes AA (2020) A GIS-based evaluation of the potential of woody short rotation coppice (SRC) in Portugal aiming at co-firing and decentralized co-generation. Biomass Bioenergy 137. https://doi.org/10.1016/j.biombioe.2020.105554

Sanchez-Lozano JM, Ramos-Escudero A, Gil-Garcia IC, Garcia-Cascales MS, Molina-Garcia A (2022) A GIS-based offshore wind site selection model using fuzzy multi-criteria decision-making with application to the case of the Gulf of Maine. Expert Syst Appl 210. https://doi.org/10.1016/j.eswa.2022.118371

Sarikaya MS, Yanalak M, Karaman H (2022) Site selection of natural gas emergency response team centers in Istanbul metropolitan area based on GIS and FAHP. ISPRS Int J Geo-Inf 11(11). https://doi.org/10.3390/ijgi11110571

Shahdabadi RS, Maleki A, Haghighat S, Ghalandari M (2021) Using multi-criteria decision-making methods to select the best location for the construction of a biomass power plant in Iran. J Therm Anal Calorim 145(4):2105–2122. https://doi.org/10.1007/s10973-020-10281-1

Shao M, Han ZX, Sun JW, Xiao CS, Zhang SL, Zhao YX (2020) A review of multi-criteria decision making applications for renewable energy site selection. Renew Energy 157:377–403. https://doi.org/10.1016/j.renene.2020.04.137

Tulun S, Arsu T, Gurbuz E (2022) Selection of the most suitable biogas facility location with the geographical information system and multi-criteria decision-making methods: a case study of Konya Closed Basin, Turkey. Biomass Convers Biorefin. https://doi.org/10.1007/s13399-022-03404-1

Waewsak J, Ali S, Gagnon Y (2020) Site suitability assessment of para rubberwood-based power plant in the southernmost provinces of Thailand based on a multi-criteria decision-making analysis. Biomass Bioenergy 137. https://doi.org/10.1016/j.biombioe.2020.105545

Wang SQ, Wei GW, Lu JP, Wu J, Wei C, Chen XD (2022) GRP and CRITIC method for probabilistic uncertain linguistic MAGDM and its application to site selection of hospital constructions. Soft Comput 26(1):237–251. https://doi.org/10.1007/s00500-021-06429-2

Wu YN, Yan YD, Wang SM, Liu FT, Xu CB, Zhang T (2019) Study on location decision framework of agroforestry biomass cogeneration project: a case of China. Biomass Bioenerg 127. https://doi.org/10.1016/j.biombioe.2019.105289

Wu YN, Liao MJ, Hu MY, Lin JW, Zhou JL, Zhang BY, Xu CB (2020) A decision framework of low-speed wind farm projects in hilly areas based on DEMATEL-entropy-TODIM method from the sustainability perspective: a case in China. Energy 213. https://doi.org/10.1016/j.energy.2020.119014

Wu YN, Deng ZQ, Tao Y, Wang LW, Liu FT, Zhou JL (2021) Site selection decision framework for photovoltaic hydrogen production project using BWM-CRITIC-MABAC: a case study in Zhangjiakou. J Clean Prod 324. https://doi.org/10.1016/j.jclepro.2021.129233

Xu D (2020) Sustainability prioritization of energy systems by developing an integrated decision support framework with hybrid-data consideration. Sustain Energy Technol Assess 39. https://doi.org/10.1016/j.seta.2020.100719

Zhang RK et al (2022) An analysis of Chinas biomass briquette fuel large-scale steam heating system. Biofuels Bioprod Biorefin 16(6):1627–1641. https://doi.org/10.1002/bbb.2398

Zhao H, Ge YY, Wang WH (2024) A study on offshore wind farm site selection based on CRITIC and CPT-TOPSIS: a case study of China. Kybernetes. https://doi.org/10.1108/k-09-2022-1267

Zheng YN, Doll CA, Qiu F, Anderson JA, Hauer G, Luckert MK (2021) Potential ethanol biorefinery sites based on agricultural residues in Alberta, Canada: a GIS approach with feedstock variability. Biosyst Eng 204:223–234. https://doi.org/10.1016/j.biosystemseng.2021.01.010

Chapter 3
Agricultural Waste Power Generation Supply Chain Network Optimization

Heng Wang

Contents

Abstract Agricultural waste power generation, as a clean and renewable energy utilization method, is of great significance for improving the utilization efficiency of agricultural resources and reducing environmental pollution. However, the current agricultural waste power generation supply chain network faces problems such as information asymmetry, multiple circulation links, and resource waste and urgently

H. Wang (✉)
College of Mechanical and Electrical Engineering, Henan Agricultural University, Zhengzhou, Henan Province, China
e-mail: dawn.wangh@henau.edu.cn

Z. Wang et al. (eds.), *Agricultural Waste Power Generation Supply Chain*, Industrial Ecology and Environmental Management 7,
https://doi.org/10.1007/978-3-032-20087-7_3

needs to be optimized. By analyzing the agricultural waste power generation supply chain network, a set of optimization strategies are finally proposed. The optimization of the agricultural waste power generation supply chain network requires starting from multiple aspects such as infrastructure construction, supply chain synergy, technological innovation, green supply chain management, and risk management to achieve the resource utilization of agricultural waste and the efficient supply of clean energy. The implementation of these optimization strategies will help promote the sustainable development of the agricultural waste power generation industry and make contributions to environmental protection and energy transformation.

Keywords Biomass power generation · Agricultural waste supply chain · Supply chain · Network optimization

3.1 Introduction

3.1.1 Background and Significance of Agricultural Waste Power Generation Supply Chain Network Optimization

Agricultural waste, encompassing crop residues, livestock manure, and processing by-products, has emerged as a critical resource for renewable energy production. With global energy demand projected to rise by 50% by 2050 (Yeston et al. 2023), converting agricultural waste into energy offers a sustainable pathway to reduce fossil fuel dependency and mitigate greenhouse gas emissions. However, the inherent complexities of agricultural waste supply chains—characterized by spatial dispersion, seasonal availability, perishability, and heterogeneous material properties—pose significant challenges to resource utilization efficiency. For instance, in countries such as India and China, the common practice of open-field burning of crop residues not only wastes energy potential but also exacerbates air pollution. Transforming these residues into bioenergy via thermochemical (e.g., gasification, pyrolysis) and biological (e.g., anaerobic digestion) processes necessitates a robust supply chain network to ensure cost-effective collection, transportation, storage, and conversion.

Optimizing agricultural waste power generation supply chains transcends mere logistical challenges, representing a multidimensional problem that integrates economic, environmental, and social objectives aligned with circular economy principles (Debnath and Sarkar 2023). Economically, the low energy density of biomass demands minimized costs in transport routes and facility locations. Environmentally, reducing carbon footprints requires emission mitigation during transportation and processing. Socially, equitable benefit distribution—such as job creation in rural areas—must be prioritized. Recent studies highlight the integration of circular economy strategies, including waste power generation conversion and material cascading, to enhance supply chain sustainability (Perdana et al. 2023; Sonu et al.

2023). For example, converting rice husks into biochar via pyrolysis for power generation, while utilizing residual ash as silicon fertilizer for soil amendment, establishes a closed-loop resource system. Such models not only mitigate pollution from open burning but also improve supply chain economics through by-product valorization (Contreras-Zarazúa et al. 2021).

3.1.2 Current Research Status and Challenges in Network Optimization

Existing literature on agricultural waste power generation supply chain network optimization has achieved substantial progress in addressing discrete facility location, inventory management, and route planning. However, a holistic framework to resolve the systemic complexities of network optimization remains lacking. International scholars have primarily focused on developing logistics models tailored to supply chain scenarios, such as variants of the Vehicle Routing Problem (VRP), including the Location-Routing Problem (LRP), Inventory-Routing Problem (IRP), and Location-Inventory Problem (LIP) (Darvish and Coelho 2018; Derigs and Pullmann 2014; Zheng et al. 2019). Despite advancements in bioenergy conversion technologies, agricultural waste supply chain networks remain suboptimally designed due to several systemic challenges:

3.1.2.1 Multimodal Logistics Complexity

The supply chain encompasses multiple stages—farm collection, intermediate storage, preprocessing (e.g., drying, pelletizing), and final conversion—each with distinct cost and energy requirements. A Two-Echelon VRP (2E-VRP) structure emerges, where primary vehicles transport bulk residues to preprocessing centers, while secondary vehicles deliver refined biomass to power plants. Although intermediate hubs (e.g., for drying or pelletizing) can reduce transportation costs by 30% through synchronized scheduling (Roudneshin and Sosa 2025), most studies focus on single-tier logistics (e.g., direct farm-to-plant routes), leaving hierarchical logistics underexplored (Rajabi-Kafshgar et al. 2023).

3.1.2.2 Lack of Reverse Logistics in Circular Economy

Reverse logistics research predominantly targets industrial remanufacturing and commercial sectors, with limited application in agricultural contexts. Closed-loop systems requiring bidirectional flows (e.g., delivering biochar to farms while collecting residues) necessitate Vehicle Routing Problems with Simultaneous Pickup and Delivery (VRPSPD). However, existing algorithms prioritize unidirectional flows, resulting in inefficient route planning.

3.1.2.3 Priority-Based Material Handling

Agricultural waste types (e.g., high-moisture manure vs. dry straw) demand differentiated treatment. Perishable or high-value materials (e.g., livestock manure for biogas, fruit pulp) require prioritized collection to prevent spoilage, while low-priority materials (e.g., dry straw) tolerate delayed processing. Due to the shelf life of by-products, priority allocation may also be required for materials stored first. Current VRP models rarely incorporate priority criteria, leading to suboptimal resource allocation and increased spoilage rates (Basile et al. 2022).

3.1.2.4 Resource Allocation Flexibility

Biomass availability fluctuates due to climatic or market factors, necessitating adaptive rerouting and inventory buffering. Traditional deterministic models lack real-time adjustment mechanisms to address dynamic supply-demand mismatches (Kayan et al. 2025).

3.1.2.5 Spatiotemporal Mismatches

The temporal and geographic dispersion of agricultural waste generation—such as postharvest peaks in rice straw availability in Southeast Asia—requires timely collection and storage to mitigate rapid decay. Existing models often overlook temporal variability, resulting in suboptimal facility utilization.

3.1.3 Toward a Novel Two-Echelon Logistics Model

To address these gaps, this chapter proposes a 2E-VRP model tailored to agricultural waste power generation supply chains. The framework integrates the following innovations:

A two-echelon VRP with simultaneous pickup and delivery for multi-category commodities, aligning with circular economy principles by enabling bidirectional flows of waste and by-products (e.g., biochar redistribution to farms). The model not only focuses on the circular economy principles of the agricultural waste supply chain but also considers the allocation priority of processing by-products while recycling waste.

Adaptive flexibility in by-product allocation, enabling resource interchange across secondary networks to enhance circularity.

Synchronized scheduling of vehicle loading and routing across tiers to address real-world operational complexities.

The model employs a hybrid metaheuristic algorithm combining improved genetic algorithms with Fuzzy C-Means Clustering (FCM). Key algorithmic enhancements include:

Simulated annealing-inspired adaptive fitness evaluation to balance exploration and exploitation.

Elite sub-path preservation and short sub-path elimination operators to maintain solution quality during multi-population evolution.

Adaptive crossover and mutation operators to dynamically adjust search parameters based on convergence trends.

This framework advances supply chain optimization theory by integrating multi-tier logistics, priority constraints, and adaptive flexibility into a unified model. Practically, it provides actionable insights for policymakers and stakeholders, bridging the gap between theoretical models and real-world complexities. By addressing hierarchical logistics, circular economy integration, and dynamic resource allocation, this research paves the way for scalable and sustainable agricultural waste power generation systems.

3.2 Basic Theory of Agricultural Waste Power Generation Supply Chain Network Optimization

3.2.1 Composition and Characteristics of the Agricultural Waste Power Generation Supply Chain Network

The system boundary of the agricultural waste power generation supply chain must be defined from a full life cycle perspective, encompassing a complete closed-loop system from raw material collection to energy output (Kundu et al. 2025). Full life cycle management must be implemented within specific spatial and temporal frameworks.

Spatially, the central place theory established by Christaller in 1933 provides a theoretical basis for facility layout. It is necessary to consider the density of regional-echelon raw material distribution, the provincial-echelon policy constraints, and the economic balance of cross-provincial collaborative networks (Belaud et al. 2019). Temporally, particular attention must be paid to seasonal fluctuation characteristics: The harvesting cycles of major crops such as rice and wheat lead to pulsed distributions of raw material supply, requiring transportation systems to have dynamic planning capabilities. A stochastic programming model confirms that accounting for seasonal fluctuations (e.g., the 3-month harvesting period for rice straw) can reduce inventory costs, a finding that complements the bullwhip effect theory proposed by Lee et al. in 1997 for agricultural product supply chains.

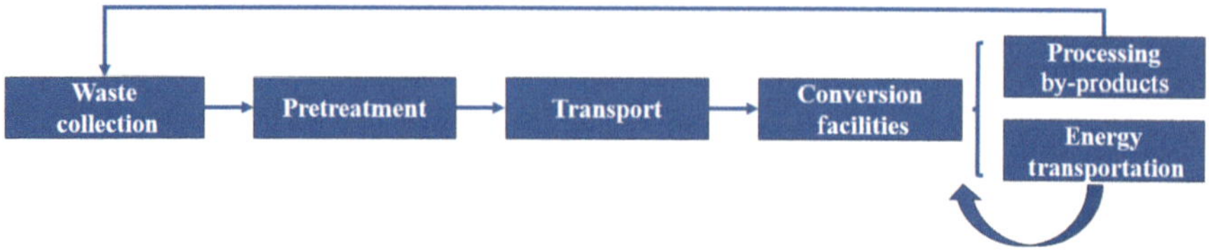

Fig. 3.1 A complete agricultural waste power generation supply chain network

In line with the "cradle-to-cradle" principle of circular economy theory, the complete system network should include six core links—raw material collection, preprocessing, transportation, conversion, energy delivery, and processing by-products—forming a closed material flow, as illustrated in Fig. 3.1.

The agricultural waste supply chain network is a complex system aimed at efficient resource utilization, consisting of collection nodes (farmers, cooperatives), preprocessing centers, resource recovery plants, distribution centers, and terminal markets. Its core components include:

Collection layer: Covering dispersed sources of agricultural waste (such as farmlands and livestock farms), a regionalized layout is required to enhance coverage. Given the widespread and highly seasonal nature of agricultural waste, a node network based on the "dispersed generation-centralized recycling" model is essential.

Processing layer: This layer encompasses technical facilities for physical sorting, biodegradation, pyrolysis gasification, etc. Processes like shredding and drying in the preprocessing stage directly impact the efficiency of subsequent resource conversion.

Distribution layer: Connecting processing plants with downstream markets, this layer involves optimizing transportation routes and multimodal transport. Establishing a region-centered logistics network is crucial to reduce transportation costs and develop emergency response mechanisms.

3.2.2 Core Objectives of Network Optimization

The core optimization objectives of the agricultural waste power generation supply chain network are efficiency, cost control, and sustainability.

Efficiency optimization: Node location selection and route planning can reduce logistics time. Models addressing vehicle routing and scheduling problems (Leng and Li 2022) demonstrate that well-designed routes can significantly enhance transportation efficiency and cut costs.

Cost control: This objective covers the entire chain of collection, transportation, and processing. Balancing equipment investment and operational expenses is necessary, such as adopting shared warehousing models or optimizing vehicle routes to reduce the number of trips and lower fixed costs.

Sustainability: Under the concepts of green supply chains and circular economy, carbon emissions and resource recycling rates must be integrated into the evaluation system.

3.2.3 Model and Methodology of Network Optimization

In modern society, with the globalization of commercial activities and the continuous development of the service industry, the demand for effective route planning and resource utilization has become particularly urgent.

The VRP is a classic issue in the field of combinatorial optimization. It involves efficiently planning the routes of a fleet of vehicles to visit multiple customer locations while satisfying various constraints. The significance of this problem extends beyond the logistics and transportation sectors; it also finds extensive applications in diverse fields, including the service industry. Practical application scenarios include, but are not limited to, logistics distribution, express delivery services, public transportation planning, urban cleaning, and emergency rescue operations. In these contexts, rational vehicle allocation, route optimization, and consideration of time and resource limitations pose critical challenges for organizations and enterprises. Solving the VRP not only reduces transportation costs and improves operational efficiency but also minimizes environmental impacts, thereby promoting sustainable urban development.

Despite its seemingly straightforward theoretical concept, the VRP's combinatorial nature leads to an exponential increase in problem size and complexity as the number of customer locations and constraints grows. Additionally, the multitude of intricate real-world issues necessitates continuous innovation in practical engineering applications, making the search for optimal solutions highly challenging. This has spurred researchers to develop a wide range of innovative models and algorithms to address various forms of real-world problems. This section will delve into all aspects of the Vehicle Routing Problem, including the definition of the basic problem, common variants, solution methods, and application domains. Through a comprehensive summary and standardization of this problem, a solid foundation is established for a profound understanding of the VRP's complexity and subsequent research endeavors.

3.2.3.1 Definition of Vehicle Routing Problem

P Problems, NP Problems, NPC Problems, and NP-Hard Problems

Polynomial (P) problems: Problems solvable by algorithms with polynomial time complexity.

Nondeterministic Polynomial (NP) problems: Problems for which a correct solution can be verified in polynomial time. It remains unknown whether all NP

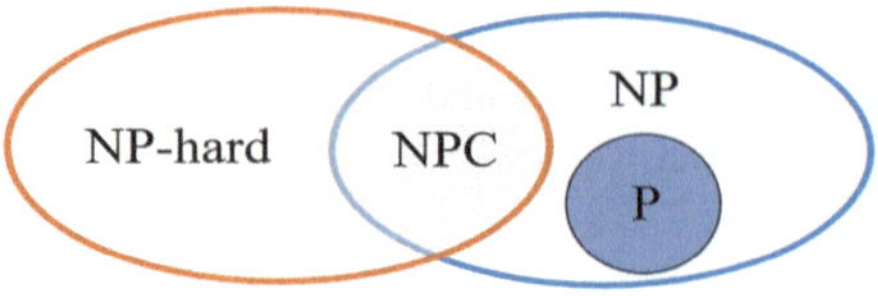

Fig. 3.2 Relationship between P problems, NP problems, NPC problems, and NP-hard problems

problems have polynomial-time algorithms, though P problems are a subset of NP problems—any problem solvable in polynomial time can always be verified in polynomial time.

Nondeterministic Polynomial Complete (NPC) problems: A class of NP problems where every other NP problem can be reduced to them.

NP-hard problems: Problems to which all NP problems can be reduced, though they may not themselves be in NP.

The relationships among these four problem classes are illustrated in Fig. 3.2.

Traveling Salesman Problem

Before defining the VRP, it is essential to understand the Traveling Salesman Problem (TSP), a renowned problem in mathematics. The TSP generally asks: How should a salesman plan a route starting and ending at their home city, visiting each target city at least once, to minimize the total travel distance? As a classic combinatorial optimization problem, the TSP is NP-complete, meaning that no polynomial-time algorithm has been found to solve it exactly.

Mathematical formulation of TSP: Given the locations of n cities, a traveler starts from one city, visits each city exactly once, and returns to the starting city. The goal is to determine the optimal visiting order to minimize the total travel distance. From a graph theory perspective, this is equivalent to finding a Hamiltonian cycle with the minimum total weight in a weighted undirected graph. In short, the problem is to find the shortest path that traverses n cities, or to find an arrangement of a natural subset (a set of n city numbers) $C = \{1, 2, 3, ..., n\}$: $\pi(X) = \{c_1, c_2, c_3, ..., c_n\}$, such that:

$$D_d = \sum_{i \in C \setminus \{n\}} d\left(c_i, c_{i+1}\right) + d\left(c_n, c_1\right) \tag{3.1}$$

The minimum value is taken, where $d\,(c_i, c_{i+1})$ represents the distance from city c_i to c_{i+1}. On this basis, a modeling method for TSP is presented. For nodes i and j, d_{ij} represents the distance from i to j, and x_{ij} is the decision variable:

$$x_{ij} \begin{cases} 1, \text{The } arc \text{ accessing } j \text{ from } i \text{ is selected} \\ 0, \text{The } arc \text{ accessing } j \text{ from } i \text{ is not selected} \end{cases} \tag{3.2}$$

The objective function is shown in Eq. (3.3):

$$\min Z = \sum_{i \in C} d_{ij} x_{ij} \tag{3.3}$$

$$s.t. \quad \sum_{j \in C} x_{ij} = 1, \forall i \in C \tag{3.4}$$

$$\sum_{i \in C} x_{ij} = 1, \forall j \in C \tag{3.5}$$

$$\sum_{i \in S} \sum_{j \in \bar{S}} x_{ij} \geq 1, \forall S \subset C, S \neq \phi \tag{3.6}$$

Constraint (3.4) indicates that each node must exit once; constraint (3.5) indicates that each node must enter once; constraint (3.6) represents the elimination of the constraint on sub-loops. S is the true subset of set C except for the empty set. If any true subset satisfies this constraint, no sub-loops will be generated.

The sub-loop phenomena in the TSP and VRP are shown in Fig. 3.3, which may occur without adding sub-loops to eliminate constraints. There are two methods to eliminate sub-loops:

Method 1:

$$\sum_{i \in S} \sum_{j \in S} x_{ij} \leq |S| - 1, \forall S \subset C, S \neq \phi \tag{3.7}$$

Method 2:

$$u_i - u_j + n x_{ij} \leq n - 1, \forall i, j \in V \setminus \{c_1\}, i \neq 1 \tag{3.8}$$

$$u_i \geq 0, \forall i \in V \tag{3.9}$$

This constraint excludes the starting node. The variable u_i serves as an auxiliary variable without practical interpretation, and Method 2 is generally considered more efficient in solving such problems.

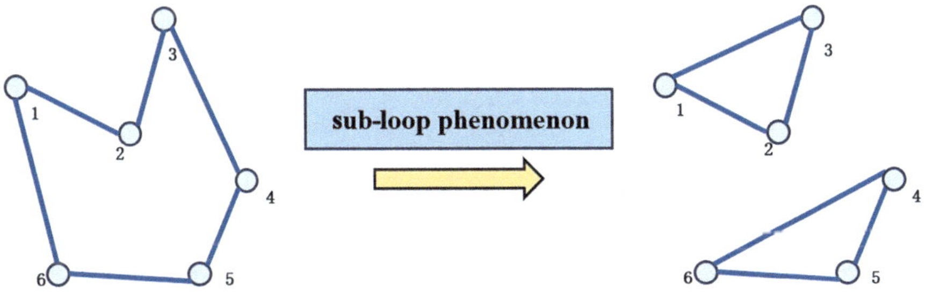

Fig. 3.3 Sub-loop phenomenon

Many other NP-complete problems can be reduced to the TSP, such as nut placement on assembly lines and production scheduling. In the context of vehicle routing, the TSP has evolved into the Multiple Traveling Salesman Problem (MTSP) due to additional constraints in real-world applications. The MTSP retains a single depot but introduces multiple salesmen, each visiting a subset of cities exactly once. Unlike the TSP, where cities have no demand and vehicles face no capacity restrictions, the MTSP focuses on minimizing total travel distance across all routes.

When cities are replaced by customers with specific demands (e.g., loading/unloading goods with volume or weight constraints) and vehicle fleets vary in capacity or type, the MTSP transitions into the VRP. Consequently, the VRP is more complex and computationally challenging than the TSP but better reflects real-world logistics scenarios.

Vehicle Routing Problem

First proposed by Dantzig and Ramser in 1959, the Vehicle Routing Problem (VRP)—also referred to as vehicle routing optimization or vehicle scheduling problem—typically involves time-related factors in the latter case. The VRP can be described as follows: A central depot serves a set of customers with distinct cargo demand quantities, using a fleet of vehicles to deliver goods. The objective is to design appropriate routing plans that satisfy customer demands while optimizing metrics (e.g., minimizing travel distance, cost, or time) under specific constraints. A schematic illustration is shown in Fig. 3.4.

The VRP is typically defined on an undirected graph $G = (N, A)$, where N is the set of nodes, including n sets of customer points $C = \{c_1, c_2, c_3, ..., c_n\}$ and a distribution center $\{d_0\}$, i.e., $N = C \cup \{d_0\}$. A is the set of arcs, $A = \{(i, j) | i, j \in N, i \neq j\}$. The vehicle set is V, and the number of vehicles owned by the distribution center is set

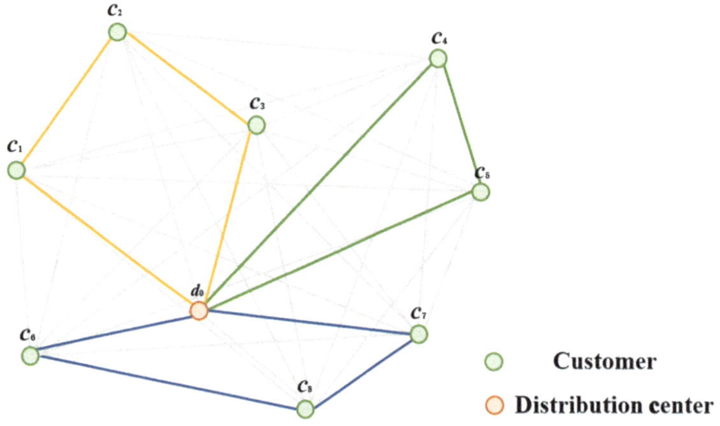

Fig. 3.4 Diagram of Vehicle Routing Problem

to *Nv*. The rated load capacity of all vehicles is *L*, and the demand q_i ($i \in C$) of all customer points is known. Let the path cost between any two nodes *i* and *j* be c_{ij}. The decision variables are x_{ijv} and y_{iv}:

$$x_{ijv} \begin{cases} 1, Arc\ ij \text{ is selected } by \text{ vehicle } v \\ 0, Arc\ ij \text{ not selected } by \text{ vehicle } v \end{cases} \tag{3.10}$$

$$y_{iv} \begin{cases} 1, \text{Customer } i \text{ is serviced } by \text{ Vehicle } V \\ 0, \text{Customer } i \text{ is serviced } by \text{ other vehicles} \end{cases} \tag{3.11}$$

The objective function is shown in Eq. (3.12):

$$\min Z = \sum_{i \in N} \sum_{j \in N} \sum_{v \in V} c_{ij} x_{ijv} \tag{3.12}$$

$$s.t. \quad \sum_{i \in N} \sum_{v \in V} x_{ikv} - \sum_{j \in N} \sum_{v \in V} x_{kjv} = 0, \forall k \in N \tag{3.13}$$

Constraint (3.13) is the traffic balance constraint; for node *k*, all traffic arriving at node *k* is equal to all traffic leaving node *k*.

$$\sum_{i \in N} x_{ijv} = y_{jv}, \forall j \in C, \forall v \in V \tag{3.14}$$

$$\sum_{j \in N} x_{ijv} = y_{iv}, \forall i \in C, \forall v \in V \tag{3.15}$$

Constraints (3.14) and (3.15) are flow direction uniqueness constraints, where all vehicles leaving customer point *i* and arriving at customer point *j* are unique, and all previous nodes *i* arriving at customer point *j* are unique.

$$\sum_{j \in C} \sum_{v \in V} x_{ijv} \le 1, \forall i \in C \tag{3.16}$$

Constraint (3.16) is the path uniqueness constraint, and in the vehicle path model, there is at most one path passing through customer point *i*.

$$\sum_{v \in V} y_{iv} = 1, \forall i \in C \tag{3.17}$$

Constraint (3.17) is the service frequency constraint, which means that each customer point has only one vehicle serving it.

$$\sum_{i \in C} y_{iv} q_i \le L, \forall v \in V \tag{3.18}$$

Constraint (3.18) is the load capacity constraint, which is one of the most common constraints in path planning. It mainly constrains that the sum of the demands of the customer points served by vehicle v must be less than the rated load capacity of the vehicle.

$$\sum_{i\in C}\sum_{v\in V} x_{d_o iv} \le N_v \tag{3.19}$$

Constraint (3.19) is the constraint on the number of vehicles, where the number of vehicles departing from the distribution center is less than the number of vehicles owned by the distribution center.

3.2.3.2 Problem Variants and Extensions

The practical application of VRP often faces diverse scenarios and demands, prompting researchers to propose various problem variants and extensions to better adapt to different scenarios, including the Vehicle Routing Problem with Time Window (VRPTW), the Multi-distribution Center Vehicle Routing Problem (MDVRP), and the Simultaneous Pickup and Delivery Vehicle Routing Problem (VRPSPD). In this section, some variations and extensions of the basic VRP are further standardized.

3.2.3.3 Vehicle Routing Problem with Time Window

In many practical scenarios, the service time of customer points is limited to a specific time window. For example, courier companies, cargo transportation companies, etc., need to ensure that delivery tasks are completed within the specified time period. The fast food and food delivery industries also need to complete delivery within the specified time window to ensure that food is delivered to customers in the freshest state. Therefore, VRPTW, as an important variant of VRP, has broad and profound research significance, affecting the optimization and efficiency of logistics, transportation, and service fields.

VRPTW introduces the concepts of vehicle travel time and time window on the basis of traditional VRP, that is, each customer point has a specified time range that can be served. This increases the complexity of the problem, requiring path planning to not only consider distance and capacity but also meet time constraints. The time window restrictions for demand points can be divided into two types. One is the hard time window, which requires vehicles to arrive within the time window. Early arrivals must wait, while late arrivals will be rejected. Another type is the soft time window, which does not necessarily have to arrive within the time window but must be penalized if the vehicle arrives outside the time window. The biggest difference between soft time window and hard time window is that penalties replace waiting and rejection.

Hard Time Window

Let t_{iv} represent the time when vehicle v arrives at node i, t_{ijv} be the time when vehicle v travels from node i to node j, and M be an infinite number used to constrain vehicle travel time. The hard time window for customer point i is $[LT_i, RT_i]$, and the supplementary constraints of VRPTW based on VRP under hard time window are as follows:

$$t_{iv} + t_{ijv} - M\left(1 - x_{ijv}\right) \le t_{jv}, \forall i, j \in N, \forall v \in V \tag{3.20}$$

Constraint (3.20) is the vehicle travel time constraint.

$$LT_i \le y_{iv} t_{iv} \le RT_i, \forall i \in C, \forall v \in V \tag{3.21}$$

Constraint (3.21) is a hard time window constraint, indicating that the vehicle needs to arrive at the customer point within the earliest and latest time requested by the customer..

Soft Time Window

VRPTW under soft time windows usually does not require additional constraints but is reflected in the form of penalty costs, as analyzed in Sect. 3.4 of this chapter.

Multi-distribution Center Vehicle Routing Problem

The Multi-distribution Center Vehicle Routing Problem (MDVRP) refers to a VRP that contains multiple distribution centers and has rich application scenarios in various fields, especially in situations where multiple distribution centers need to be coordinated for optimized path planning. For example, in large logistics and distribution networks, there are multiple distribution centers, each responsible for serving a specific area. E-commerce and retail companies typically have multiple warehouses or distribution centers. The manufacturing industry involves multiple suppliers, production centers, and distribution centers. The schematic diagram of MDVRP is shown in Fig. 3.5.

The model construction method for MDVRP is as follows: Assuming that there are m distribution centers, the set of distribution center nodes is changed from $\{d_0\}$ to $D = \{d_1, d_2, d_3, \cdots, d_m\}$, and each distribution center has the same number of vehicles as N_v. The remaining assumptions are consistent with traditional VRP. The flow balance constraint, flow direction uniqueness constraint, path uniqueness constraint, service frequency constraint, and load capacity constraint remain unchanged. The constraints for changes and additions are as follows:

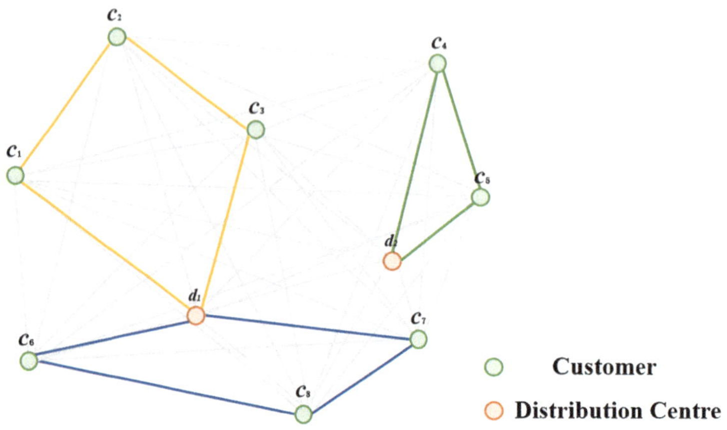

Fig. 3.5 Diagram of multi-distribution center VRP

$$\sum_{i \in C} \sum_{v \in V} x_{jiv} \leq N_v, \forall j \in D \tag{3.22}$$

Constraint (3.22) is a constraint on the number of vehicles, where the number of vehicles departing from each distribution center is less than the number of vehicles they own.

$$\sum_{i \in D} \sum_{j \in D} x_{ijv} = 0, \forall v \in V \tag{3.23}$$

Constraint (3.23) is the distribution center path constraint, indicating that there is no path between each distribution center.

The vehicle paths in the above models all belong to closed paths, that is, the vehicle starts from the distribution center and returns to the distribution center at the time of departure after completing the task, which can be classified as closed VRP. MDVRP can still be expanded into open VRP and semi-open VRP based on specific application scenarios.

Open VRP: Vehicles depart from the distribution center and stop at the last customer point after completing the task, without returning to the distribution center.

Semi-open VRP: Vehicles depart from the distribution center and return to the distribution center closest to their last customer point of service after completing the task, without being required to return to the distribution center at the time of departure.

Simultaneous Pickup and Delivery Vehicle Routing Problem

In the above VRP model, vehicles only involve one type of demand business when serving customer points. However, in the actual express and cargo transportation industries, delivery vehicles need to handle both pickup and delivery tasks simultaneously to plan the vehicle path in the optimal way, ensuring that goods can efficiently arrive at their destination from the delivery point. In food delivery services, delivery personnel also need to receive food orders and deliver them to customers simultaneously. Therefore, modeling and solving VRPSPD that involves both pickup and delivery services is of great practical significance. The schematic diagram of VRPSPD is shown in Fig. 3.6.

The model construction of VRPSPD adds the following variables and constraints on the basis of traditional VRP:

P_i: Customer demand for pickup at point i
D_i: Delivery demand at customer point i
P_{ijv}: Real-time pickup volume of the vehicle leaving node i when it travels on the ij arc
D_{ijv}: Real-time load capacity of the delivery goods leaving node i when vehicle i is traveling on the ij arc

$$x_{ijv} P_{ijv} = x_{kiv} \left(P_{kiv} + P_i \right), \forall i \in C, \forall j, k \in N, \forall v \in V \tag{3.24}$$

$$x_{ijv} D_{ijv} = x_{kiv} \left(D_{kiv} - D_i \right), \forall i \in C, \forall j, k \in N, \forall v \in V \tag{3.25}$$

Constraints (3.24) and (3.25) are load variation constraints, representing changes in the vehicle's load capacity after serving customer point i.

$$P_{ijv} + D_{ijv} \le L, \forall i, j \in N, \forall v \in V \tag{3.26}$$

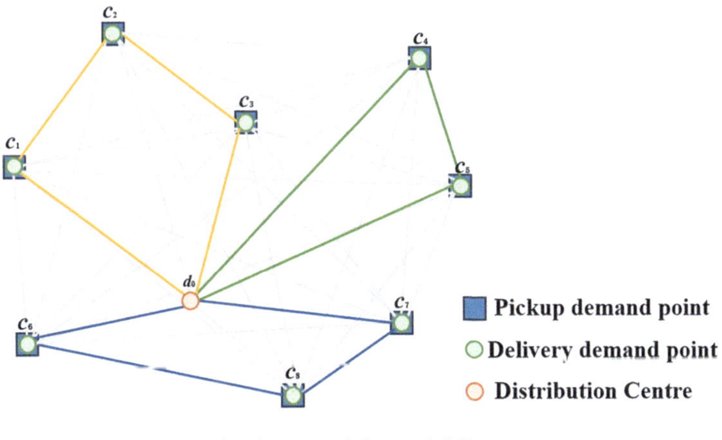

Fig. 3.6 Diagram of VRP with simultaneous pickup and delivery

Constraint (3.26) is the updated load capacity constraint, which means that the real-time load capacity of the vehicle cannot exceed its rated load capacity at any time.

$$P_{ijv} \geq 0, \forall i, j \in N, \forall v \in V \tag{3.27}$$

$$D_{ijv} \geq 0, \forall i, j \in N, \forall v \in V \tag{3.28}$$

Constraints (3.27) and (3.28) define the properties of the decision variables.

3.2.3.4 Solution Algorithm

VRP belongs to NP-hard problems. With the gradual deepening of research in this field, scholars have proposed models that are more in line with real transportation behavior, which has made VRP models increasingly complex. Although traditional exact algorithms can accurately obtain the absolute optimal solution of the problem, they cannot be well applied to practical engineering due to their limited solution scale. Artificial intelligence optimization algorithms are more adaptable and flexible in solving large-scale or structurally complex problems and are widely used in this field. Below are several commonly used basic optimization algorithms for VRP solving.

Genetic Algorithm

Genetic algorithm (GA) was first designed and proposed by John Holland in the 1970s based on the evolutionary laws of organisms in nature. GA is currently the most widely used artificial intelligence optimization algorithm (Özlük et al. 2025). The algorithm follows the principle of "survival of the fittest" in nature and is a randomized search algorithm that draws on the natural selection and genetic mechanisms of the biological world. The algorithm flowchart is shown in Fig. 3.7.

The basic idea of the algorithm is to simulate the process of natural evolution, by performing genetic operations on individuals with certain structural forms in a population, such as selection, crossover, and mutation to generate a population representing a new solution set. The selection operator refers to the selection of individuals in a population based on their fitness, following the principle of survival of the fittest. The roulette wheel method is generally used for screening. The physical model of roulette wheel method is to allocate regions of different sizes on the turntable based on the fitness of individuals. Each time the pointer rotates and stops in a certain area, the individuals in that area are selected as the reserved individuals. The final result is that the larger the area and fitness of the individuals, the higher the probability of being selected. The crossover operation refers to encoded crossover recombination, similar to chromosome crossover recombination. There are several common crossover operators, including Partial-Mapped Crossover (PMX), Order

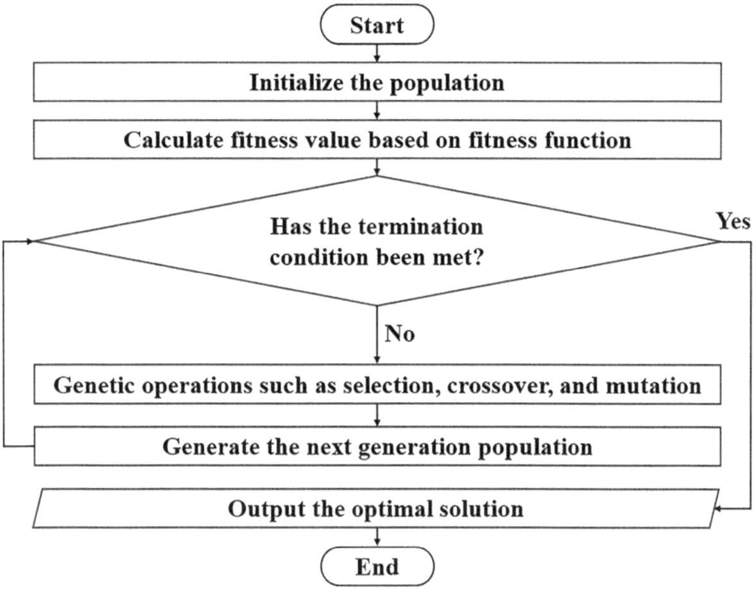

Fig. 3.7 Flow diagram of GA

Crossover (OX), Position-Based Crossover (PBX), and Order-Based Crossover (OBX). Mutation is similar to genetic mutations, involving encoding changes produced by small-probability perturbations. During the solving process, each individual in the population represents a possible solution to the problem and gradually approaches the optimal solution.

Comparison between GA and Particle Swarm Optimization (PSO): According to algorithm principles, it can be known that GA and PSO have many similarities. Both randomly initialize populations and use fitness values to evaluate systems, conducting certain random searches based on fitness values. However, PSO does not have genetic operations such as crossover and mutation but rather determines the search based on its own speed. At the same time, PSO also has an important feature of memory. In genetic algorithms, chromosomes share information with each other, so the movement of the entire population is relatively uniform toward the optimal region. In PSO, only individual optimal values and community optimal values provide information to other particles, which is a one-way information flow. The entire search and update process follows the current optimal solution.

Particle Swarm Optimization

Particle Swarm Optimization (PSO) was proposed by Eberhart and Kennedy in the 1990s. It originated from the study of bird hunting behavior, and its algorithm flowchart is shown in Fig. 3.8. The algorithm was initially inspired by the regularity of

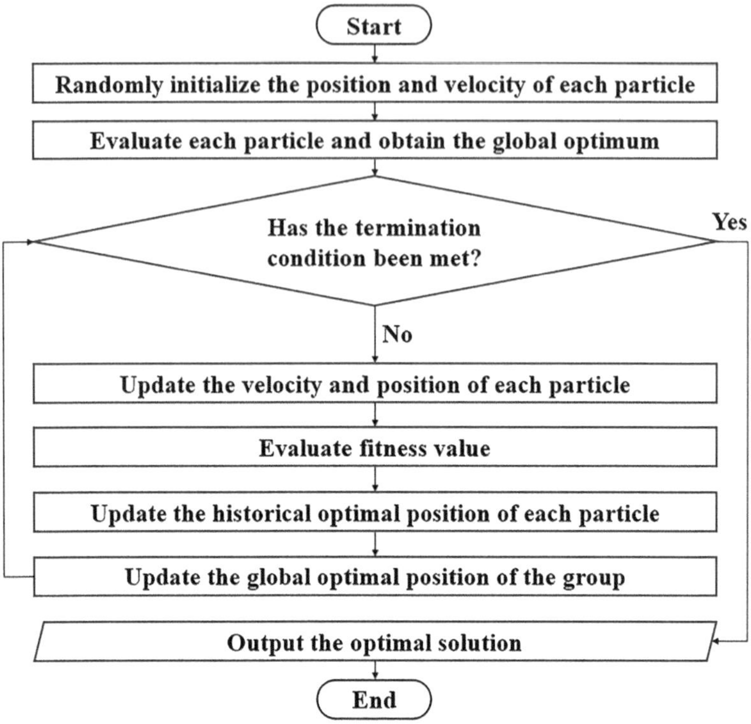

Fig. 3.8 Flow diagram of PSO

bird clustering activities, and then a simplified model was established using swarm intelligence. Based on the observation of animal swarm activity, the particle swarm algorithm utilizes information sharing among individuals in the group to generate an evolutionary process from disorder to order in the solving space, thereby obtaining the optimal solution.

Assuming there is a scenario in which a group of birds is searching for food in a fixed area, where there is only one piece of food and none of the birds knows where the food is. However, they can perceive how far they are from the food and constantly approach it through communication with each other. In algorithms, each solution can be seen as a bird searching for food, called a "particle," and the food represents the optimal solution.

The particle swarm algorithm has the advantages of a simple principle, few parameters that need to be adjusted, and simultaneous use of individual local information and group global information for collaborative search, thus having fast convergence speed. However, the PSO algorithm is a probabilistic algorithm that cannot guarantee the search for the global optimal solution. There are mainly two aspects: Sometimes, Particle Swarm Optimization may miss the global optimal solution during the diving process, and when applying PSO algorithm to deal with high-echelon complex problems, the algorithm may converge too early. In addition, the algorithm has poor local search ability and low search accuracy, so it is often combined with other algorithms for use.

Simulated Annealing Algorithm

Simulated annealing (SA) is a probabilistic algorithm proposed by Metropolis in the 1950s for optimizing problems. Its inspiration comes from the annealing process in solid-state physics, where the arrangement of atoms in crystal materials reaches the lowest energy state during cooling. SA introduces this physical process into solving optimization problems by simulating the processes of "heating" and "cooling," allowing the solution to explore a larger search space at high temperatures and gradually refine the search to find the global or near-optimal solution at low temperatures. The core mechanism of the algorithm is probabilistic decision-making for receiving new solutions, decreasing probability of receiving worse solutions as the temperature cools, and generating new solutions through random perturbations of the current solution. The algorithm flowchart of SA is shown in Fig. 3.9.

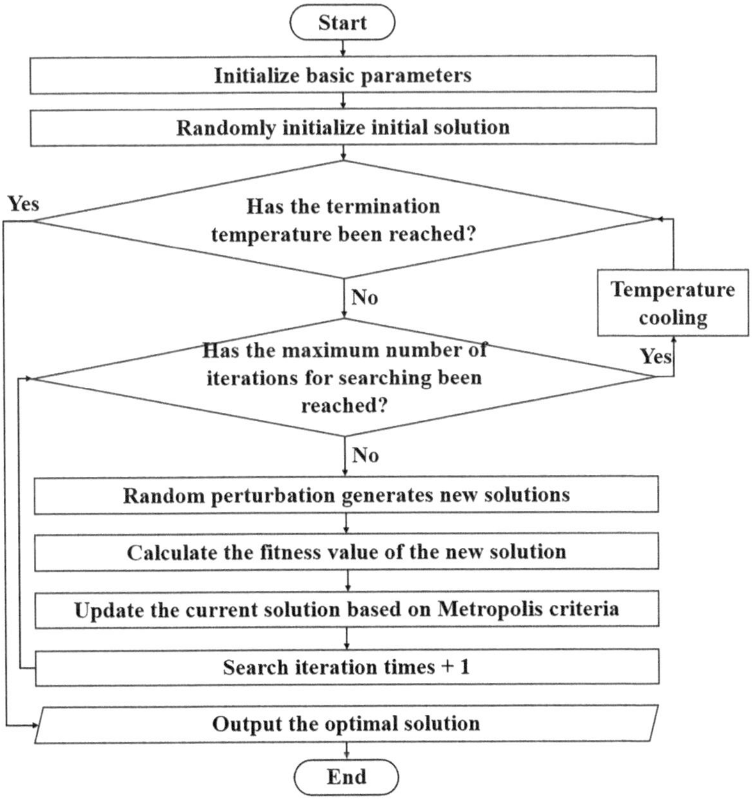

Fig. 3.9 Flow diagram of SA

3.3 Analysis of Problems in Agricultural Waste Supply Chain

According to the above discussion, considering that the agricultural waste power generation supply chain network contains a 2E-VRP, it is necessary to optimize the network using this variant model of VRP. Meanwhile, in the Vehicle Routing Problem of agricultural waste transportation, there is a problem of spoilage of processed by-products after long-term storage. It is necessary to allocate them according to the principle of using the waste stored first, that is, considering the priority of allocation for transportation, to reduce the risk of pollution or loss caused by decay and improve the efficiency of circular economy utilization. Additionally, since collecting waste while returning compost or recycled resources to farmers can promote resource recycling, various business operations such as transporting processing by-products, collecting waste, and flexibly allocating processing by-products in the actual network transportation scenario also need to be comprehensively considered in the network optimization. Finally, innovating and making up for the deficiencies of 2E-VRPSPD also has significant practical value (Hernández-Pérez et al. 2021). According to the literature review, many studies simplify transportation synchronization by only considering simultaneous pickup and delivery within the second-echelon problem, which may lead to increased transportation costs in the first-echelon problem, while studies on VRPSPD usually assume that vehicles leave the distribution center fully loaded or at a certain load ratio, which may result in potential load redundancy and waste.

Given all the above problems to be solved, this section first describes the real problem and abstracts it into a special Two-Echelon Vehicle Routing Problem with Simultaneous Pickup and Delivery for Multi-commodity (2E-VRPSPDM). Then, it analyzes the supply chain network transportation cost, which is the optimization objective of this mathematical problem, and constructs a two-layer optimization model to solve the 2E-VRPSPDM. Finally, a hybrid intelligent optimization algorithm is designed, which first clusters and decouples and then solves the optimization.

3.3.1 Two-Echelon Agricultural Waste Power Generation Supply Chain Network

In the complete agricultural waste supply chain network shown in Fig. 3.1, the supply chain network system adopting the transportation mode of farmers—collection points—processing center is illustrated in Fig. 3.10.

The by-products generated by this supply chain network are divided into three categories based on the distance of production batches: newly produced products that can be stored recently A, intermediate products that have been produced in the past but still meet the usage standards B, and agricultural waste and spoiled products that do not meet the usage standards C. The positive numbers in parentheses in the figure represent the surplus of type A products that have been determined at each

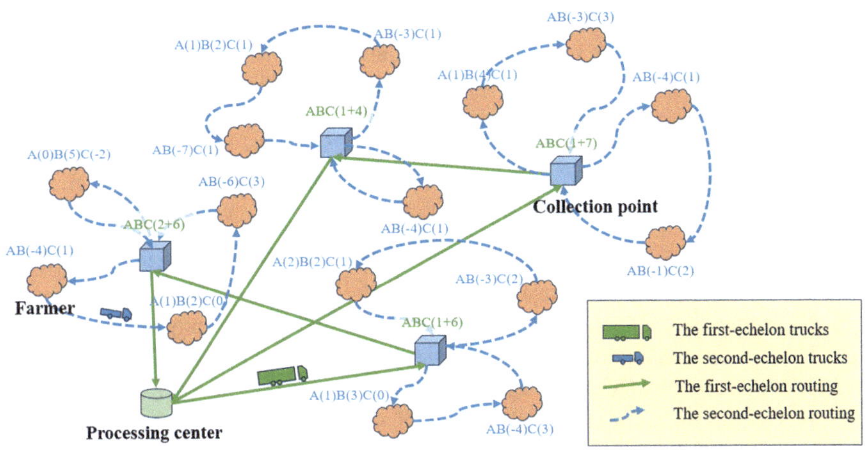

Fig. 3.10 Schematic diagram of the 2E-VRPSPDM

node, as well as the pickup demand for type *B* products and type *C* goods. Negative numbers represent the confirmed delivery demand. Assuming that the rated load capacities of the first- and second-echelon trucks in the diagram are 20 and 10 units, respectively, in the second-echelon transportation network between farmers and collection points, small vehicles depart from each collection point. To reduce production waste, it is necessary to prioritize the distribution and allocation of product *B* from the departure warehouse and the farmers it serves. When the total amount of product *B* is insufficient to meet the needs of farmers, product *A* is distributed and allocated as needed, while goods *C* are collected from various locations and finally returned to the departure warehouse. In the first-echelon transportation network between collection points and processing centers, large vehicles depart from the processing center and prioritize the allocation of product *B* from each collection point based on the second-echelon transportation situation. At the same time, goods *C* brought by reverse logistics are recovered on the way, and product *A* from each warehouse is replenished according to the return and allocation situation and finally returned to the processing center. The vehicles of the two-echelon system need to complete all the transportation services mentioned above in one go according to the optimal transportation route. This problem can be abstracted as a two-echelon vehicle routing problem with simultaneous pickup and delivery of multiple types of goods (2E-VRPSPDM).

3.3.2 Mathematical Description and Basic Assumptions

The 2E-VRPSPDM is defined in the directed delivery network graph $G = (N, A_E)$, where N is the set of nodes, $N = N_f \cup N_w \cup N_s$, and A_E is the set of arcs, $A_E = A_1 \cup A_2$. The first-echelon node set is defined as $N_1 = N_f \cup N_w$, and the second-echelon node set is defined as $N_2 = N_w \cup N_s$. $N_f = \{f_0\}$ represents the processing center node, and the node sets of m collection points and n farmers are represented as $N_w = \{w_1, w_2, w_3, \ldots, w_m\}$ and $N_s = \{s_1, s_2, s_3, \ldots, s_n\}$, respectively. The first-echelon arc set is represented as $A_1 = \{(i, j) | i, j \in N_f \cup N_w, i \neq j\}$, and the second-echelon arc set is represented as $A_2 = \{(i, j) | i, j \in N_w \cup N_s, i \neq j, (i, j) \notin N_w \times N_w\}$. The product category set is defined as K, where $K = \{A, B, C\}$. To simplify and standardize the model, some basic assumptions are made as follows:

1. As shown in Fig. 3.10, the demand for pickup and delivery of various types of waste and processing by-products by farmers has been obtained in advance.
2. The vehicles at all echelons are homogeneous and limited in quantity.
3. There is no division of vehicle compartments at all echelons, and various types of goods can be mixed together.
4. The requirements of each node cannot be separated, that is, each node can only be serviced by one vehicle once.
5. Vehicles of the same echelon travel at the same average speed.
6. The real-time cargo capacity of various types of goods in a vehicle shall not exceed the rated cargo capacity of the vehicle at any time and shall not be less than 0.
7. When a vehicle service has both pickup and drop-off requirements, delivery service should be provided first, and then pickup service should be provided.
8. Each warehouse node has an adequate inventory of type A products.
9. Type A and type B products are commonly referred to as compliant by-products, and there is no difference in meeting the usage needs of farmers.
10. Both echelons are closed vehicle path problems, where the vehicle returns to the departure node after completing the transportation task.

The optimization objective of 2E-VRPSPDM is to plan the sequence of vehicle service nodes in the two-echelon system, namely the macroscopic transportation path, so that all transportation behaviors mentioned above can be achieved and the requirements and constraints of all nodes can be met, while minimizing the total transportation cost.

3.4 Optimization of Agricultural Waste Supply Chain Network

3.4.1 Symbol Interpretation

According to the need to construct an optimization model for 2E-VRPSPDM abstracted from the agricultural waste power generation supply chain, other mathematical symbols used are summarized in Tables 3.1 and 3.2. Among them, Table 3.1 provides the definitions of decision variables, while Table 3.2 provides the definitions of sets and parameters.

Table 3.1 Description of the decision variables

Symbol	Interpretation
x_{ijv}	If vehicle v travels from node i to node j, it is 1; otherwise, it is 0
y_{ij}	If farmer j is served by vehicles in collection point i, it is 1; otherwise, it is 0
z_v	If product B in the service path of vehicle v is insufficient to meet the needs of farmers, it is 1; otherwise, it is 0
z_{iv}	If the product B in vehicle v at node i is insufficient to meet the need of the farmer, it is 1; otherwise, it is 0
$p^K{}_i$	Actual pickup quantity of K-type products at node i
$d^K{}_i$	Actual delivery quantity of K-type products at point i

3.4.2 Modeling Framework

In the 2E-VRP, intermediate nodes serve as both customer points in the first-echelon problem and distribution centers in the second-echelon problem. The solution to the first-echelon problem is based on the second-echelon problem, while the solution to the second-echelon problem is influenced by the first-echelon problem. Therefore, the two-echelon VRP system usually brings about a huge solution space due to the mutual coupling relationship. Meanwhile, as the 2E-VRPSPDM proposed in this chapter involves numerous transportation businesses and functions, it is rather complex, posing significant challenges to the construction of the mathematical model and the solution of the problem. Therefore, to enhance the constructability and solvability of the 2E-VRPSPDM, this study adopts a two-echelon programming approach to model the 2E-VRPSPDM.

3.4.3 Objective Function Analysis

The objective function for optimizing the model is the total transportation cost, denoted as C_T. Considering the practical engineering application scenarios of 2E-VRPSPDM and in response to the "dual carbon" policy, the final analysis of C_T includes several components, namely fixed cost, travel cost, time window penalty cost, and carbon tax cost, denoted as C_1, C_2, C_3, and C_4, respectively.

Fixed cost (C_1) refers to the depreciation of transportation vehicles, drivers' wages, and other factors directly related to the number of vehicle departures. In this study, the fixed cost comes from two echelons of transportation fleets, so C_1 can be expressed as:

$$C_1 = F_1 \sum_{v \in V_1} \sum_{i \in N_w} x_{f_o iv} + F_2 \sum_{v \in V_2} \sum_{w_m \in N_w} \sum_{i \in N_S} x_{w_m iv} \tag{3.29}$$

Table 3.2 Description of the sets and parameters

Symbol	Interpretation
V_E	Vehicles of the *E*-echelon assemble
F_E	Fixed cost of a single departure for vehicles in the *E*-echelon
M_E	Driving cost per unit distance of a single vehicle for vehicles in the *E*-echelon
L_{rE}	Rated cargo capacity of vehicles in the *E*-echelon
L_{ijv}	Actual cargo capacity of vehicle *v* when it is on arc *Ij*
$L^K{}_{ijv}$	Actual cargo load of type *K* goods when vehicle *v* is on arc *Ij*
$d^{AB}{}_i$	Distribution need for compliant by-products at farmer *i*
$s^K{}_i$	Storage volume of type *K* goods at node *i* (which may include r_i)
$r^C{}_i$	Return need for type *C* goods at node *i*
$S^B{}_{i0}$	Initial storage quantity of type *B* goods at collection point *i*
$S^B{}_i$	Real-time storage volume of type *B* goods at collection point *i*
N_i	Maximum number of vehicles that can depart from collection point *i*
P_i	Equivalent node of the pickup demand node *i*
D_i	Equivalent node of the distribution demand node *i*
D_{ij}	Traveling distance from node *i* to node *j*
δ_1	Early arrival penalty factor
δ_2	Delayed arrival penalty factor
t_{iv}	Time when vehicle *v* arrives at node *i*
T_{Li}	Left time window of acceptable service at node *i*
T_{Ri}	Right time window of acceptable service at node *i*
T_{li}	Satisfactory left time window of node *i*
T_{ri}	Satisfactory right time window of node *i*
M	Huge fine cost for violating the time window
μ	Carbon dioxide emission coefficient
θ	Carbon tax per unit mass
$\rho^E{}_0$	Fuel consumption per unit distance of *E*-echelon vehicles when empty
$\rho^E{}_f$	Fuel consumption per unit distance of *E*-echelon vehicles when fully loaded

Operating costs (C_2) mainly include expenses incurred by factors related to the driving distance during vehicle operation, such as energy consumption, which are also produced by two-echelon transportation fleets. C_2 can be expressed as:

$$C_2 = M_1 \sum_{v \in V_1} \sum_{ij \in A_1} x_{ijv} D_{ij} + M_2 \sum_{v \in V_2} \sum_{ij \in A_2} x_{ijv} D_{ij} \tag{3.30}$$

Demand points usually have certain requirements for the time nodes of services within a certain range, such as the designated small-scale time window for takeout and express packages, as well as the time window caused by the timeliness of the collected and transported waste in the problem of this article. If the service time is earlier or later than the time window required by the demand point, it will cause certain losses to the service quality of the transportation company. To avoid this situation as much as possible, VRP usually sets corresponding hard or soft time window penalty costs according to the actual problem. To accurately describe the real scenarios, this study sets a time window penalty cost (C_3) combining software and hardware, which comes from the penalty costs generated by all nodes. The penalty cost $C_{iv}(t_{iv})$ generated at a single node is related to the time t_{iv} for vehicle v to reach the demand point i, as shown in Fig. 3.11.

C_3 can be expressed as:

$$C_3 = \sum_{v\in V_1}\sum_{ji\in A_1} x_{ji} C_{iv}\left(t_{iv}\right) + \sum_{v\in V_2}\sum_{ji\in A_2} x_{ji} C_{iv}\left(t_{iv}\right) \tag{3.31}$$

Here, the calculation formula of $C_{iv}(t_{iv})$ is:

$$C_{iv}\left(t_{iv}\right) = \begin{cases} M, & t_{iv} \in \left[0, T_{Li}\right) \cup \left(T_{Ri}, +\infty\right) \\ 0, & t_{iv} \in \left[T_{li}, T_{ri}\right] \\ \left(T_{li} - t_{iv}\right)\delta_1, & t_{iv} \in \left[T_{Li}, T_{li}\right) \\ \left(t_{iv} - T_{ri}\right)\delta_2, & t_{iv} \in \left(T_{ri}, T_{Ri}\right] \end{cases} \tag{3.32}$$

Since road transportation currently accounts for a large proportion of carbon emissions, considering the cost of carbon tax and including it in the total transportation cost is particularly important for reducing carbon emissions (Velimirovic et al. 2023). The carbon tax cost (C_4) is related to the distance traveled by the vehicle and its fuel consumption and is derived from the carbon emissions generated by the transportation behavior of the two-echelon fleet. The calculation formula of C_4 is as follows:

$$C_4 = \mu\theta\left(\sum_{v\in V_1}\sum_{ij\in A_1} x_{ijv} D_{ijv} \rho_{ij}^1 + \sum_{v\in V_2}\sum_{ij\in A_2} x_{ijv} D_{ijv} \rho_{ij}^2\right) \tag{3.33}$$

Among them, the variables $\rho 1\ ij$ and $\rho 2\ ij$, respectively, represent the actual fuel consumption per unit distance of the first-echelon and second-echelon vehicles, and

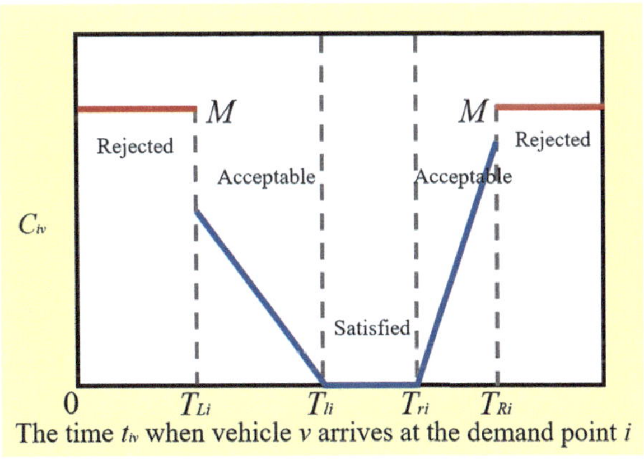

Fig. 3.11 Schematic diagram of the time window penalty cost generated at a single node

depend on the actual cargo load. According to relevant work (Chen et al. 2020), the calculation formula for the actual fuel consumption per unit distance $\rho E\ ij$ of vehicles of E-echelon is as follows:

$$\rho_{ij}^{E} = \rho_{0}^{E} + \frac{\left(\rho_{f}^{E} - \rho_{0}^{E}\right)}{L_{rE}} L_{ijv}, E = \{1,2\} \tag{3.34}$$

3.4.4 Optimization Model

Based on the above analysis, the 2E-VRPSPDM single-objective mixed-integer linear programming model is established as follows:

$$\min C_T = C_1 + C_2 + C_3 + C_4 \tag{3.35}$$

Equation (3.35) represents the objective function oriented toward minimizing the total transportation cost of the two-echelon system.

$$s.t.: \sum_{v \in V_2} \sum_{j \in N_s} x_{ijv} \le N_i, \forall i \in N_w; \tag{3.36}$$

$$\sum_{v \in V_2} \sum_{j \in N_w} x_{ijv} = 0, \forall i \in N_w; \tag{3.37}$$

$$\sum_{j \in N_2} x_{jiv} = \sum_{j \in N_2} x_{ijv}, \forall i \in N_2, \forall v \in V_2; \tag{3.38}$$

$$\sum_{v \in V_2} \sum_{j \in N_2} x_{jiv} = 1, \forall i \in N_s; \tag{3.39}$$

$$x_{ijv}\left(L_{ijv}^{A} + L_{ijv}^{B} + L_{ijv}^{C}\right) \le L_{r2}, \forall ij \in A_2, \forall v \in V_2; \tag{3.40}$$

$$x_{ijv} L_{ijv}^{A} = x_{kiv} L_{kiv}^{A} + z_{iv} x_{kiv}\left(L_{kiv}^{B} - d_{i}^{AB}\right) \ge 0, \forall i,j,k \in N_2, i \ne N_w, \forall v \in V_2; \tag{3.41}$$

$$x_{ijv} L_{ijv}^{B} = x_{kiv} L_{kiv}^{B} + x_{kiv}\left(s_{i}^{B} - d_{i}^{AB}\right) \ge 0, \forall i,j,k \in N_2, i \notin N_w, \forall v \in V_2; \tag{3.42}$$

$$x_{ijv} L_{ijv}^{C} = x_{kiv} L_{kiv}^{C} + x_{kiv} r_{i}^{C}, \forall i,j,k \in N_2, i \notin N_w, \forall v \in V_2; \tag{3.43}$$

$$\sum_{v \in V_2} x_{ijv} = 0, \forall i \in P_k, \forall j \in D_k, \forall k \in N_s; \tag{3.44}$$

$$z_v x_{w_m kv} L^A_{w_m kv} = \sum_{i \in N_s} y_{w_m i} x_{jiv} \left(d_i^{AB} - s_i^B - p_i^A\right) - S^B_{w_m}, \forall j,k \in N_2, \forall w_m, \forall v \in V_2; \tag{3.45}$$

$$x_{w_m kv} L^B_{w_m kv} = \left(1 - z_v\right) \sum_{i \in N_s} y_{w_m i} x_{jiv} \left(d_i^{AB} - s_i^B\right) + z_v S^B_{w_m}, \forall j,k \in N_2, \forall w_m, \forall v \in V_2; \tag{3.46}$$

Equations (3.36)–(3.46) are the constraint conditions of the second-echelon pickup and delivery network. Constraint (3.36) indicates that the number of vehicles departing from each collection point cannot exceed the maximum number of vehicles at that collection point. Constraint (3.37) indicates that in the second-echelon pickup and delivery network, there is no path between each collection point. Constraint (3.38) indicates that the number of vehicles entering and exiting at each node of the second-echelon is balanced. Constraint (3.39) represents the indivisibility of the demands of each farmer, that is, each farmer node is served and can only be served once by one vehicle. Constraint (3.40) represents the capacity constraint of the second-echelon vehicle during the service process. Constraint (3.41) represents the variation of the cargo volume of type *A* products when the second-echelon vehicle serves a certain farmer node. When type *B* products are insufficient to meet the demands of farmers, the distribution of type *A* products begins. Constraint (3.42) represents the change in the cargo capacity of a vehicle when it gives priority to delivering type *B* products. Constraint (3.43) represents the variation of the cargo volume in the reverse logistics of type *C* goods. Constraint (3.44) indicates that when simultaneous delivery and collection are required at the farmer's location, the vehicle delivers first and then collects the waste. Constraints (3.45) and (3.46), respectively, represent the on-demand load conditions of type *A* goods and type *B* products when the second-echelon vehicle departs.

$$d^A_{w_m} = \sum_{v \in V_2} \sum_{k \in N_s} x_{w_m kv} L^A_{w_m kv} + \sum_{v \in V_2} \sum_{k \in N_s} x_{w_m kv} L^B_{w_m kv} - d^B_{w_m} + r_{w_m}, \forall w_m; \tag{3.47}$$

$$s^B_{w_m} = S^B_{w_m 0} - \sum_{v \in V_2} \sum_{k \in N_s} x_{w_m kv} L^B_{w_m kv}, \forall w_m; \tag{3.48}$$

$$r^C_{w_m} = r_{w_m} + \sum_{i \in N_S} y_{w_m i} r_i^C, \forall w_m; \tag{3.49}$$

Equations (3.47)–(3.49) represent the cargo flow constraints between the two-echelon pickup and delivery networks, that is, the demand for various types of goods at each collection point is determined by the transportation plan of the second-echelon.

$$\sum_{v \in V_1} \sum_{i \in N_w} x_{f_0 iv} \le N_{f_0}; \tag{3.50}$$

$$\sum_{j\in N_1} x_{jiv} = \sum_{j\in N_1} x_{ijv}, \forall i \in N_1, \forall v \in V_2; \tag{3.51}$$

$$\sum_{v\in V_1}\sum_{j\in N_1} x_{jiv} = 1, \forall i \in N_w; \tag{3.52}$$

$$x_{ijv}\left(L_{ijv}^A + L_{ijv}^B + L_{ijv}^C\right) \le L_{r1}, \forall ij \in A_1, \forall v \in V_1; \tag{3.53}$$

$$x_{ijv}L_{ijv}^A = x_{kiv}L_{kiv}^A - x_{kiv}d_i^A \ge 0, \forall i,j,k \in N_1, i \notin N_f, \forall v \in V_1; \tag{3.54}$$

$$x_{ijv}L_{ijv}^B = x_{kiv}L_{kiv}^B + x_{kiv}\left(p_i^B - d_i^B\right) \ge 0, \forall i,j,k \in N_1, i \notin N_f, \forall v \in V_1; \tag{3.55}$$

$$x_{ijv}L_{ijv}^C = x_{kiv}L_{kiv}^C + x_{kiv}r_i^C, \forall i,j,k \in N_1, i \notin N_f, \forall v \in V_1; \tag{3.56}$$

$$\sum_{v\in V_1} x_{ijv} = 0, \forall i \in P_k, \forall j \in D_k, \forall k \in N_w; \tag{3.57}$$

$$x_{f_0kv}L_{f_0kv}^A = \sum_{i\in N_1} x_{jiv}d_i^A, \forall j,k \in N_w, \forall v \in V_1. \tag{3.58}$$

Equations (3.50)–(3.58) are the constraint conditions of the first-echelon pickup and delivery network. Constraint (3.50) indicates that the number of vehicles departing from the factory cannot exceed the maximum number of vehicles in this processing center. Constraint (3.51) indicates that the number of vehicles entering and exiting each node at the first level is balanced. Constraint (3.52) represents the indivisibility of the requirements at each collection point. Constraint (3.53) represents the capacity constraint of the first-echelon vehicle during the service process. Constraint (3.54) represents the variation of the cargo volume of type A products when the first-echelon vehicle serves a certain collection point. Constraint (3.55) shows the changes in the cargo capacity of vehicles when they give priority to delivering type B products. Constraint (3.56) represents the variation of the cargo volume in the reverse logistics of type C goods. Constraint (3.57) represents the operation sequence constraint when goods need to be delivered and picked up simultaneously at the collection point. Constraint (3.58) represents the on-demand load situation of the first-echelon vehicle at departure.

3.4.5 Design of Hybrid Intelligent Optimization Algorithm

From Sect. 3.4.2, the first-echelon problem of 2E-VRPSPDM is the Vehicle Routing Problem with Simultaneous Pickup and Delivery for Multi-commodity (VRPSPDM). The second-echelon problem is the Multi-depot Vehicle Routing Problem with Simultaneous Pickup and Delivery for Multi-commodity (MVRPSPDM). Due to the coupling of the two-level problem and the complexity of the proposed 2E-VRPSPDM function, the feasibility of directly using the algorithm for overall encoding and solution is relatively low. If a reasonable clustering algorithm is designed to decompose the second-echelon problem into m relatively simple subproblems, there will be a total of $m + 1$ VRPSPDM subproblems. Then, each of the decomposed subproblems is optimized respectively. At this time, the solution space of the problem can be decomposed from the product of the solution spaces of each subproblem to the sum of the solution spaces of each subproblem. Therefore, based on the characteristics of 2E-VRPSPDM, this study adopts a strategy of clustering decomposition first and then solving for optimization, which can greatly reduce the solution space of the problem, limit the algorithm to a smaller and better area for search, shorten the search time of the algorithm, and improve the search performance of the intelligent algorithm.

The 2E-VRPSPDM model constructed in this chapter is a hybrid extension of the 2E-VRP and VRPSPD models, and the subproblem VRPSPDM, after adopting the clustering decomposition strategy, all originate from the classical VRP model that has been proven to be an NP-hard problem. Therefore, they both fall within the category of NP-hard problems. According to Chap. 2, when solving NP-hard problems, the exact algorithm has limited feasibility in practical engineering applications because its computing space increases exponentially with the increase of the problem scale. However, when dealing with NP-hard problems, intelligent optimization algorithms can obtain approximate solutions of higher quality in a relatively short time and usually have more practical application significance. Furthermore, due to the frequent pickup and delivery behaviors of vehicles for multiple types of goods in VRPSPDM, combined with the condition that the real-time cargo capacity of various goods in the vehicle, as stipulated by the model, cannot exceed the rated cargo capacity of the vehicle at any time and cannot be less than 0, it is difficult for the algorithm to construct a feasible solution to this problem. Considering that the coding method of the genetic algorithm (GA) can allow constraints to be temporarily violated during the process of constructing feasible solutions and impose significant penalties on such violations to eventually obtain a truly feasible solution, compared with other intelligent optimization algorithms, the coding method of the genetic algorithm is more suitable for solving this problem. Finally, in view of the practical problems in the algorithm optimization process, in order to further improve the solution speed and quality of the algorithm, this study adopts the combination of Variable Neighborhood Search (VNS). The improved Enhanced Multi-population Genetic Algorithm (EMPGA) of VNS is used to solve and optimize the problem.

Based on the above design concept, the overall flowchart of the hybrid intelligent optimization algorithm for solving the mathematical model in this chapter is shown in Fig. 3.12.

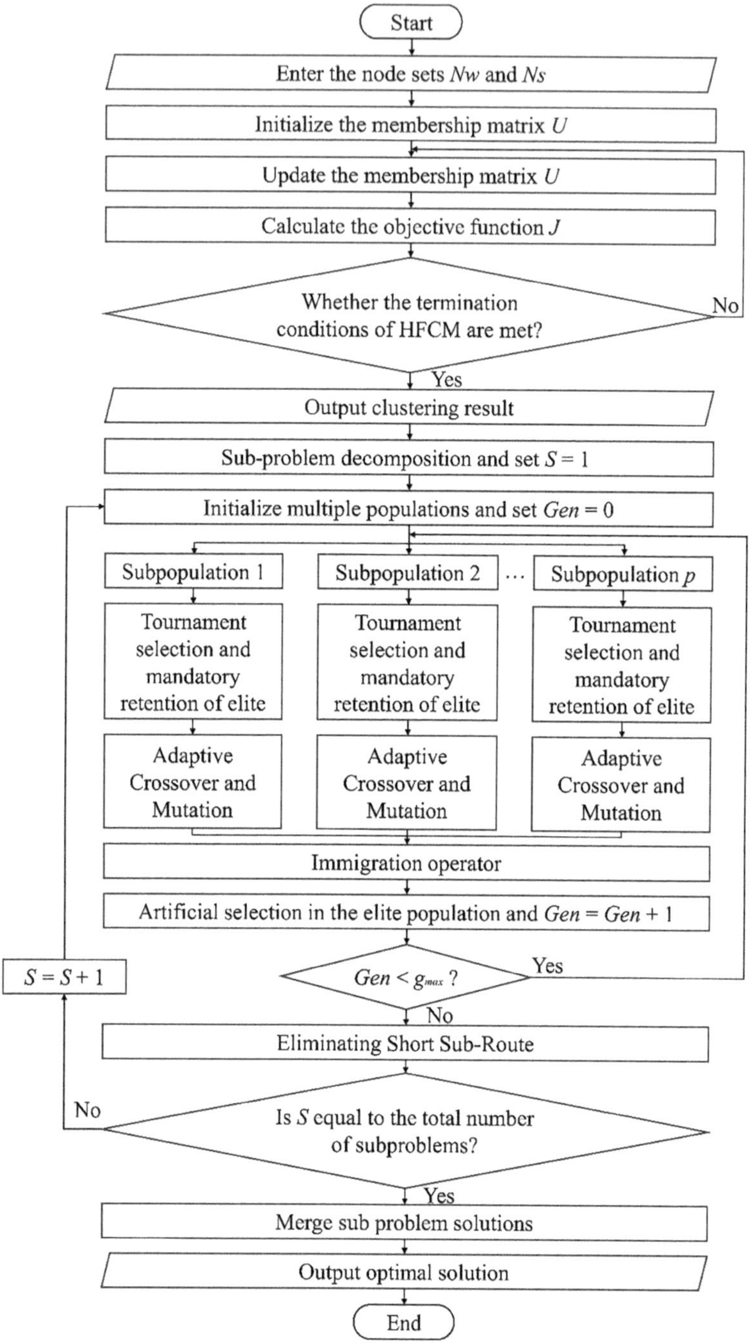

Fig. 3.12 Flow diagram of the hybrid intelligent optimization algorithm

3.4.5.1 Fuzzy C-Means Clustering

Typically, there are no clear boundaries among numerous chain stores in cities, and the same situation exists among farmers in the supply chain network studied. Therefore, the distribution between farmers and collection points has inherent uncertainties. Compared with commonly used hard clustering algorithms such as K-means, soft clustering algorithms allow data points to have different degrees of membership in multiple clusters, that is, the concept of membership degree. This is particularly useful when the boundaries between clusters are not clear or when there is noise in the data. Fuzzy C-Means Clustering (FCM) represents a kind of soft clustering algorithm, which alleviates the shortcomings of K-means by introducing the concept of membership degree and increasing the number of iterations. Therefore, in this study, FCM based on membership degree was adopted to cluster all chain stores for allocation to each warehouse.

Given the position information of the farmer node i ($i \in N_s$) and the collection node j ($j \in N_w$), the membership function from node i to node j is defined as u_{ij}, where $u_{ij} \in [0,1]$. For ∀i, the sum of the membership functions of all j is equal to 1. The membership matrix is represented as U. The objective function of FCM is defined by the following equation:

$$J(U,m) = \sum_{j \in N_w} \sum_{i \in N_s} u_{ij}^{f} \left\| i - j \right\|^2 \tag{3.59}$$

Here, J represents the objective function, and the purpose of clustering is to minimize J. m represents the number of elements in N_w, corresponding to the number of clusters. The parameter f represents the ambiguity coefficient, where $f \in [0, \infty]$. Assuming the position of the cluster center has been determined, the formula for updating the membership function derived by the Lagrange multiplier method is as follows:

$$u_{ij} = \left(\sum_{k \in N_w} \left(\frac{\left\| i - j \right\|}{\left\| i - k \right\|} \right)^{\frac{2}{f-1}} \right)^{-1} \tag{3.60}$$

The termination condition of FCM is defined as:

$$\max_{ij} \left\{ \left| u_{ij}^{(t+1)} - u_{ij}^{(t)} \right| \right\} < \varepsilon \tag{3.61}$$

Among them, t represents the iteration step size, and ε is a small constant representing the error threshold. Through iterative refinement, FCM outputs the clustering results when the maximum change of u_{ij} between consecutive iterations does not exceed the error threshold ε.

3.4.5.2 Construction Method of Fitness for Simulated Annealing

In GA and MPGA, fitness evaluates the excellence of an individual and plays a crucial role. For optimization problems that minimize the objective function, it is usually set as the reciprocal of the objective function. When solving the problems raised in this chapter, in order to enhance the adaptive optimization ability of MPGA, the concept of simulated annealing was introduced in the fitness calculation of the hybrid intelligent optimization algorithm, and the fitness function was reconstructed as shown in the following equation:

$$F(X) = \frac{e^{\{T(g)\cdot[1+C_T(X)]\}^{-1}}}{\sum_{X\in P} e^{\{T(g)\cdot[1+C_T(X)]\}^{-1}}} \tag{3.62}$$

Here, P is the current population and X represents one of the individuals, which is a solution in the problem of this chapter. $F(X)$ represents the fitness value of individual X, and $T(g)$ represents the annealing temperature function, which is related to the total number of iterations g and is defined as:

$$T(g) = t_0 a_t^g, g = 1,2,3,\ldots, g_{\max} \tag{3.63}$$

Here, $g_{\max}$ represents the maximum number of genetic iterations. t_0 is the initial annealing temperature, which is generally set as $2g_{\max}$. ag t is the annealing temperature coefficient, usually taking a value slightly less than 1.

Through this fitness calculation, the EMPGA proposed in this paper shows a relatively slow convergence behavior in the early stage of operation. Especially at higher annealing temperatures, individuals with similar fitness levels have comparable probabilities of generating offspring to promote the generation of more diverse individuals and prevent falling into local optima. As the annealing temperature gradually decreases, the fitness differences of individuals are magnified, so that the algorithm can quickly obtain better individuals.

3.4.5.3 Adaptive Crossover and Mutation Strategy

Although MPGA has addressed, to a certain extent, the drawback of GA's reliance on crossover and mutation operators through multiple population dimensions, in order to further improve the algorithm performance and enable it to generate diverse individuals as much as possible in the early iterations to prevent local convergence, the focus will shift to retaining the excellent individuals in the population in the later iterations to accelerate the convergence speed. Combined with the fitness function constructed in the previous subsection, this paper introduces adaptive crossover and mutation strategies in EMPGA.

1. First, the adaptive crossover probability p_c and the mutation probability p_m are, respectively, defined as:

$$p_c = \begin{cases} \lambda_1 - (\lambda_1 - \lambda_2)\dfrac{F_h - F_a}{F_{\max} - F_a}, F_h \geq F_a \\ \lambda_1 \quad , F_h < F_a \end{cases} \tag{3.64}$$

$$p_m = \begin{cases} \lambda_3 - (\lambda_3 - \lambda_4)\dfrac{F(X) - F_a}{F_{\max} - F_a}, F(X) \geq F_a \\ \lambda_3 \quad , F(X) < F_a \end{cases} \tag{3.65}$$

Here, $F_{\max}$ represents the maximum fitness of the population, F_a represents the average fitness of the population, F_h represents the higher fitness between paired individuals, λ_1 and λ_2 are crossover parameters, and λ_3 and λ_4 are mutation parameters.

2. Next, the adaptive cross-mixing operator is constructed.

In the initial stage of the algorithm, the traditional Order Crossover (OX) operator is called to promote the generation of new individuals and ensure diversity of individuals within the population. The OX operator refers to first selecting two different parent individuals, randomly choosing a starting point and an endpoint index, replicating the genes of the two parents from the starting point to the endpoint at the same position into the two offspring individuals, and then placing the other genes of the other parent into the offspring in sequence. Eventually, two new offspring individuals are obtained.

As mentioned above, since it is relatively difficult to construct feasible solutions for VRPSPDM, during the algorithm iteration process, when individuals with significantly high fitness appear within the population, it is considered to construct a protective sub-path line operator to quickly protect the optimal sub-path of the individual with the highest fitness value in the current population, preventing the excellent sub-path from being randomly damaged by the OX operator. To improve the efficiency of optimization in later stages, the principle of protecting the excellent sub-route operator is shown in Fig. 3.13. Meanwhile, this operator also solves the problem that the traditional OX operator cannot generate new offspring when encountering the same parent individual in later stages.

3.4.5.4 Eliminate Overly Short Sub-paths

In the feasible solutions obtained by the algorithm, there sometimes exist sub-paths with fewer service nodes. Vehicles only serve one demand point and then return to the distribution center, which greatly wastes vehicle resources. In order to further optimize the optimal solution and improve the vehicle loading efficiency, this study introduces VNS into the elite selection operator to eliminate the overly short sub-routes in the elite individuals where the number of path nodes is less than 4. Among

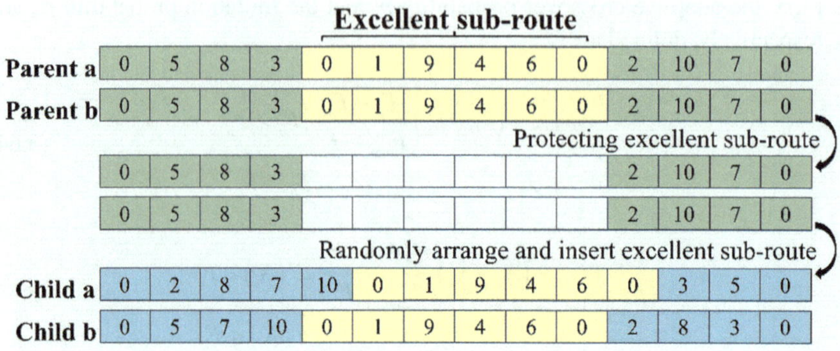

Fig. 3.13 Schematic diagram of the protective excellent sub-route operator

them, taking other sub-paths except for the overly short sub-paths as the neighborhood structure, and taking the demand points left after inserting and deleting the overly short sub-paths at different positions in the sub-path selection as the perturbation operator, the pseudocode for eliminating the overly short sub-path operator using VNS is shown in Table 3.3.

3.4.5.5 Algorithm Complexity Analysis

Let the population size of the EMPGA be defined as N_c, and let the number of neighborhood search iterations be defined as N_v (where N_v is a constant related to the number of nodes in the subproblem). The population number of EMPGA is denoted as p. The complexity T of the designed intelligent hybrid optimization algorithm is composed of the complexities of FCM and EMPGA. Specifically, the complexity of FCM is $O(\mathrm{nm}^2)$. For the first-echelon pickup and delivery network, the complexity of EMPGA is $O(m + pN_c + g_{\max}p(N_c + N_c + N_c p_c) + N_v)$, and for the second-echelon pickup and delivery network, its complexity is $O(n + mpN_c + mg_{\max}p(N_c + N_c + N_c p_c) + mN_v)$. To sum up, the complexity of the hybrid intelligent optimization algorithm designed in this chapter is expressed as:

$$
\begin{aligned}
T &= O\big(nm^2 + m + pN_c + g_{\max} p\left(N_c + N_c + N_c p_c\right) + N_v \\
&\quad + n + mpN_c + mg_{\max} p\left(N_c + N_c + N_c p_c\right) + mN_v\big) \\
&= O\left(nm^2 + \left(pN_c g_{\max} + N_v\right)m + pN_c g_{\max} p_c + N_v + n\right)
\end{aligned}
\tag{3.66}
$$

Table 3.3 Eliminating short sub-route operator

Algorithm 1: Eliminating Short Sub Route Operator

Input the elite individual X_{elite} and the collection of sub routes R $[R_1, R_2, R_3, \cdots, R_r]$
  a set of neighborhood structures N_{is} for $s = 1, 2, 3, \cdots, s_{\max}$ for shaking
  a set of neighborhood structures N_{isl} for $l = 1, 2, 3, \cdots, l_{\max}$ for local search
for all $i = 1$ to r **do**
  Calculate the sub path length Len_i for the R_i
  if $Len_i \geq 4$ **then** $i = i+1$
  else Delete the sub route R_i record the shop involved s and update R
    Nis = the collection A_j of all arcs in the other sub-route R_j
    N_{isl} = the collection R_{re} of remaining sub-routes $R \backslash R_i$
    $X = X_0$ /* Generate the initial solution */
    for all $s = 1$ to s_{max} **do**
      Shaking: pick a random solution X' from the s-th neighborhood $N_{is}(X)$
      for all $l = 1$ to l_{max} **do**
        VND: Local search the optimal neighborhood X'' of X' in $N_{isl}(X')$
      end for
      Calculate the fitness values of X'' based on Eq. (33)
      if $F(X'') < F(X)$ **then** $s = s+1$
      else $X = X''$
      end if
    end for
  end if
  Update the elite individual based on $X_{elite} = X$
  Calculate the optimal distribution cost C_{min}
end for
Output X_{elite} and C_{min}

3.5 Conclusion

Due to the deep-rooted circular economy, building an efficient reverse logistics and circular economy system in the agricultural waste-to-power generation supply chain can effectively enhance the sustainable benefits of agricultural production and transportation and promote the further development of the social economy. The generation of agricultural waste is characterized by wide distribution and large scale. Meanwhile, considering the passage restrictions on vehicles of different specifications in urban areas, the supply chain network mostly adopts a multi-echelon transportation model. Furthermore, numerous processing by-products produced from agricultural waste recycling have a certain shelf life. The inventory management principle of using goods that were stored first has played a significant role in reducing the risk of by-product spoilage and resource waste. Therefore, it is of great significance to construct a paradigm transportation network in the supply chain network that can scientifically and flexibly solve the current problems faced in this field. This study is oriented toward reducing the production and transportation costs of the supply chain network. It considers the problems of transportation based on the priority allocation of production batches of processing by-products, the flexible allocation of compliant by-products among each node of farmers and collection points, as well as the problems of simultaneous pickup and delivery at both levels and on-demand loading of vehicles. An optimization model is established with the goal of minimizing the transportation costs of enterprises, and an improved multi-population genetic algorithm combined with a clustering algorithm is designed to form a hybrid intelligent optimization algorithm for solution. The main tasks are as follows:

1. Research on the Optimization of Multi-Echelon Transportation Systems in Combination with the Actual Transportation Behaviors of the Supply Chain

At present, the agricultural waste supply chain has adopted a variety of practical measures to save production and transportation resources, such as different priority transportation plans based on the production and warehousing batches of goods and flexible allocation of compliant by-products at each node in the transportation system. This study comprehensively considered all the abovementioned actual transportation behaviors, successfully implemented them in the constructed agricultural waste transportation system, and proposed a flexible allocation model for compliant by-products. It not only proves the feasibility of these transportation behaviors in reducing the transportation costs of network production but also enables the supply chain to comprehensively analyze the impact of these actual behaviors and provide decision-making guidance and optimization suggestions when optimizing transportation costs.

2. Research on the Theoretical Expansion of Two-Echelon Vehicle Routing for Integrated Simultaneous Pickup and Delivery

When considering the multi-echelon transportation mode of the agricultural waste power generation supply chain, due to the requirements of actual demands, a comprehensive study is conducted on the combination of the two-level vehicle routes and the simultaneous pickup and delivery vehicle routes. In the existing studies, the two-level vehicle routing problem integrating simultaneous pickup and delivery only considers the transportation activities of a single commodity. At the same time, it does not take into account the influence of dynamic demand and the simplification problems existing in the two-echelon pickup and delivery network, which hinders practical application in this field. There is an urgent need for theoretical expansion to enhance the practical application significance. This study combines the actual transportation behavior of “first-in, first-out” inventory management, expands the problem to the dimension of multiple types of commodities, and enables the two-echelon network to pick up and deliver goods completely simultaneously and vehicles to load and start as needed. Finally, it dynamically expands the model and constructs the corresponding integer programming model, extending the research of the theoretical model and better guiding practical engineering applications.

3. Research on the Design of Solution Schemes for Complex Two-Echelon Vehicle Routing Problems

The two-echelon vehicle routing has the inherent problem of solving spatial coupling. Considering various actual transportation behaviors and integrating the simultaneous pickup and delivery of multiple goods makes the solution and optimization of this two-echelon vehicle routing problem more difficult and cumbersome. Referring to relevant studies and the characteristics of the problem, this paper proposes a solution idea of clustering first and then optimizing. Meanwhile, the multi-population genetic algorithm is selected and the corresponding operators are designed for improvement based on the adaptive idea and the problems that occur during the solution. Combined with the Fuzzy C-Means Clustering algorithm, a hybrid intelligent optimization algorithm is constructed, successfully forming the solution framework of this complex vehicle routing problem.

The conclusions of this study are as follows: The research provides a paradigm model for optimizing the power generation supply chain network from agricultural waste and designs an optimization algorithm for decoupling the supply chain network. Combining the analysis of algorithm complexity and experimental comparison, compared with the multi-population genetic algorithm before improvement, the designed improved multi-population genetic algorithm has significantly improved in terms of optimization effect. Not only has the optimal solution been obtained within a relatively short running time but also the running time and complexity have not significantly increased compared with those before improvement. It provides a theoretical basis and ideas for the optimization of the supply chain network of power generation from agricultural waste.

References

Basile F, Pilotti L, Ugolini M, Lozza G, Manzolini G (2022) Supply chain optimization and GHG emissions in biofuel production from residues in Sweden. Renew Energy 196:405–421. https://doi.org/10.1016/j.renene.2022.06.095

Belaud JP, Prioux N, Vialle C, Sablayrolles C (2019) Big data for agri-food 4.0: application to sustainability management for by-products supply chain. Comput Ind 111:41–50. https://doi.org/10.1016/j.compind.2019.06.006

Chen JM, Dan B, Shi J (2020) A variable neighborhood search approach for the multi-compartment vehicle routing problem with time windows considering carbon emission. J Clean Prod 277. https://doi.org/10.1016/j.jclepro.2020.123932

Contreras-Zarazúa G, Martin M, Ponce-Ortega JM, Segovia-Hernández JG (2021) Sustainable design of an optimal supply chain for furfural production from agricultural wastes. Ind Eng Chem Res 60(40):14495–14510. https://doi.org/10.1021/acs.iecr.1c01847

Darvish M, Coelho LC (2018) Sequential versus integrated optimization: production, location, inventory control, and distribution. Eur J Oper Res 268(1):203–214. https://doi.org/10.1016/j.ejor.2018.01.028

Debnath A, Sarkar B (2023) Effect of circular economy for waste nullification under a sustainable supply chain management. J Clean Prod 385. https://doi.org/10.1016/j.jclepro.2022.135477

Derigs U, Pullmann M (2014) Solving multitrip vehicle routing under order incompatibilities: a VRP arising in supply chain management. Networks 64(1):29–39. https://doi.org/10.1002/net.21554

Hernández-Pérez H, Landete M, Rodríguez-Martín I (2021) The single-vehicle two-echelon one-commodity pickup and delivery problem. Comput Oper Res 127. https://doi.org/10.1016/j.cor.2020.105152

Kayan RR, Jauhar SK, Kamble SS, Belhadi A (2025) Optimizing bio-hydrogen production from agri-waste: a digital twin approach for sustainable supply chain management and carbon neutrality. Comput Ind Eng 204. https://doi.org/10.1016/j.cie.2025.111021

Kundu P, Vineetha SV, Mohan A, Ravikumar A (2025) Bio-hydrogen production from various waste resources through circular economy: current technologies and future perspective. J Mater Cycles Waste Manag 27(3):1263–1282. https://doi.org/10.1007/s10163-025-02183-x

Leng KJ, Li SH (2022) Distribution path optimization for intelligent logistics vehicles of urban rail transportation using VRP optimization model. IEEE Trans Intell Transp Syst 23(2):1661–1669. https://doi.org/10.1109/TITS.2021.3105105

Özlük R, Aydin F, Yildiz Y (2025) Artificial intelligence algorithms, simulation tools and software for optimization of adaptive facades: a systematic literature review. J Build Eng 106. https://doi.org/10.1016/j.jobe.2025.112566

Perdana T, Kusnandar K, Perdana HH, Hermiatin FR (2023) Circular supply chain governance for sustainable fresh agricultural products: minimizing food loss and utilizing agricultural waste. Sustainable Prod Consumption 41:391–403. https://doi.org/10.1016/j.spc.2023.09.001

Rajabi-Kafshgar A, Gholian-Jouybari F, Seyedi I, Hajiaghaei-Keshteli M (2023) Utilizing hybrid metaheuristic approach to design an agricultural closed-loop supply chain network. Expert Syst Appl 217. https://doi.org/10.1016/j.eswa.2023.119504

Roudneshin M, Sosa A (2025) A novel approach for optimisation of bioenergy supply chain: integrating mathematical programming, geographic information system, and analytic hierarchy process. Bioresour Technol 418. https://doi.org/10.1016/j.biortech.2024.131827

Sonu R et al (2023) Agro-waste to sustainable energy: a green strategy of converting agricultural waste to nano-enabled energy applications. Sci Total Environ 875. https://doi.org/10.1016/j.scitotenv.2023.162667

Velimirovic LZ, Janjic A, Vranic P, Velimirovic JD, Petkovski I (2023) Determining the optimal route of electric vehicle using a hybrid algorithm based on fuzzy dynamic programming. IEEE Trans Fuzzy Syst 31(2):609–618. https://doi.org/10.1109/TFUZZ.2022.3205045

Yeston JS, Vinson V, Thorp HH (2023) Energy in 2023. Science 379(6627):7–7. https://doi.org/10.1126/science.adg4111

Zheng XJ, Yin MX, Zhang YX (2019) Integrated optimization of location, inventory and routing in supply chain network design. Transp Res Part B-Methodol 121:1–20. https://doi.org/10.1016/j.trb.2019.01.003

Chapter 4
Agricultural Waste Power Generation Supply Chain Synergetic Development Research

Zhanwu Wang, Zhenfeng Wang, Nadeem Tahir, Heng Wang, Jin Li, and Guangyin Xu

Contents

Abstract Power generation from agricultural waste not only helps to meet energy demands but also provides an effective way to dispose of agricultural waste. However, in order to boost agricultural waste power generation industry, many factors such as willingness of farmers, high collection cost, and uncoordinated supply chain are the major issues that need to be solved. In the present work, synergetic development of the agricultural waste power supply chain has been studied by establishing collaborative evolution model, and the marginal utility of government subsidy has been defined. Our results showed that in the farmer subsystem, the selling price of agricultural waste is the order parameter of the agricultural waste sup-

Z. Wang · Z. Wang (✉) · N. Tahir · H. Wang · J. Li · G. Xu
College of Mechanical and Electrical Engineering, Henan Agricultural University, Zhengzhou, Henan Province, China
e-mail: zfking@henau.edu.cn

Z. Wang et al. (eds.), *Agricultural Waste Power Generation Supply Chain*, Industrial Ecology and Environmental Management 7,
https://doi.org/10.1007/978-3-032-20087-7_4

ply chain and has positive feedback in straw supply chain system, which can lead the development of the system. Increase in selling price can promote the synergetic development of the agricultural waste supply chain. The results indicate that subsidies affect the synergetic development of the agricultural waste supply chain, and an increase in government subsidies will promote the development of agricultural waste power generation system. The findings suggest that a government subsidy of 53 CNY/t would optimize utility, given the existing synergetic system. The results can be helpful for government policymakers to support agricultural waste power generation projects.

Keywords Biomass power generation · Supply chain · Synergetic development · Government subsidies · Marginal utility

4.1 Introduction

Economy of the world depends heavily on the consumption of energy produced from fossil fuels, and depletion of fossil fuel reserves is a major threat to energy security. At the same time, burning of fossil fuels is piling up pollutants to our environment, which eventually affects not only human health but also the whole ecosystem. Therefore, it is the responsibility of every country in the world to explore renewable energy resources to alleviate the energy crisis and environmental pollution. Biomass has emerged as a clean renewable energy resource and the fourth largest energy reserve to replace fossil fuels (Saxena et al. 2009). However, current biomass energy utilization is still below par in minimizing our dependence on fossil fuel. According to China's 13th Five-Year Plan for comprehensive utilization of straw, in 2015 only 100 million tons out of 720 million tons were employed for energy conversion, which was only 11.4%. Compared with 43.2% (390 million tons of straw) used for fertilizer, energy utilization of straw was minimal (Golecha and Gan 2016). It is suggested that the main reason for this difference is the complex straw collection mechanism, low straw density, high cost of straw collection (An and Searcy 2012; Wang and Watanabe 2017; Zhang et al. 2014), and lack of effective straw supply chain management and collaboration ability (Zhao and Yan 2012). Straw supply chain is an important part of the straw power generation industry. The efficient operation of the straw supply chain not only helps reduce costs but also ensures the supply of straw, so it is worth studying the straw supply chain.

The straw supply chain comprises three parts: farmers, straw brokers, and straw power plants, where each part plays a vital role (Wen and Zhang 2015). Study of straw supply chain requires understanding of individual participants. It has been observed that farmer's decisions play a vital role in affecting the collection and supply chain of biomass (Sherrington et al. 2008; Wang and Watanabe 2016). Farmer's enthusiasm can be increased by providing market security and land production subsidy policies from the government (Glithero et al. 2013). It has also been suggested that the government can sign contracts with farmers to guarantee their income and subsidies (Sherrington and Moran 2010; Spiegel et al. 2018). Not only government subsidy but also awareness of benefits to farmers can improve the supply chain of the straw. By analyzing the comprehensive mode of local biomass fuel production, it has been observed that comprehensive production can provide extra income to farmers, which is conducive to their participation in biomass fuel production (Cohen et al. 2018). Keeping in mind China's traditions, Zhang et al. (2016) introduced the concept of organization of the village committee to study the influence of the committee on the behavior of farmers and their willingness to participate. It has been seen that farmers' willingness to supply straw can be improved after farmers develop trust in intermediaries and awareness of supply chain risks from the perspectives of education, income, and trust (Wang and Watanabe 2017). Straw brokers have the responsibility to complete the logistics activities (straw collection, storage, processing, transportation, and other processes) of straw from farmers to power plant. An efficient supply chain requires strong cooperation between these processes; however, various straw characteristics and the scale of power plants may lead to several problems in each process. It has been proposed that mixed storage methods can overcome issues associated with seasonal straw harvesting (Rentizelas et al. 2009). Ebadian et al. (2013) analyzed three different storage scenarios with minimum transportation cost by considering the number and location of farms and establishing a straw storage model to allocate straw to different storage points. Zhang et al. (2012) optimized biomass fuel location by simulating the impact of extreme weather, such as snow, on the straw supply chain in northern part of China. Sun et al. (2017) analyzed different transportation modes of straw brokers to reduce transportation cost.

Straw power plant is at the end of straw supply chain system, which converts biomass energy into electrical energy. The operation of power plants requires sufficient straw, but the equipment operational efficiency of many straw power plants in China is no more than 60% (Zhao and Yan 2012). In order to improve the current situation, Lv et al. (2014) considered the straw supply chain and conducted a site selection optimization study of straw power plant. Zhao et al. (2017) proposed a spatially explicit optimization model to optimize the site selection of power plants. Zhang et al. (2013) suggested that acquisition contracts can guarantee the sustainable supply of raw materials.

The straw supply chain is like a bridge to link production areas and power plants of straw. Breakages in the supply chain can interrupt the flow of straw. Many scholars have studied the straw supply chain in order to build an efficient and sustainable supply chain. Wang et al. (2015) showed that scientific design and planning can ensure an adequate supply of straw. Singh (2016) showed that the economy of power generation systems can be improved through biomass management system. Mafakheri and Nasiri (2014) studied the operation and management model of biomass supply chains by considering the variability of biomass supply and sources. Ng and Maravelias (2017) proposed a multi-period mixed-integer model to study the design and planning of biomass supply chains. Yu et al. (2012) and Zhao and Li (2016) established straw collection cost model combined with Geographic Information System (GIS) model to study the planning and allocation of biomass power plants. By analyzing the impact of biomass species on the supply chain through multiscale modeling, Yue et al. (2014) showed that the straw supply chain had low operating efficiency due to unbalanced profit distribution within the system. Cao et al. (2016) constructed a straw supply system of regional network based on uneven straw distribution in rural areas. It has been observed that the development of the renewable energy power generation industry depends on power generation costs and government investment in the industry (Reddy 2018). Liu et al. (2018) designed an incentive mechanism for the straw power supply chain by combining system dynamics and simulation. A dynamic incentive mechanism based on game theory was developed by Wang et al. (2017) and Wen and Zhang (2015) by considering the relationships between supply chain participants. Zhao et al. (2017) also implemented game theory to design a synergetic mechanism for stakeholders. Wolde et al. (2017) studied incentive factors of farmers based on socioeconomic methods to improve their willingness to promote the supply of biomass. Reddy (2018) showed that economic compensation is an important factor to encourage farmers to supply straw for biomass-to-energy conversion. From the limited available literature, it is clear that, being an external factor, government has a great impact on the straw supply chain. External policy incentives can effectively improve the development of the straw supply chain system.

To sum up, research on the straw supply chain mainly focuses on the analysis of supply chain factors, optimization of supply chain models, network construction, policy incentives, etc. In the present work, we have studied the synergetic development of the straw power supply chain by employing synergetic theory combined with government subsidy policy, as little work has been done in this area. In order to fill this gap, the present study includes straw price and government subsidy as incentives to carry out research on the synergetic development of the supply chain. We provide a method and theoretical basis for the synergetic development of the straw power supply chain.

4.2 Methodology

4.2.1 Straw Power Supply Chain System

The straw supply chain system is a complex system, and based on the different functions of participants, it is divided into three subsystems: supply subsystem, transport subsystem, and power generation subsystem. The model of the straw supply chain is shown in Fig. 4.1.

At present, there are two main modes of straw supply chain. The first mode is direct transportation, whereas the second mode is indirect transportation through straw brokers. In China, however, scattered farmlands increase the cost of straw power plants due to the transportation. The second mode can reduce the working intensity and improve the operating efficiency by adding straw brokers as straw purchasing agents. It has been observed that adoption of the agency model can

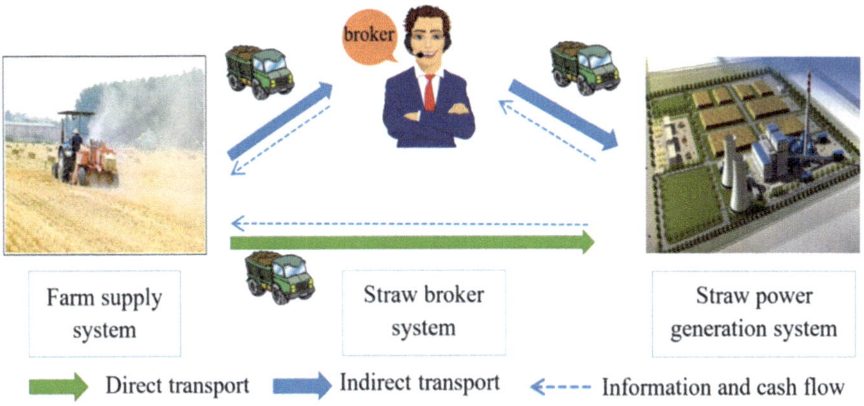

Fig. 4.1 Transportation model of straw supply chain

reduce the cost of the supply chain (Golecha and Gan 2016). Therefore, we have selected the second mode as the research object to study the synergetic development of the supply chain composed of farmers, brokers, and power plants.

4.2.2 Characteristics of Straw Supply Chain

The efficient operation of the straw supply chain requires cooperative efforts from all the participants. Based on the different purposes and activities of participants, the supply chain system has the following characteristics.

1. Dynamic development characteristics: During the operation of the straw supply chain from farmland to the power plant, it passes through exchanges of material flow, information flow, and cash flow. These dynamic characteristics make each participant have the right of independent choice.
2. Independent characteristics: Each subsystem of the straw supply chain has its own characteristics. During actual operation of the supply chain, the participants pursue the maximization of their own benefits. For example, farmers are the owners of straw resources, and they try to sell straw only at high prices.
3. Nonlinear correlation: Each subsystem depends on, restricts, and interacts with each other. The nonlinear correlation between subsystems makes the system have complex characteristics such as diversity and unpredictability.
4. Self-organize characteristic: The self-organization of the system urges the three subsystems to develop a stable supply chain system.
5. Open features: The straw supply chain system is permanently associated with the external environment and is affected by the outside environment. For example, the supply chain system can develop harmoniously under the action of policy, and straw use will change under competition in market straw resources.

The above-described complex characteristics make straw supply chain development difficult. The function of the system is not only to add subsystem functions but also to analyze the influencing factors of the system. The development of the straw supply chain system is dominated by some factors, and the changes in rules of the system can be found after defining these leading factors.

4.2.3 Model Construction

According to the synergetic theory established by Haken, it is well established that the subsystems within a complex system are in competition and collaboration, where their relationships help the subsystems to form an organization with certain functions. Synergetic theory points out that the system has states of order and disorder, which define the system direction. The concept of ordered degree is introduced to show the system's degree of order (Perez-Fortes et al. 2012). In the study of coevolution systems, order parameters represent the order degree of the system. Synergetic theory holds that the structural change of a system determined by a few order parameters. The evolution trend of the system can be studied by analyzing the order parameters. In the present work, a synergetic development model of the straw supply chain system has been proposed by employing synergetic theory (Wang et al. 2020). The main steps are as follows:

Step 1: Selection of system state variables

The subsystems of the straw supply chain play an important role, as they are affected by multiple factors. The main factors need to be selected as the state variables of the system and are set to x_i. The subsystems and their influencing factors of the straw supply chain are summarized in Table 4.1.

Table 4.1 Influencing factors of straw power supply chain system

Subsystem	Influencing factors
Farm supply system	Farmers' willingness to supply
	Sale price of straw
	Labor force level
Straw broker system	Cost of purchasing straw
	Transportation cost of straw
	Straw storage cost
	Cost of straw processing
Straw power generation system	Straw purchase price
	Straw purchase amount

Step 2: Establishment of the synergetic evolution equation of the system

Langevin's evolution equation of synergetic theory is adopted in the present work. The synergetic evolution equation of the straw supply chain can be obtained by improving the idea of Al-Agtash and Su (2002), as shown in Eq. (4.1).

$$\frac{d_{x_i}}{dt} = -\partial_i x_i + \sum_{j \neq i}^{m} \left(\gamma_j x_j + \beta_j x_j^2\right) + f_i(t) \tag{4.1}$$

where $-\partial_i x_i$ is the development of variables in the evolution process of the straw supply chain system, ∂_i is the damping coefficient of the variables, $\gamma_j x_j + \beta_j x_j^2$ is the nonlinear effect of variable x_j on variable x_i, $\gamma_j x_j$ is driving effect, and $\beta_j x_j^2$ is the obstruction effect. γ_j is the driving factor, β_j is hindrance factor, $f_i(t)$ is the interference of the external environment on the straw supply chain system, and m represents the number of selected parameters.

Step 3: Identification of system order parameters

According to actual survey data, EViews software has been employed to fit the system state variables. The variables ∂_i, γ_j, and β_j in Eq. (4.1) are estimated, and the coevolution model of the system is obtained. Fast and slow variables in the system are identified based on the size of ∂_i. For fast variables, larger damping coefficient ∂_i leads to rapid changes in parameter. Conversely, variables with smaller damping coefficients are characterized as "slow variables," which dominate the evolution of the system and are therefore considered to be the "order parameters."

Step 4: Establishment of the evolution equation of order parameter

The adiabatic method is used to solve the evolution equation of order parameters, supposing x_n, x_k, and x_s. In the system of synergetic evolution equations, the method of eliminating fast variables in the evolution equation as shown in Eq. (4.2).

$$\begin{cases} \dfrac{d_{x_1}}{dt} = F_1\left(g_1\left(x_n, x_k, x_s\right), g_2\left(x_n, x_k, x_s\right), \cdots, g_n\left(x_n, x_k, x_s\right)\right) = G_1\left(x_n, x_k, x_s\right) \\ \dfrac{d_{x_2}}{dt} = F_2\left(g_1\left(x_n, x_k, x_s\right), g_2\left(x_n, x_k, x_s\right), \cdots, g_n\left(x_n, x_k, x_s\right)\right) = G_2\left(x_n, x_k, x_s\right) \\ \vdots \\ \dfrac{d_{x_m}}{dt} = F_m\left(g_1\left(x_n, x_k, x_s\right), g_2\left(x_n, x_k, x_s\right), \cdots, g_n\left(x_n, x_k, x_s\right)\right) = G_m\left(x_n, x_k, x_s\right) \end{cases} \tag{4.2}$$

where $g_i(x_n, x_k, x_s)$ is the i-th state variable in terms of order parameters x_n, x_k, and x_s. $G_m(x_n, x_k, x_s)$ represents the evolution function of order parameters. One can study changes in the system with multiple order parameters by obtaining a comprehensive order parameter. Let r be the comprehensive order parameter, and the calculation method is shown in Eq. (4.3).

$$r = \sum_{i=1}^{m} \omega_i x_i \tag{4.3}$$

where ω_i is the weight coefficient of the i-th order parameter, which satisfies the conditions $\sum_{i=1}^{m} \omega_i = 1, 0 \leq \omega_i \leq 1$. The evolution equation of the comprehensive order parameter can be obtained as shown in Eq. (4.4).

$$\frac{dr}{dt} = a + br + cr^2 + \delta r^3 \cdots er^n \tag{4.4}$$

Step 5: Calculation of the potential function of straw supply chain

According to mutation theory, every system has a certain potential function (potential "energy" of the system). The increase of the potential function leads to the instability of the system. The potential function and the evolution equation of the order parameter are described by the following relationship, as shown in Eq. (4.5).

$$\frac{dr}{dt} = -\frac{\partial_v}{\partial r} \tag{4.5}$$

The potential function of the straw supply chain can be obtained by combining Eqs. (4.4) and (4.5). By analyzing the extreme value of potential function, we can estimate the running state of the straw supply chain system. The trend of the potential function can be adjusted by changing the value of the order parameter with internal or external excitation.

4.2.4 Method for Solving the Synergetic Evolution Equation

In the process of system coevolution, different factors have different effects. The force of each factor on the system can be expressed by a specific function. The coevolution process of the system can be described by a set of equations through the analysis of multiple factors. The synergetic evolution equations of the system can be solved by the following methods:

Step 1: Dimensionless processing of data

Set the original data of the system as $x_i(t)$ and process it using the method described in Eq. (4.6), and set the processed data as $x_i^0(t)$.

$$x_i^0(t) = \frac{x_i(t) - x_i(t_{\min})}{x_i(t_{\max}) - x_i(t_{\min})} \tag{4.6}$$

where $x_i(t_{max})$ is the maximum value of parameter x_i, and $x_i(t_{min})$ is the minimum value.

- Step 2: Data accumulation

According to grey theory, the change rule of the original data is not obvious. However, after the accumulation of the data, it starts growing exponentially. Therefore, $x_i^0(t)$ needs to be accumulated

$$x_i^1(t) = \sum_{k=1}^{t} x_i^0(t) \tag{4.7}$$

- Step 3: Parameter solution

If $f_i(t) = 0$ and $-\partial_i = \gamma_{ii}$, Eqs. (4.8) and (4.9) can be obtained.

$$\frac{dx_i^1}{dt} = x_i^1(t) - x_i^1(t-1) = x_i^0(t), t = 2, 3, \cdots n \tag{4.8}$$

$$x_i^0(t) = \sum_{j=1}^{m} \gamma_{ij} x_j^1(t) + \sum_{j \neq i}^{m} \beta_{ij} \left(x_j^1(t)\right)^2 \tag{4.9}$$

Take $t = 2, 3, \ldots n$ into Eq. (4.9); when $i = 1$, we can get Eq. (4.10):

$$\begin{cases} x_1^0(2) = \sum_{j=1}^{m} \gamma_{1j} x_j^1(2) + \sum_{j \neq i}^{m} \beta_{1j} \left(x_j^1(2)\right)^2 \\ x_1^0(3) = \sum_{j=1}^{m} \gamma_{1j} x_j^1(3) + \sum_{j \neq i}^{m} \beta_{1j} \left(x_j^1(3)\right)^2 \\ \vdots \\ x_1^0(\mathrm{n}) = \sum_{j=1}^{m} \gamma_{1j} x_j^1(n) + \sum_{j \neq i}^{m} \beta_{1j} \left(x_j^1(n)\right)^2 \end{cases} \tag{4.10}$$

Equation (4.10) can be converted into a matrix form as shown in the following: $y = X\vartheta$.

$$y = [x_1^0(2) \quad x_1^0(3) \cdots \quad x_1^0(n)]^T \tag{4.11}$$

$$\vartheta = [\gamma_{11} \quad \gamma_{12} \cdots \quad \gamma_{1m} \; \beta_{11} \quad \beta_{12} \cdots \beta_{1m}]^T \tag{4.12}$$

$$X = \begin{bmatrix} x_1^1(2) & x_2^1(2) \cdots & x_m^1(2) & \left(x_2^1(2)\right)^2 & \left(x_3^1(2)\right)^2 \cdots & \left(x_m^1(2)\right)^2 \\ x_1^1(3) & x_2^1(3) \cdots & x_m^1(3) & \left(x_2^1(3)\right)^2 & \left(x_3^1(3)\right)^2 \cdots & \left(x_m^1(3)\right)^2 \\ \vdots & \vdots \cdots & \vdots & \vdots & \vdots \cdots & \vdots \\ x_1^1(n) & x_2^1(n) \cdots & x_m^1(n) & \left(x_2^1(n)\right)^2 & \left(x_3^1(n)\right)^2 \cdots & \left(x_m^1(n)\right)^2 \end{bmatrix} \quad (4.13)$$

The result of coefficient matrix ϑ can be obtained by employing EViews software to estimate the parameters of the multivariate linear regression model, as shown in Eq. (4.10). Similarly, the synergetic evolution equations of the system can be extracted by solving the evolution equations of each parameter with $i = 2, 3, \ldots, m$.

4.2.5 *Data Sources*

Shandong is one of the major agricultural provinces in China, which is rich in straw resources. In the present work, we studied the incentive effect of government subsidies on the synergetic development of the straw supply chain by considering a biomass power plant in Shandong as a case study. Based on the previous study by Wolde et al. (2017), we stipulate the dimensionless representations of straw selling price, transportation cost, and straw purchase price as x_1, x_2, and x_3, respectively. The data are summarized in Table 4.2.

Table 4.2 Statistical data of main parameters of biomass power generation system

Year	Sale price of straw (CNY/t)	Transportation cost of straw (CNY/t)	Straw purchase price (CNY/t)
2007	0.0000	0.0833	0.0667
2008	0.0889	0.0000	0.0000
2009	0.0667	0.1667	0.1429
2010	0.2222	0.2917	0.2667
2011	0.4222	0.3750	0.4571
2012	0.6667	0.4583	0.5524
2013	0.6889	0.5833	0.7238
2014	0.8222	0.7917	0.9143
2015	1.0000	1.0000	1.0000

4.3 Results and Discussion

4.3.1 Calculation of Results

EViews software was used to solve synergetic evolution equations of the straw supply chain system. The results are shown in Eq. (4.14).

$$\begin{cases} \frac{d_{x_1}}{dt} = 0.0039 + 0.5034x_1 - 0.3195x_2 + 0.4956x_3 + 0.8762x_2^2 \\ \frac{d_{x_2}}{dt} = -0.0017 + 0.5293x_2 + 0.7753x_3 + 0.3285x_1^2 - 0.4796x_3^2 \\ \frac{d_{x_3}}{dt} = 0.6579x_2 + 0.5277x_3 + 0.0823x_1^2 - 0.0249 \end{cases} \quad (4.14)$$

It can be seen from the synergetic evolution Eq. (4.14) that the damping coefficients of the three subsystems are 0.5034, 0.5293, and 0.5277, respectively. The damping coefficient of state variable x_1 (sale price of straw) is the minimum, so x_1 is the order parameter of the system. The evolution equation of the order parameter is described as shown in Eq. (4.15).

$$\frac{d_{x_1}}{dt} = -0.0419 + 0.5034x_1 - 0.0856x_1^2 + 0.9253x_1^4 \quad (4.15)$$

The potential function V of the order parameter x_1 of the straw supply chain can be concluded as shown in Eq. (4.16).

$$V = 0.0419x_1 - 0.2517x_1^2 + 0.0285x_1^3 - 0.1851x_1^5 \quad (4.16)$$

where V is the potential function of the straw power supply chain, and x_1 is the order parameter of the straw supply chain system.

4.3.2 Result Analysis

By analyzing the change in potential function and its extreme values, we estimated the running situation of the straw supply chain system. The change in potential function has the following four scenarios:

1. When $dv/dx = 0$, $d^2v/dx^2 > 0$, potential function is at the minimum point, whereas the straw supply chain system is in a stable state, and the system is in good running state.

2. When $dv/dx = 0$, $d^2v/dx^2 = 0$, the system is in a critical condition, where straw supply chain system is under mutation, which leads to an unstable state.
3. When $dv/dx = 0$, $d^2v/dx^2 < 0$, the potential function has a maximum point, and the straw supply chain system has the greatest potential "energy," which leads to an unstable system state. The order parameter of the system may change.
4. When $dv/dx \neq 0$, there are two trends: toward increasing price if dv/dx is negative and toward lower price if dv/dx is positive.

By selecting $dv/dx_1 = 0$, that is, $0.0419 - 0.5034x_1 + 0.0855x_1^2 - 0.9255x_1^4 = 0$, we get $x_1 = 0.0844$. By solving the second derivative of Eq. (4.16), we get Eq. (4.17):

$$\frac{d^2v}{d(x)^2} = -0.503 + 0.1712x_1 - 3.7012x_1^3 \tag{4.17}$$

By inserting the value of $x_1 = 0.0844$ into Eq. (4.17), we get $d^2v/d(x)^2 = -0.4912 < 0$. As described earlier, under this criterion, the straw supply chain system has a maximum point, and the system is unstable. Since $d^2v/d(x)^2 < 0$, the potential function is a convex function. When $x > 0.0844$, all functional values are less than the maximum value point; therefore, no mutation occurs in the potential function of the straw power supply chain system. It means that the order parameter of the system does not change. The gradual decrease in potential function of the system shows that the adjustment of the selling price of straw can effectively promote the stable development of the straw supply chain, which is very important for the synergetic development of the straw supply chain.

4.3.3 Discussion

4.3.3.1 Current Situation Analysis of Straw Supply Chain

According to the evolution function, the positive coefficient of x_1 leads to an increase in the straw sales price and has a positive feedback effect on the synergetic development of the straw supply chain. The negative coefficient of x_2 shows an increase in transportation price of straw for straw brokers and has a negative feedback effect on the system. When coefficient of x_3 is positive, purchase price of straw in power plant has a positive feedback effect on the synergetic development of the system.

It has been observed that, in the farmer subsystem, the sale price of straw and transport price are growing slowly (see Fig. 4.2). The growth trend of the purchase price in power plants is relatively large. This suggests that power plant enterprises are more willing to strengthen the synergetic development of the straw supply chain. However, the price of straw sold by farmers grows slowly, with a lower growth rate. The selling price of straw is the order parameter and leads the development of the system. The synergetic development of the supply chain system is promoted with an

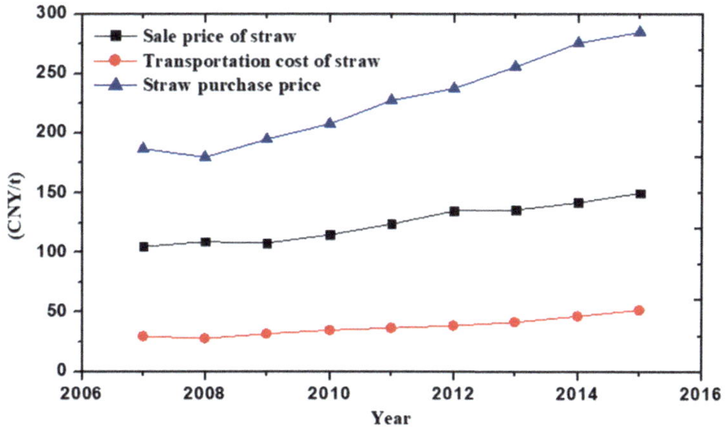

Fig. 4.2 Annual statistics of straw supply chain system

increase in the selling price of farmers' straw. Therefore, external forces are useful to promote the synergetic development of the system, and the government can serve as the main driving force.

4.3.3.2 The Impact of Government Subsidies on the Supply Chain

By incorporating the data from Table 4.2, it is assumed that the government started to subsidize the selling prices of farmers' straw in 2012. In order to observe the impact of government financial subsidies on the straw supply chain, we set the price of government subsidies at a gradient of 10 CNY/t. The potential functions of the straw supply chain under different subsidy prices are summarized in Table 4.3.

Every running system has a potential function. A higher potential function of the system leads to deviation from its current situation, and, eventually, the system enters a relatively disordered state of operation. On the contrary, the system is in a relatively orderly state of operation. The potential function evolution diagram under different subsidy prices is shown in Fig. 4.3.

It is clear that, with an increase in government financial subsidies, the potential function value of the straw supply chain gradually decreases (see Fig. 4.3). The findings suggest that the straw supply chain system can be in a stable state with a high potential for synergetic development under government financial support. Government subsidies to farmers affect their willingness to supply straw. The higher government subsidies can encourage farmers to collect more straw, which ensures the operation of the supply chain. At the same time, subsidies also affect the straw power generation system, and sufficient straw can improve the production efficiency of the power plants. Furthermore, government subsidies may help to increase the enthusiasm of participants and eventually improve the efficient operation of the straw power supply chain.

Table 4.3 Supply chain potential functions under different subsidy prices

Subsidized price	Potential function
10 CNY/t	$V = -0.02128x_1 - 0.47458x_1^2 - 0.22752x_1^3 - 0.00056x_1^4$
20 CNY/t	$V = -0.03425x_1 - 0.52245x_1^2 - 0.23361x_1^3 + 0.00154x_1^4$
30 CNY/t	$V = -0.04939x_1 - 0.53822x_1^2 - 0.23281x_1^3 + 0.00263x_1^4$
40 CNY/t	$V = -0.06027x_1 - 0.55399x_1^2 - 0.23243x_1^3 + 0.00323x_1^4$
50 CNY/t	$V = -0.07185x_1 - 0.56901x_1^2 - 0.23214x_1^3 + 0.00344x_1^4$
60 CNY/t	$V = -0.08286x_1 - 0.57706x_1^2 - 0.22671x_1^3 + 0.00344x_1^4$
70 CNY/t	$V = -0.08909x_1 - 0.58485x_1^2 - 0.22665x_1^3 + 0.00339x_1^4$
80 CNY/t	$V = -0.09549x_1 - 0.59265x_1^2 - 0.22662x_1^3 + 0.00328x_1^4$

When $x_1 = 1$, we get nine points of a, b, c, d, e, f, g, m, and p (from Fig. 4.3). It has been observed that a continuous increase in government subsidy price leads to a gradual decrease in potential function of the straw supply chain system. By analyzing these nine points, it is observed, for subsidy price of 0–20 CNY/t, the potential function value of the straw supply chain system decreases rapidly with abrupt changes. When the financial subsidy is 20–50 CNY/t, the decay rate of the potential function value tends to decrease. However, at a subsidy of 50–80 CNY/t, there is a slight decrease in the value of the potential function. In order to clarify the promotion effect of government subsidies on the straw supply chain, the abovementioned nine points from a to p require further insight. In order to find the optimal subsidy price, the change in the potential function of the system is deeply studied for different subsidy prices.

4.3.3.3 Government Optimal Subsidy Price Decision

The government provides subsidies to encourage straw utilization for the reduction of greenhouse gas (GHG) emissions. Therefore, the study assumes that government decision-makers are independent economic agents. This manuscript studies the effect of government subsidies on the straw power supply chain based on utility theory. Marginal utility refers to addition of satisfaction gained from consuming units of goods or services. According to the law of diminishing marginal utility, an increase in consumption leads to an increase in marginal utility, and consequently, total utility increases by a large margin. When consumption reaches a certain level, marginal utility decreases gradually, whereas total utility continues to increase. When marginal utility decreases to zero, there is no further increase in total utility. The marginal utility theory provides a scientific method for studying the utility of government subsidies.

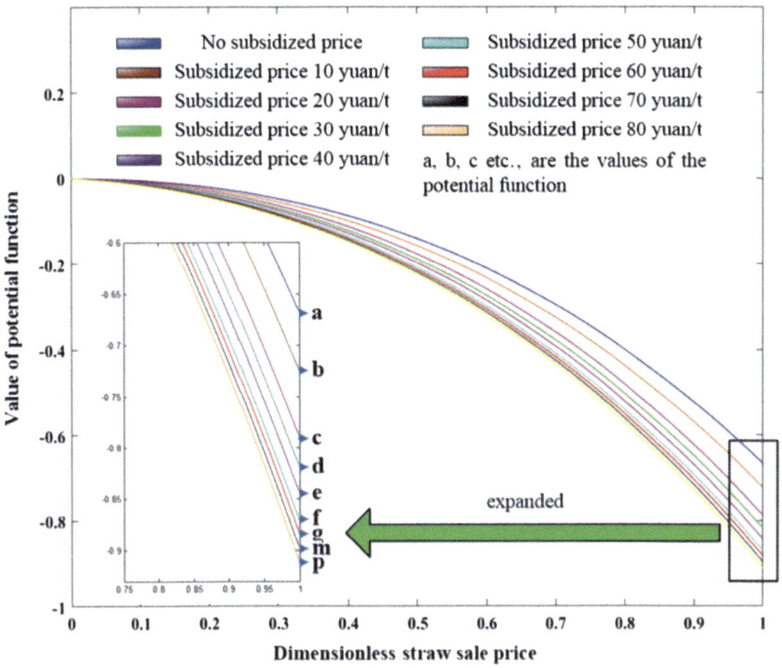

Fig. 4.3 Supply chain potential function under different subsidy prices

The purpose of government subsidies is to encourage the energy utilization of straw. From field research, we know that increasing subsidies can improve the energy utilization of straw, but the government cannot provide unlimited subsidies due to limited ability. Therefore, it is necessary to study the optimal subsidy price under the government's expectations. Based on the study of the potential function (see Fig. 4.3), we found that increasing subsidies can reduce the potential function. Therefore, we assume that the changes in the potential function as the utility of government subsidies. With reference to utility theory, the utility of government subsidies, the marginal utility of government subsidies, and the minimum expected utility of fiscal subsidies are defined as follows.

Definition 1: The utility of government subsidies refers to the total change in the potential function caused by government financial subsidies.

Definition 2: The marginal utility of fiscal subsidies (marked as h) refers to the change in the potential function of increasing the unit subsidy.

Definition 3: The minimum expected utility of fiscal subsidies (marked as H) refers to the minimum change in the potential function after the government provides unit subsidy.

We assume H and h as the minimum expected utility of government financial subsidy and the marginal utility of the subsidy price, respectively. If $h > H$, the government will increase the subsidy price. If $h < H$, the increase in subsidy prices will have only a small effect, and the government will stop increasing the subsidy price. The minimum expected utility of government subsidies (*H*) can be calculated by Eq. (4.18).

$$H = \frac{\sum_{i=1}^{t} f(t) - f(\max) - f(\min)}{t} \tag{4.18}$$

where *f*(*t*) is the change in the potential function of the straw supply chain system after the subsidy price is increased by one gradient, $\sum_{i=1}^{t} f(t)$ is the sum of the changes in the potential function of the system, *f*(*max*) is the maximum change in the potential function, *f*(*min*) is the minimum change in the potential function, and *t* is the number of gradients.

According to Fig. 4.3 (expanded in the window), when $x = 1$ is postulated, the values of the potential function under different government subsidies can be obtained as shown in Table 4.4.

By analyzing the data in Table 4.4, the effect of incentive from the government subsidies on the synergetic development of the straw supply chain can be clearly observed. By analyzing the variation in the potential function, the trend of data is shown in Fig. 4.4. It can be seen that, with the increase in subsidy price, the change in potential function of the straw supply chain gradually decreases and stabilizes. In order to analyze the change law of potential function change amount, a quadratic fitting method is chosen. The fitting results are shown in Fig. 4.4.

The grey system theory highlights that the original data sequence of system elements has no obvious variation law. However, after data accumulation, the new sequence follows an exponential growth law. The changes in the potential function were accumulated and dimensionless. The utility graph of the subsidized prices (the relationship between the government subsidy price and the change of potential function) is obtained, as shown in Fig. 4.5.

Table 4.4 Values and variations of potential functions under different subsidy prices

Subsidized price	Function value	Amount of change	Cumulative change
0	−0.66689	0	0
10 CNY/t	−0.72394	0.05705	0.05705
20 CNY/t	−0.78876	0.06482	0.12187
30 CNY/t	−0.81778	0.02902	0.15089
40 CNY/t	−0.84343	0.02565	0.17654
50 CNY/t	−0.86955	0.02612	0.20266
60 CNY/t	−0.88318	0.01363	0.21629
70 CNY/t	−0.89720	0.01402	0.23031
80 CNY/t	−0.91149	0.01429	0.24460

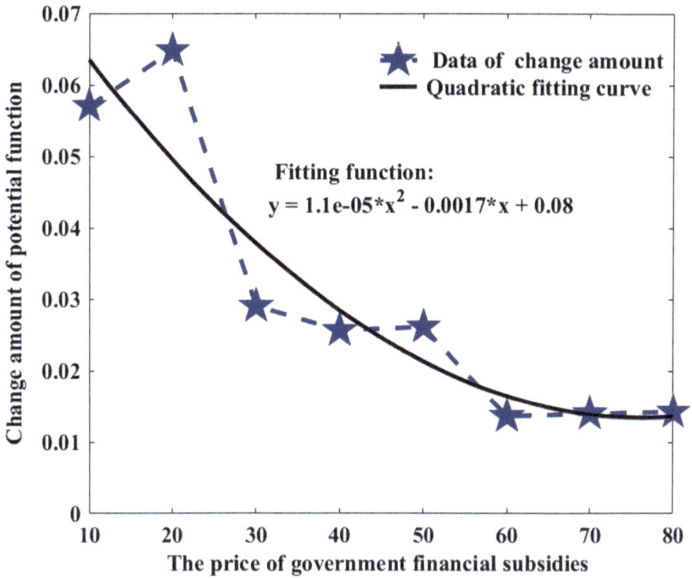

Fig. 4.4 Gradient variation of potential function in straw supply chain system

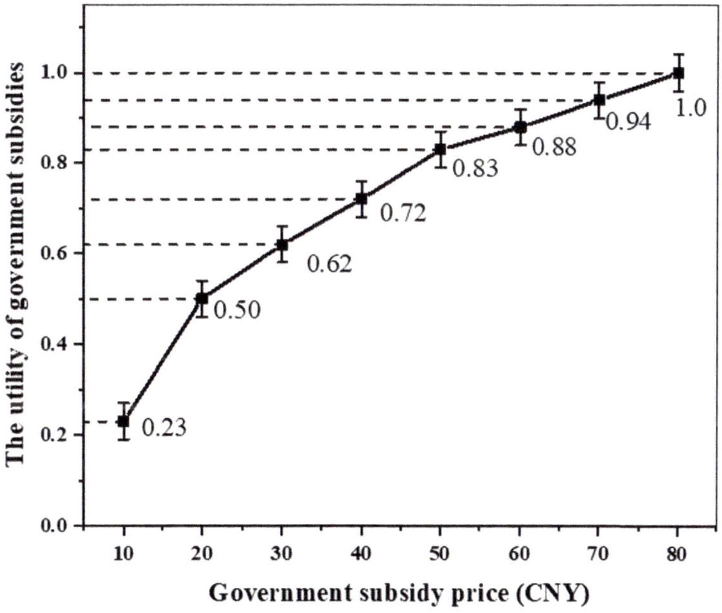

Fig. 4.5 Utility of government subsidies

Table 4.5 Effect of government minimum expected utility on the optimal subsidy price

Change in minimum expected utility	−20%	−10%	0	+10%	20%
Minimum expected utility	0.0166	0.0187	0.0208	0.0229	0.0249
Optimal subsidy price (CNY)	63	57	53	49	46

An increase in subsidy price leads to an increase in utility (see Fig. 4.5). This confirms that government subsidy policy plays a significant role in promoting the synergetic development of the straw supply chain system. However, when the straw subsidy price increases to a certain value, the utility of government subsidies has a very small change, which suggests that the increase in subsidy price has minor effect on the straw supply chain.

The minimum expected utility H of government fiscal subsidies is calculated using Eq. (4.18), which is 0.0208. This shows that change in potential function cannot be lower than 0.0208; otherwise, the government will not provide more subsidies. The optimal subsidy price can be calculated at this time.

Suppose the change in the potential function is the marginal utility of the unit subsidy price (h), then the relationship between marginal utility and subsidy price can be calculated through the fitting function in Fig. 4.4, as shown in Eq. (4.19).

$$h = \frac{1.1}{10^5}x^2 - 0.0017x + 0.08 \tag{4.19}$$

When $h = H = 0.0208$, the optimal subsidy price is 53 CNY. Therefore, when the government subsidy price of straw to farmers is 53 CNY/t, the straw supply chain can achieve no better synergetic development.

4.3.3.4 Sensitivity Analysis of Optimal Subsidies

From above analysis, we know that the minimum expected utility of government subsidies affects the optimal subsidy price. In order to analyze the impact of the minimum expected utility on the optimal subsidy, a sensitivity analysis was conducted. The results are shown in Table 4.5.

It can be seen that the larger the minimum expected utility of the government subsidy, the smaller the optimal subsidy price (Table 4.5). The results of the sensitivity analysis of the optimal subsidy price can provide assistance to government decision-makers to choose a reasonable subsidy price.

4.4 Conclusions and Policy Implications

This study investigated the synergetic development of the straw supply chain and showed that the selling price of farmers' straw dominates the development of the straw supply chain. By analyzing the coevolution equations of the supply chain, it was found that an increase in selling price of straw can promote the synergetic development of the supply chain. Government financial support can promote the synergetic development of the straw supply chain. According to our findings, the government subsidy of up to 53 CNY/t can provide synergetic benefit for the existing straw supply chain system.

The present work suggests that development of the straw power generation industry can be promoted by building a straw supply chain with synergetic development, improving the operation efficiency, and enhancing the enthusiasm of supply chain participants. Based on the above observations, the following suggestions are made:

First, the supply of straw is the beginning of supply chain system and requires more attention. The sufficient supply of straw can provide power for the development of supply chain. In China, the farmer's planting is dispersed, and the amount of straw from per farmer is small, so the collection of straw is difficult. According to the results of this study, it is suggested that the government should actively assume social responsibility, implement the central government's policy against burning straw, and innovate the method of "turning blockage into thinning." Government financial subsidies can increase farmers' income, which can stimulate farmers to reduce the open burning of straw and provide more straw raw materials.

Second, it is necessary to innovate the benefit distribution mechanism among the members of the straw supply chain, improve the coordination relationship among the participants, and build a synergetic development supply chain. Unfair distribution of benefits will reduce the enthusiasm of participants. If a participant in the supply chain slackens or stops working, it will reduce the operational efficiency of the straw supply chain. The government allocates the financial subsidy (raised by the government and the straw power plant) to farmers, which actually distributes income from the power plant to farmers. This promotes farmers' willingness to supply and the participation of brokers and makes the straw power supply chain develop steadily and coordinately.

Third, to build a stable straw trading market, straw as a raw material can be used not only for power generation but also for other purposes. This leads to competition for straw resources. In order to compete for resources, enterprises often raise the price of straw, which easily leads to confusion in the straw market and increases production pressure on enterprises. Therefore, it is necessary to build a stable trading market through contracts and reasonably allocate straw resources. The government provides subsidies for straw power generation projects, which can increase the selling price of straw, improve competitiveness, and bring the synergetic benefits to the straw power supply chain system.

Finally, government price subsidies help attract business investment and promote the coordinated development of supply chains. Straw power generation is a labor-intensive industry, which requires a lot of manpower and material resources during the straw harvest season, which results in higher costs.

Acknowledgments This chapter has been reprinted from Z. Wang, Z. Wang, N Tahir, et al. *Study of Synergetic Development in Straw Power Supply Chain: Straw Price and Government Subsidy as Incentive* [J]. Energy Policy, 2020, 146: 111788. Copyright (2020), with permission from Elsevier.

References

Al-Agtash S, Su RJ (2002) Economic efficiency of coordinated multilateral trades in electricity markets. Int J Electr Power Energy Syst 24(10):843–850. https://doi.org/10.1016/S0142-0615(01)00090-4

An H, Searcy SW (2012) Economic and energy evaluation of a logistics system based on biomass modules. Biomass Bioenergy 46:190–202. https://doi.org/10.1016/j.biombioe.2012.09.002

Cao J, Pang B, Mo X, Xu F (2016) A new model that using transfer stations for straw collection and transportation in the rural areas of China: a case of Jinghai, Tianjin. Renew Energy 99:911–918. https://doi.org/10.1016/j.renene.2016.07.061

Cohen M, Lepesant G, Lamari F, Bilodeau C, Benyei P, Angles S et al (2018) Biomolecules from olive pruning waste in sierra Magina—engaging the energy transition by multi-actor and multidisciplinary analyses. J Environ Manag 216:204–213. https://doi.org/10.1016/j.jenvman.2017.03.067

Ebadian M, Sowlati T, Sokhansanj S, Townley-Smith L, Stumborg M (2013) Modeling and analysing storage systems in agricultural biomass supply chain for cellulosic ethanol production. Appl Energy 102:840–849. https://doi.org/10.1016/j.apenergy.2012.08.049

Glithero NJ, Wilson P, Ramsden SJ (2013) Prospects for arable farm uptake of short rotation coppice willow and miscanthus in England. Appl Energy 107:209–218. https://doi.org/10.1016/j.apenergy.2013.02.032

Golecha R, Gan J (2016) Biomass transport cost from field to conversion facility when biomass yield density and road network vary with transport radius. Appl Energy 164:321–331. https://doi.org/10.1016/j.apenergy.2015.11.070

Liu Y, Zhao R, Wu K-J, Huang T, Chiu ASF, Cai C (2018) A hybrid of multi-objective optimization and system dynamics simulation for straw-to-electricity supply chain management under the belt and road initiatives. Sustainability 10(3). https://doi.org/10.3390/su10030868

Lv H, Ding H, Zhou D, Zhou P (2014) A site selection model for a straw-based power generation plant with CO2 emissions. Sustainability 6(10):7466–7481. https://doi.org/10.3390/su6107466

Mafakheri F, Nasiri F (2014) Modeling of biomass-to-energy supply chain operations: applications, challenges and research directions. Energy Policy 67:116–126. https://doi.org/10.1016/j.enpol.2013.11.071

Ng RTL, Maravelias CT (2017) Design of biofuel supply chains with variable regional depot and biorefinery locations. Renew Energy 100:90–102. https://doi.org/10.1016/j.renene.2016.05.009

Perez-Fortes M, Lainez-Aguirre JM, Arranz-Piera P, Velo E, Puigjaner L (2012) Design of regional and sustainable bio-based networks for electricity generation using a multi-objective MILP approach. Energy 44(1):79–95. https://doi.org/10.1016/j.energy.2012.01.033

Reddy BS (2018) Economic dynamics and technology diffusion in Indian power sector. Energy Policy 120:425–435. https://doi.org/10.1016/j.enpol.2018.05.044

Rentizelas AA, Tolis AJ, Tatsiopoulos IP (2009) Logistics issues of biomass: the storage problem and the multi-biomass supply chain. Renew Sustain Energy Rev 13(4):887–894. https://doi.org/10.1016/j.rser.2008.01.003

Saxena RC, Adhikari DK, Goyal HB (2009) Biomass-based energy fuel through biochemical routes: a review. Renew Sustain Energy Rev 13(1):167–178. https://doi.org/10.1016/j.rser.2007.07.011

Sherrington C, Moran D (2010) Modelling farmer uptake of perennial energy crops in the UK. Energy Policy 38(7):3567–3578. https://doi.org/10.1016/j.enpol.2010.02.034

Sherrington C, Bartley J, Moran D (2008) Farm-level constraints on the domestic supply of perennial energy crops in the UK. Energy Policy 36(7):2504–2512. https://doi.org/10.1016/j.enpol.2008.03.004

Singh J (2016) Identifying an economic power production system based on agricultural straw on regional basis in India. Renew Sustain Energy Rev 60:1140–1155. https://doi.org/10.1016/j.rser.2016.02.002

Spiegel A, Britz W, Djanibekov U, Finger R (2018) Policy analysis of perennial energy crop cultivation at the farm level: short rotation coppice (SRC) in Germany. Biomass Bioenergy 110:41–56. https://doi.org/10.1016/j.biombioe.2018.01.003

Sun Y, Cai W, Chen B, Guo X, Hu J, Jiao Y (2017) Economic analysis of fuel collection, storage, and transportation in straw power generation in China. Energy 132:194–203. https://doi.org/10.1016/j.energy.2017.05.077

Wang L, Watanabe T (2016) A Stackelberg game theoretic analysis of incentive effects under perceived risk for China's straw-based power plant supply chain. Energies 9(6). https://doi.org/10.3390/en9060455

Wang L, Watanabe T (2017) Influence of trust on biomass supply decision-making in China. Energies 10(11). https://doi.org/10.3390/en10111749

Wang C, Zhang L, Chang Y, Pang M (2015) Biomass direct-fired power generation system in China: An integrated energy, GHG emissions, and economic evaluation for Salix. Energy Policy 84:155–165. https://doi.org/10.1016/j.enpol.2015.04.025

Wang Y, Ebadian M, Sokhansanj S, Webb E, Lau A (2017) Impact of the biorefinery size on the logistics of corn Stover supply—a scenario analysis. Appl Energy 198:360–376. https://doi.org/10.1016/j.apenergy.2017.03.056

Wang Z, Wang Z, Tahir N, Wang H, Li J, Xu G (2020) Study of synergetic development in straw power supply chain: straw price and government subsidy as incentive. Energy Policy 146. https://doi.org/10.1016/j.enpol.2020.111788

Wen W, Zhang Q (2015) A design of straw acquisition mode for China's straw power plant based on supply chain coordination. Renew Energy 76:369–374. https://doi.org/10.1016/j.renene.2014.11.062

Wolde B, Lal P, Burli P (2017) Forestland owners' willingness to consider multiple ways of supplying biomass simultaneously: implications for biofuel incentive policies. Energy Policy 105:183–190. https://doi.org/10.1016/j.enpol.2017.02.040

Yu H, Wang Q, Ileleji KE, Yu C, Luo Z, Cen K, Gore J (2012) Design and analysis of geographic distribution of biomass power plant and satellite storages in China. Part 2: road delivery. Biomass Bioenergy 46:785–792. https://doi.org/10.1016/j.biombioe.2012.06.028

Yue D, You F, Snyder SW (2014) Biomass-to-bioenergy and biofuel supply chain optimization: overview, key issues and challenges. Comput Chem Eng 66:36–56. https://doi.org/10.1016/j.compchemeng.2013.11.016

Zhang F, Johnson DM, Johnson MA (2012) Development of a simulation model of biomass supply chain for biofuel production. Renew Energy 44:380–391. https://doi.org/10.1016/j.renene.2012.02.006

Zhang Q, Zhou D, Zhou P, Ding H (2013) Cost analysis of straw-based power generation in Jiangsu Province, China. Appl Energy 102:785–793. https://doi.org/10.1016/j.apenergy.2012.08.032

Zhang H, Li L, Zhou P, Hou J, Qiu Y (2014) Subsidy modes, waste cooking oil and biofuel: policy effectiveness and sustainable supply chains in China. Energy Policy 65:270–274. https://doi.org/10.1016/j.enpol.2013.10.009

Zhang X, Luo K, Tan Q (2016) A feedstock supply model integrating the official organization for China's biomass generation plants. Energy Policy 97:276–290. https://doi.org/10.1016/j.enpol.2016.07.027

Zhao X-g, Li A (2016) A multi-objective sustainable location model for biomass power plants: case of China. Energy 112:1184–1193. https://doi.org/10.1016/j.energy.2016.07.011

Zhao Z-y, Yan H (2012) Assessment of the biomass power generation industry in China. Renew Energy 37(1):53–60. https://doi.org/10.1016/j.renene.2011.05.017

Zhao X, Ma X, Wang K, Long Y, Zhang D, Xiao Z (2017) A spatially explicit optimization model for agricultural straw-based power plant site selection: a case study in Hubei Province, China. Sustainability 9(5). https://doi.org/10.3390/su9050832

Chapter 5
Research on Supply Chain Risk Control for Agricultural Waste Power Generation

Zhanwu Wang

Contents

Abstract Biomass power generation can reduce pollutant emissions and alleviate the energy crisis. However, the development of straw power generation supply chain is affected by various risk factors. Therefore, this study analyzed the causal relationship of risk factors, clarified the internal structure and feedback mechanisms of the risk factor system, and constructed the system dynamics model of the agricultural waste power generation supply chain to explore the impact of risk factors on the supply chain in terms of cost. Subsequently, simulations were carried out using the Vensim PLE software. The results indicate that environmental risk has the greatest impact on the total cost of agricultural waste power generation supply chain, followed by market risk and information risk, with technical risk having the least impact. Risk control strategies were then proposed for the most influential risks, and

Z. Wang (✉)
College of Mechanical and Electrical Engineering, Henan Agricultural University, Zhengzhou, Henan Province, China
e-mail: wzw6997@henau.edu.cn

Z. Wang et al. (eds.), *Agricultural Waste Power Generation Supply Chain*, Industrial Ecology and Environmental Management 7,
https://doi.org/10.1007/978-3-032-20087-7_5

the system dynamics simulation was carried out. The results show that the effect of information risk control strategies is the most effective, followed by market risk and environmental risk. Finally, the corresponding risk control strategy recommendations are proposed to reduce the cost of agricultural waste power generation. This study provides a theoretical foundation for risk control in straw power generation supply chain and contributes to the sustainable development of the agricultural waste power generation industry.

Keywords Straw power generation · Supply chain · Risk control · System dynamics

5.1 Introduction

As an important biomass energy source, straw has the characteristics of being clean and renewable (Nguea and Fotio 2024). Its use in power generation and utilization can not only reduce the consumption of fossil fuels but also alleviate environmental pollution problems (Noor et al. 2023; Zahraee et al. 2020). However, the operation of the straw power generation supply chain involves numerous links, uneven distribution of straw resources, and low level of informatization (Zhang et al. 2020). It is affected by multiple risk factors during the operation process, which causes problems such as insufficient supply of raw materials, high operating costs, and uncoordinated supply chains. Therefore, it is necessary to study the risks of straw power generation supply chain to reduce the probability of risk occurrence and decrease the costs associated with straw power generation.

The straw power generation supply chain is an essential component of the straw power generation industry. Numerous scholars have conducted research on the supply chain of straw power generation. In the economic aspect, Wang et al. (2023) constructed a straw power generation benefit distribution model to study the promotion effect of supply chain revenue coordination on straw power generation. Ebadian et al. (2013), according to the influence of different-scale storage systems on the cost of supply chain participants, planned the supply radius of agricultural biomass and the location and quantity of storage sites. Wang et al. (2024a) optimized the economic benefits of the straw supply chain by establishing a model to determine the location of facilities, the reasonable resource allocation, and the best transportation route. In terms of environmental benefits, Wang et al. (2024b) used the life cycle assessment (LCA) method to analyze the pollutant emissions at different stages of agricultural waste briquettes, and the carbon emission reduction of the briquette heating system was evaluated. Sun et al. (2020) used life cycle assessment to compare and analyze the environmental impacts of six wheat straw utilization methods. Wang et al. (2020), considering environmental and economic indicators, constructed a comprehensive sustainable evaluation model of straw direct combustion power generation. A comparison with coal-fired power generation indicated that straw-fired power generation had excellent environmental sustainability. In the straw power supply chain location and optimization aspects, Cao et al. (2021) established the integer programming model of the Location-Routing Problem of Biomass Supply Chain (LRP-BSC) and studied the two key decision-making problems of facility location and vehicle routing in biomass supply chain optimization. Gao et al. (2024), by using the comprehensive method of Multi-criteria Decision-Making and Geographic Information System integration, completed the research on the site selection of biomass cogeneration projects. Zhang et al. (2023) evaluated and optimized five straw energy utilization modes by constructing the EmLCA-II model. Sun et al. (2017) compared the cost of artificial models and mechanical models, and the results indicated that the mechanical collection model was more economical.

Supply chain risk control constitutes another significant research direction, characterized by a variety of research methods, among which system dynamics method has been widely employed. Ricardo Saavedra et al. (2018) summarized studies related to the application of system dynamics in renewable energy supply chains (RESC) and showed the possibility of systematically studying renewable energy supply chains. Nakandala et al. (2016), according to the characteristics of the relevant supply chain, established a total cost model to identify risks, help enterprises control costs, and reduce losses caused by product deterioration risks. Chen (2021), by constructing the system dynamics model of fresh agricultural products supply chain risk in Shandong Province, analyzed the main risk factors and risk control strategies. These findings indicate that the system dynamics method can be effectively used to relevant research on the system.

Currently, research on risk control in the supply chain of straw power generation is limited. In order to study the law of change of the total cost of straw power generation supply chain under the condition of risk occurrence, and to reveal the control effect of the risk control strategy on each risk factor, the system dynamics model can be employed to simulate the relevant factors. This paper aims to research risk control in the supply chain of straw power generation, constructing a system dynamics model for risk control in the supply chain, exploring the impact of four types of risks—environmental, market, technical, and information risks—on the total cost of straw power generation supply chain, and putting forward effective control measures for major risks. It provides a scientific basis for the risk control in straw power generation supply chain and ensures the sustainable development of the straw power generation industry.

5.2 Methods

5.2.1 The Risk Factors of Straw Power Generation Supply Chain System

Risk analysis and evaluation are the basis of risk control. According to the relevant literature, the risk factors are divided into four first-level risks: environmental risk, market risk, technical risk, and information risk (Pan and Lin 2023; Zhong et al.

Table 5.1 Risk factors of straw power generation supply chain

Level 1 risk factors	Level 2 risk factors
Environmental risk	Natural disasters, policy risks, storage environment, traffic environment, sudden accidents, inadequate safety supervision, imperfect rules and regulations, farmers' willingness to purchase, subsidized electricity prices, environmental protection policies
Market risk	Straw price fluctuations, supply-demand changes, straw quality and safety, default risk, straw inventory fluctuations, peer competition, fuel price fluctuations
Technical risk	Design technology, mechanical equipment maintenance, stacking technology, transportation route planning, environmental protection exceeding emissions, maintenance technology
Information risk	Information delay, information asymmetry, information inaccuracy, information security risk, low level of information, information management system instability, imperfect information sharing mechanism

2024). Among these, environmental risk refers to the external risk of straw from the growth to the utilization stage (Zhao et al. 2020). Market risk refers to the instability in straw power generation supply chain and the increase in raw material price caused by the change of market environment for farmers, brokers, and power plants (Lo et al. 2021). Technical risk refers to the risk associated with the use of machinery, stackers, boiler combustion equipment, and other technologies in the operation of straw power generation supply chain. Information risk refers to the uncoordinated operation of the supply chain caused by the lack of public information exchange platform or the interruption of market information transmission among the major players in the supply chain. The risk factors of straw power generation supply chain are shown in Table 5.1.

5.2.2 *Construction of Supply Chain Risk Factors System Dynamics Model*

5.2.2.1 System Dynamics Method

System dynamics is a discipline that studies information feedback systems. It is also an interdisciplinary and comprehensive discipline that understands and addresses systemic issues. By utilizing computer simulation technology, according to the relationship between structure, function, and behavior, the system architecture function is associated with the system dynamic behavior, the feedback mechanism is clarified, the dynamic relationship between system factors is studied, and the dynamic complexity problem is solved. The main steps of system dynamics modeling include (Garbolino et al. 2016) (1) determining the system boundary, (2) drawing a causality diagram, (3) constructing the system dynamics flowchart, (4) determining the model parameters and equations, and (5) analyzing the system dynamics simulation.

5.2.2.2 Determining the Boundary of Straw Power Generation Supply Chain System

Since the study of system dynamics focuses on the internal system factors, the system boundaries are determined based on the research objective. The operation process of straw power generation supply chain includes four links: straw collection, storage, transportation, and power generation utilization. So the system boundary starts from straw collection to the end of power generation utilization. According to the analysis of the risk factors of the straw power generation supply chain (Abdel-Basset and Mohamed 2020), combined with expert opinions and literature analysis (Sarkar and Guchhait 2023), the more impactful risk factors are selected to carry out the system dynamics research. Consequently, natural disasters and policy risks from environmental risks, straw price fluctuations and supply-demand changes in straw from market risks, transportation route planning, mechanical equipment maintenance, and stacking technology from technical risks, and information delay and information asymmetry from information risks are selected.

5.2.2.3 Constructing a Causality Diagram

In the analysis of the relationship between the risk factors and the system results of the straw power generation supply chain, considering the impact of environmental risk, market risk, technical risk and information risk on the supply chain system, the causal relationship between the risk factors and the total cost is determined. The causal relationship diagram of the system dynamics model is constructed, as illustrated in Fig. 5.1.

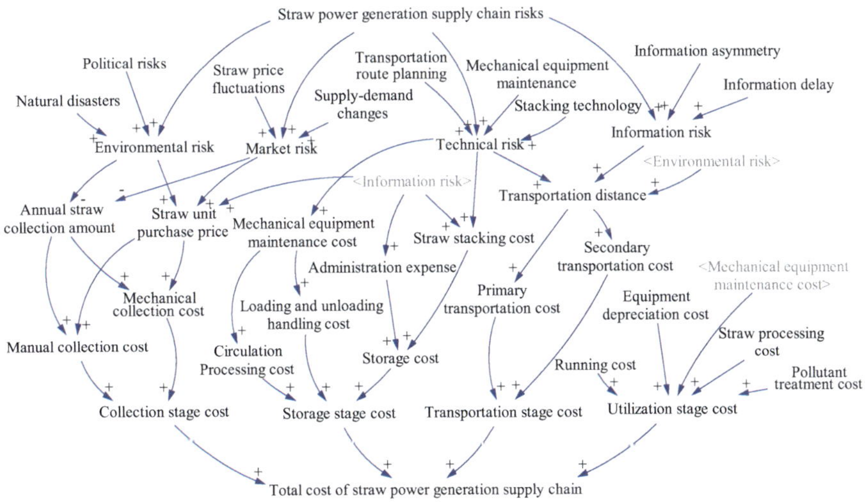

Fig. 5.1 Causal relationship diagram of risk factors in straw power generation supply chain

5.2.2.4 Constructing the System Dynamics Flowchart

Based on the different operation links of straw power generation supply chain, the total cost of straw power generation is divided into collection cost, storage cost, transportation cost, and utilization cost. Due to the impact of supply chain risks on the costs at different stages, the appropriate state variables, rate variables, auxiliary variables, and constants are selected (Jahani et al. 2023). According to the causal relationship between risk factors and supply chain system, Vensim PLE software is used to construct the system dynamics flowchart of straw power generation supply chain, as shown in Fig. 5.2.

From Fig. 5.2, the system dynamics flowchart of straw power generation supply chain involves 57 variables, including 8 state variables, 8 rate variables, 14 auxiliary variables, and 27 constants. Among these, environmental risk, market risk, technical risk, information risk, collection stage cost, storage stage cost, transportation stage cost, and utilization stage cost are state variables; incidence rate 1, incidence rate 2, and cost variation 1 are rate variables; and the others are auxiliary variables and constants.

5.2.2.5 Determining the Main Equation of the Model

Based on the influence of risk factors of straw power generation supply chain on the costs of different operational links, a system dynamics model of risk factors affecting the total cost of straw power generation supply chain is constructed. The main equations in the model are listed in Table 5.2.

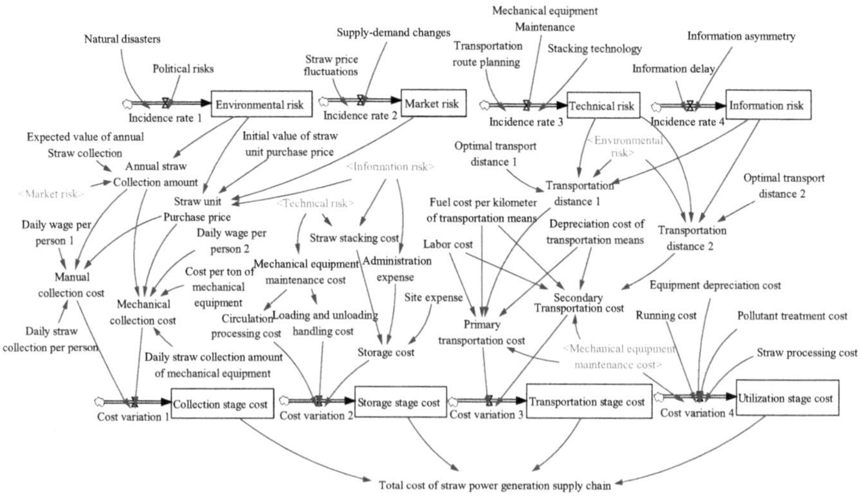

Fig. 5.2 Straw power generation supply chain system dynamics flowchart

Table 5.2 Main equations of straw power generation supply chain system dynamics model

Number	Mathematical expression
1	Environmental risk = INTEG (incidence rate 1,0)
2	Market risk = INTEG (incidence rate 2,0)
3	Technical risk = INTEG (incidence rate 3,0)
4	Information risk = INTEG (incidence rate 4,0)
5	Straw unit purchase price = initial value of straw unit purchase price + initial value of straw unit purchase price * (environmental risk + market risk + information risk)
6	Manual collection cost = daily wage per person 1* (annual straw collection amount/daily straw collection per person) + straw unit purchase price * annual straw collection amount
7	Mechanical collection cost = daily wage per person 2* annual straw collection amount/daily straw collection amount of mechanical equipment + straw unit purchase price * annual straw collection amount + cost per ton of mechanical equipment * annual straw collection amount
8	Storage cost = site expense + straw stacking cost + administration expense
9	Primary transportation cost = labor cost + mechanical equipment maintenance cost + depreciation cost of transportation means + fuel cost per kilometer of transportation means * transportation distance 1
10	Secondary transportation cost = labor cost + mechanical equipment maintenance cost + depreciation cost of transportation means + fuel cost per kilometer of transportation means * transportation distance 2
11	Collection stage cost = INTEG (cost variation 11,746)
12	Storage stage cost = INTEG (cost variation 22,349)
13	Transportation stage cost = INTEG (cost variation 33,087)
14	Utilization stage cost = INTEG (cost variation 44,871)

5.3 Case Study

5.3.1 Setting Model Parameters

The feedback behavior of system dynamics is insensitive to the initial value of the parameters, and its model behavior mainly depends on the model structure (Zheng et al. 2022). Therefore, with reference to relevant literature and research data (Sun et al. 2017; Wang et al. 2018), the parameter settings of the system dynamics model are indicated in Table 5.3.

5.3.2 System Initial State Setting

1. The operational lifespan of a straw power plant is 20 years. Therefore, Vensim PLE software is used for simulation, with the initial time of the system set to 0, the end time set to 20, and the simulation step set to 1, in units of years.

Table 5.3 Main parameters of system dynamics model (Wu et al. 2021)

Name	Values	Unit
Daily wage per worker	100	CNY
Daily straw collection per person	1	Ton/day
Daily straw collection amount of mechanical equipment	10	Ton/day
Storage space rental cost	15,000	CNY/hm^2
Straw storage height	5	m
Straw land occupation coefficient	0.8	/
Storage management personnel wage	120	CNY/day
Transport staff wage	200	CNY/day
Diesel price	7.5	CNY/L
Distance from field to transfer station	10	km
Distance from transfer station to power generation plant	30	km
Electricity price	0.6	CNY/kWh
Annual straw collection amount	250,000	Ton

2. A 30 MW straw-fired power plant consumes 250,000 tons of straw annually.
3. It is assumed that the cost of straw power generation supply chain in year 0 is the initial value under the risk-free state.
4. When Vensim PLE software is used for simulation, the values of some auxiliary variables do not change with the increase in years.

5.4 Results and Discussion

5.4.1 Simulation Results

The equations and main parameters are input into the system dynamics flowchart of straw power generation supply chain. Under the compound simulation state of Vensim PLE software, different probability scenarios of environmental risk, market risk, technical risk, and information risk are set up to calculate the total cost of straw power generation supply chain. When the occurrence probabilities of various risk factors in the supply chain system are each set to 0.1, the total cost of straw power generation under different risk scenarios can be obtained. The results are depicted n Fig. 5.3.

From Fig. 5.3, it can be seen that when the probability of risk occurrence is 0.1, the total cost of the straw power generation supply chain increases over time under different risk scenarios. The largest power generation cost is observed in the environmental risk scenario, while the lowest cost is seen in the technical risk scenario.

To clarify the influence of risk factors on the total cost of straw power generation supply chain, the total cost of straw power generation is calculated when the probability of each risk is 0.2. Additionally, the cost at the end of the period with a risk probability of 0.1 and 0.2 is selected for comparison. The results are shown in Table 5.4.

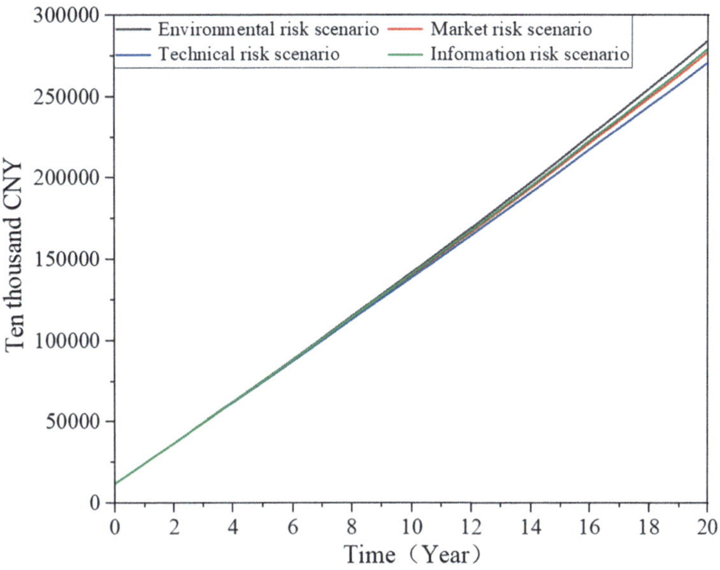

Fig. 5.3 Total cost of straw power generation supply chain in different scenarios

Table 5.4 Cost changes in different scenarios

Scenarios	Probability 0	Probability 0.1	Probability 0.2	Cost change rate with probability 0.1 (%)	Cost change rate with probability 0.2 (%)
Environmental risk	260,280	284,821	311,337	9.43	19.62
Market risk	260,280	277,554	296,803	6.64	14.03
Technical risk	260,280	271,335	282,390	4.25	8.49
Information risk	260,280	279,681	299,081	7.45	14.91

The results in Table 5.4 demonstrate that when the probability of risk occurrence is 0.1, the maximum cost change of straw power supply chain under environmental risk scenario is 9.43% and the minimum cost change under technical risk scenario is 4.25%. When the probability of risk occurrence is 0.2, the cost variation rate is highest under the environmental risk scenario at 19.62% and lowest under the technical risk scenario at 8.49%. The simulation results reveal that an increase in the probability of environmental risk leads to the most obvious upward trend in the total cost of the straw power generation supply chain, followed by the cost variation caused by market risk and information risk, while the increase in technical risk probability results in the smallest cost change.

5.4.2 *Analysis of Results Under Risk Control*

According to the simulation results, when the probability of environmental risk, market risk, and information risk increases, the total cost of straw power generation supply chain gradually increases. Risk control strategies need to be implemented to reduce the impact of risk factors on the system of straw power generation supply chain. However, the risk control process requires enterprises to invest in relevant technical equipment. To further analyze the impact of risk control strategies on the total cost of the supply chain, it is assumed that the cost of system risk control accounts for 15% of the total cost (Yang et al. 2017). System dynamics is used to analyze the changes in the total cost of straw power generation supply chain under risk control.

5.4.2.1 Construction of System Dynamics Model with Risk Control Strategies

The sustainable operation of the straw power generation supply chain requires the joint efforts of stakeholders, and corresponding control strategies can be proposed according to different risk factors to reduce the probability of risk occurrence. For environmental risks, joint management (Assa et al. 2021) and emergency management measures (Su et al. 2023) are applied. In response to market risks, the government should implement preferential strategies (Cai et al. 2021) and agreement management (Ji et al. 2024). For information risk, contract management (Konovalenko and Ludwig 2019), supervision management (Gao et al. 2020), and information management (Arif et al. 2024) are used to reduce the risk occurrence. The system dynamics model of straw power generation supply chain with risk control is shown in Fig. 5.4.

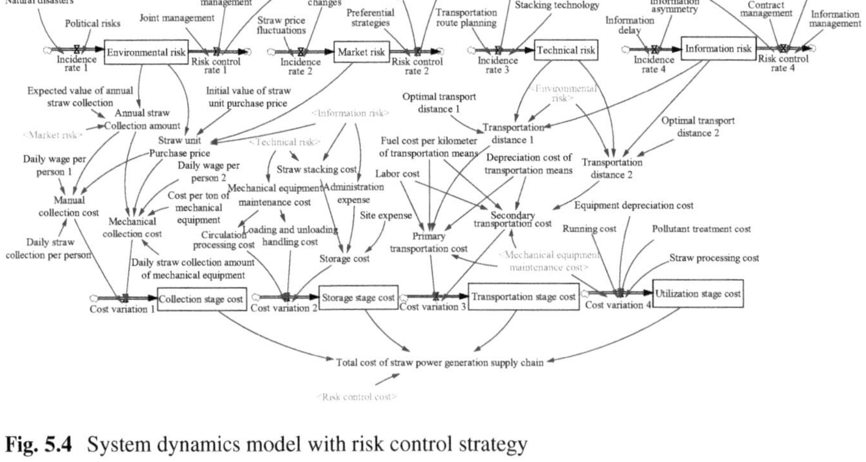

Fig. 5.4 System dynamics model with risk control strategy

5.4.2.2 Impact of Risk Control on the Probability of Occurrence

Control strategies aimed at environmental, market, and information risks can reduce the probability of risk occurrence. Scenarios with a risk probability of 0.1 and the risk control scenarios were set, and Vensim PLE software was used to plot the probability variation curve of risk occurrence, as depicted in Fig. 5.5.

Since the entry into force of the risk control strategy has a certain lag, it can be observed from Fig. 5.5 that after the introduction of the risk control strategies, compared with the scenario when the probability of risk occurrence is 0.1, the probability curves of environmental risk, market risk, and information risk show a downward trend between 2 and 4 years. Between 4 and 20 years, these curves exhibit a gradually declining trend with fluctuations. Of these, the decrease in the probability of information risk is greater than that of environmental and market risks, indicating that the impact of risk control is more significant on information risk.

5.4.2.3 Impact of Risk Control on Total Cost

To explore the impact of changes in the probability of risk occurrence on the total cost of the straw power generation supply chain, the total cost under different scenarios of risk factors is calculated. The results are shown in Fig. 5.6.

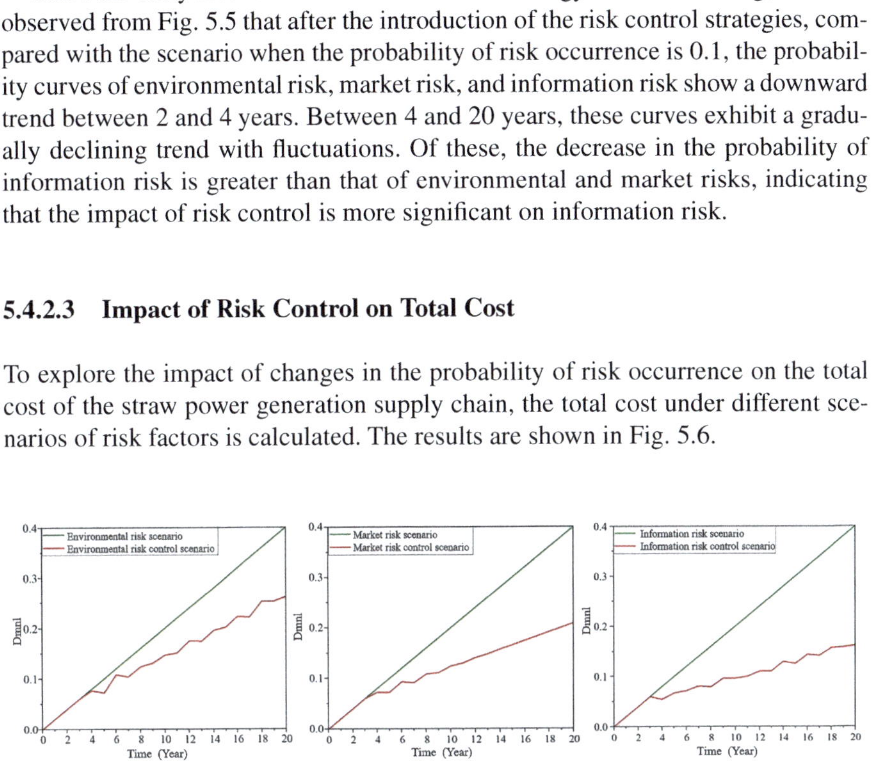

Fig. 5.5 Impact of risk control strategy on the probability of occurrence

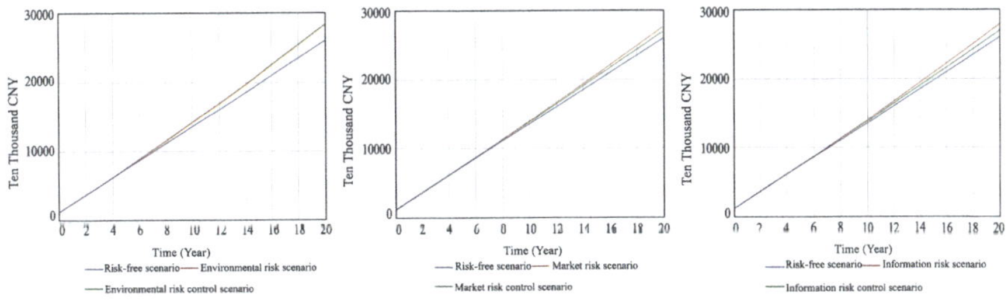

Fig. 5.6 Impact of risk control strategy on total cost

As illustrated in Fig. 5.6, the total cost curve of the straw power generation supply chain under environmental risk control experiences a slight decline between 6 and 10 years after the risk control is implemented, compared to a risk probability of 0.1. Under market risk control, the total cost exhibits a minor downward trend in 6–10 years, with the decline trend gradually increasing from 10 to 20 years. Under information risk control, the total cost shows a significant and continuous decrease between 10 and 20 years. Therefore, the downward trend in the total cost curve is most pronounced with the implementation of information risk control strategies, followed by market risk control, while the decline is least under environmental risk control.

5.4.3 Risk Control Strategy Recommendations

Through the implementation of the risk control strategy, the risk probability of the straw power generation supply chain can be reduced, thereby decreasing the total cost of the supply chain. According to the simulation results, the most optimal effect is observed with information risk control, followed by market risk control, with environmental risk control having the least impact. Therefore, reasonable and appropriate control strategies should be proposed based on the effectiveness of risk control. The following control strategy recommendations are suggested for information risk, market risk, and environmental risk.

5.4.3.1 Information Risk Control Strategy

By incorporating contract management, information management, and supervision management control strategies, an information management platform can be established to enable comprehensive information sharing within the straw power generation supply chain. This allows for effective communication and coordination among all stages and participants. Farmers, brokers, and straw power generation enterprises can sign information sharing contracts to share information in the stages of collection, storage, transportation, and utilization. This reduces the risk probability of information asymmetry. In terms of information management, Radio-Frequency Identification (RFID), Global Positioning System (GPS), and other Internet of Things (IoT) technologies can be applied to the straw power generation supply chain. The information can be transmitted to the information management platform, which can enhance the degree of informatization and reduce the risk of information delay. Local governments can introduce policies to standardize the management of network information security in all aspects of the supply chain, improve the level of information management, and reduce the occurrence of information risks.

5.4.3.2 Market Risk Control Strategy

The market risk control strategies primarily involve contract management and preferential policies. By signing contracts, farmers, brokers, and straw power generation enterprises can improve the profit distribution system, avoid imbalanced profit distribution, and promote the stable development of the straw supply chain. The government should respond to the preferential policies issued by the straw utilization industry and provide preferential measures or cost reductions to power generation enterprises in terms of taxation. Additionally, promoting the benefits of straw utilization can encourage farmers to sell straw, alleviate supply shortages, and reduce straw acquisition costs.

5.4.3.3 Environmental Risk Control Strategy

Environmental risk control strategies include joint management and emergency management. Given the sudden and inevitable characteristics of natural disasters, natural disaster warning mechanisms are incorporated based on regional climate, weather, and geographic conditions. Various preventive measures and emergency management control strategies are implemented in advance. For example, fire control and temperature and humidity control facilities are added at storage areas, monitoring devices are installed during transportation, risk prevention and emergency management preparations are made beforehand, and supply chain operations are supervised in real time. In response to operational environmental risks, government departments can introduce agricultural production insurance specifically for the straw power generation supply chain, providing protection for the operation of the agricultural product supply chain.

5.5 Conclusion

The operation of the straw power generation supply chain is affected by multiple risk factors. In order to explore the losses caused by different risk factors to the system. In this paper, a system dynamics model of the straw power generation supply chain was constructed to investigate the risk factors affecting the supply chain. The impact of changes in risk occurrence probabilities on the total cost of the straw power generation supply chain under different scenarios was analyzed. The results indicate that environmental risk, market risk, and information risk significantly influence the cost of the straw power generation supply chain, whereas technical risk has a relatively smaller impact. A risk control scenario was constructed to study the change in the total cost of straw power generation supply chain under risk control. The findings reveal that the probability of information risk occurrence decreases more rapidly under risk control, demonstrating greater sensitivity to the implementation of strategies, followed by market risk and, finally, environmental risk. The reason for this phenomenon is that the environmental risk control measures are more challenging to implement and the control strategies are slow to take effect.

References

Abdel-Basset M, Mohamed R (2020) A novel plithogenic TOPSIS- CRITIC model for sustainable supply chain risk management. J Clean Prod 247. https://doi.org/10.1016/j.jclepro.2019.119586

Arif J, Samadhiya A, Naz F, Kumar A (2024) Exploring the application of ICTs in decarbonizing the agriculture supply chain: a literature review and research agenda. Heliyon 10(8):e29564. https://doi.org/10.1016/j.heliyon.2024.e29564

Assa H, Sharifi H, Lyons A (2021) An examination of the role of price insurance products in stimulating investment in agriculture supply chains for sustained productivity. Eur J Oper Res 288(3):918–934. https://doi.org/10.1016/j.ejor.2020.06.030

Cai Z, Ye F, Xie Z, Zhang L, Cui T (2021) The choice of cooperation mode in the bioenergy supply chain with random biomass feedstock yield. J Clean Prod 311. https://doi.org/10.1016/j.jclepro.2021.127587

Cao J, Zhang Z, Zhou Y (2021) A location-routing problem for biomass supply chains. Comput Ind Eng 152. https://doi.org/10.1016/j.cie.2020.107017

Chen Z (2021) Research on risk control of fresh agricultural products supply chain based on system dynamics. Market Modernizat 19:9–12. https://doi.org/10.14013/j.cnki.scxdh.2021.19.003

Ebadian M, Sowlati T, Sokhansanj S, Townley-Smith L, Stumborg M (2013) Modeling and analysing storage systems in agricultural biomass supply chain for cellulosic ethanol production. Appl Energy 102:840–849. https://doi.org/10.1016/j.apenergy.2012.08.049

Gao Q, Guo S, Liu X, Manogaran G, Chilamkurti N, Kadry S (2020) Simulation analysis of supply chain risk management system based on IoT information platform. Enterp Inf Syst 14(9–10):1354–1378. https://doi.org/10.1080/17517575.2019.1644671

Gao J, Wang C, Wang Z, Lin J, Zhang R, Wu X, Wang Z (2024) Site selection decision for biomass cogeneration projects from a sustainable perspective: a case study of China. Energy 286. https://doi.org/10.1016/j.energy.2023.129518

Garbolino E, Chery J-P, Guarnieri F (2016) A simplified approach to risk assessment based on system dynamics: an industrial case study. Risk Anal 36(1):16–29. https://doi.org/10.1111/risa.12534

Jahani H, Gholizadeh H, Hayati Z, Fazlollahtabar H (2023) Investment risk assessment of the biomass-to-energy supply chain using system dynamics. Renew Energy 203:554–567. https://doi.org/10.1016/j.renene.2022.12.038

Ji C, Tan Z, Chen B, Zhou D-C, Qian W-Y (2024) The impact of environmental policies on renewable energy investment decisions in the power supply chain. Energy Policy 186. https://doi.org/10.1016/j.enpol.2024.113987

Konovalenko I, Ludwig A (2019) Event processing in supply chain management – the status quo and research outlook. Comput Ind 105:229–249. https://doi.org/10.1016/j.compind.2018.12.009

Lo SLY, How BS, Leong WD, Teng SY, Rhamdhani MA, Sunarso J (2021) Techno-economic analysis for biomass supply chain: a state-of-the-art review. Renew Sustain Energy Rev 135. https://doi.org/10.1016/j.rser.2020.110164

Nakandala D, Lau H, Zhang J (2016) Cost-optimization modelling for fresh food quality and transportation. Ind Manag Data Syst 116(3):564–583. https://doi.org/10.1108/imds-04-2015-0151

Nguea SM, Fotio HK (2024) Synthesizing the role of biomass energy consumption and human development in achieving environmental sustainability. Energy 293. https://doi.org/10.1016/j.energy.2024.130500

Noor M, Khan D, Khan A, Rasheed N (2023) The impact of renewable and non-renewable energy on sustainable development in South Asia. Environ Dev Sustain 26:14621–14638. https://doi.org/10.1007/s10668-023-03210-3

Pan W, Lin M (2023) Dynamics and risk assessment of a remanufacturing closed-loop supply chain system using the internet of things and neural network approach. J Supercomput 79(4):3878–3901. https://doi.org/10.1007/s11227-022-04727-6

Ricardo Saavedra MM et al (2018) Sustainable and renewable energy supply chain: a system dynamics overview. Renew Sustain Energy Rev 82:247–259. https://doi.org/10.1016/j.rser.2017.09.033

Sarkar B, Guchhait R (2023) Ramification of information asymmetry on a green supply chain management with the cap-trade, service, and vendor-managed inventory strategies. Electron Commer Res Appl 60. https://doi.org/10.1016/j.elerap.2023.101274

Su X, Shi X, Geng L, Yu R (2023) Complex evolution game and system dynamics simulation on the impact of coevolution on the stability of the fresh agricultural products green supply chain. Int J Bifurc Chaos 33(07). https://doi.org/10.1142/s0218127423500815

Sun Y, Cai W, Chen B, Guo X, Hu J, Jiao Y (2017) Economic analysis of fuel collection, storage, and transportation in straw power generation in China. Energy 132:194–203. https://doi.org/10.1016/j.energy.2017.05.077

Sun M, Xu X, Wang C, Bai Y, Fu C, Zhang L, Wang Y (2020) Environmental burdens of the comprehensive utilization of straw: wheat straw utilization from a life-cycle perspective. J Clean Prod 259. https://doi.org/10.1016/j.jclepro.2020.120702

Wang X, Li K, Song J, Duan H, Wang S (2018) Integrated assessment of straw utilization for energy production from views of regional energy, environmental and socioeconomic benefits. J Clean Prod 190:787–798. https://doi.org/10.1016/j.jclepro.2018.04.191

Wang Z, Wang Z, Xu G, Ren J, Wang H, Li J (2020) Sustainability assessment of straw direct combustion power generation in China: from the environmental and economic perspectives of straw substitute to coal. J Clean Prod 273. https://doi.org/10.1016/j.jclepro.2020.122890

Wang Z, Hou T, Wang Z (2023) Revenue coordination of straw power generation supply chain under carbon trading environment. Waste Biomass Valorizat 14(5):1769–1785. https://doi.org/10.1007/s12649-022-01960-1

Wang X, Cao J, Lv Y (2024a) Resource allocation and route planning under the collection price uncertainty for the biomass supply chain. Biosyst Eng 241:68–82. https://doi.org/10.1016/j.biosystemseng.2024.03.011

Wang Z, Huang W, Wang H, Gao J, Zhang R, Xu G, Wang Z (2024b) Research on the improvement of carbon neutrality by utilizing agricultural waste: based on a life cycle assessment of biomass briquette fuel heating system. J Clean Prod 434. https://doi.org/10.1016/j.jclepro.2023.140365

Wu J, Zhang J, Yi W, Cai H, Su Z, Li Y (2021) Economic analysis of different straw supply modes in China. Energy 237. https://doi.org/10.1016/j.energy.2021.121594

Yang Y, Yang X, Yu Q (2017) A system dynamics-based study on the international cold chain logistics operational risk control of fresh agro products: take the international cold chain logistics of fresh vegetables in Yunnan Province as an example. J Beijing Jiaotong Univer (Soc Sci Edition) 16(03):119–128. https://doi.org/10.16797/j.cnki.11-5224/c.20170619.007

Zahraee SM, Shiwakoti N, Stasinopoulos P (2020) Biomass supply chain environmental and socio-economic analysis: 40-years comprehensive review of methods, decision issues, sustainability challenges, and the way forward. Biomass Bioenergy 142. https://doi.org/10.1016/j.biombioe.2020.105777

Zhang X, Sun P, Xu J, Wang X, Yu J, Zhao Z, Dong Y (2020) Blockchain-based safety management system for the grain supply chain. IEEE Access 8:36398–36410. https://doi.org/10.1109/access.2020.2975415

Zhang Y, Wang Q, Tian S, Xu Y, Yuan X, Ma Q, Liu C (2023) Evaluation and optimization of five straw energy utilization modes based on the improved emergy-based life cycle assessment (EmLCA-II) method. Energy Convers Manag 298. https://doi.org/10.1016/j.enconman.2023.117764

Zhao G, Liu S, Lopez C, Chen H, Lu H, Mangla SK, Elgueta S (2020) Risk analysis of the Agri-food supply chain: a multi-method approach. Int J Prod Res 58(16):4851–4876. https://doi.org/10.1080/00207543.2020.1725684

Zheng R, Huang Z, Wu Z (2022) Risk control simulation of the closed-loop supply chain of waste electrical and electronic equipment based on system dynamics. Front Energy Res 10. https://doi.org/10.3389/fenrg.2022.963211

Zhong J, Cheng H, Jia F (2024) Supply chain resilience capability factors in agri-food supply chains. Oper Manag Res. https://doi.org/10.1007/s12063-024-00470-8

Chapter 6
Emergy Analysis of Agricultural Waste Power Generation Supply Chain

Yufeng Sun, Yapeng Wang, Bin Yang, Zipeng Zheng, Chun Wang, Bo Chen, Suiliang Li, Jilai Ying, Xinping Liu, Liang Chen, and Wenlong Mu

Contents

Abstract The efficient utilization of agricultural waste resources lies in a modern and sound agricultural waste supply chain system for crops, but due to the problems of low density and strong seasonality of agricultural waste, the collection, storage, and transportation of agricultural waste are difficult, which in turn affects the utilization of agricultural waste for power generation. Therefore, a systematic analysis

---

Y. Sun (✉) · Y. Wang · B. Yang · Z. Zheng · C. Wang · B. Chen · S. Li · J. Ying · X. Liu · L. Chen · W. Mu
College of Mechanical and Electrical Engineering, Henan Agricultural University, Zhengzhou, Henan Province, China
e-mail: hansyf9397@henau.edu.cn

Z. Wang et al. (eds.), *Agricultural Waste Power Generation Supply Chain*, Industrial Ecology and Environmental Management 7,
https://doi.org/10.1007/978-3-032-20087-7_6

of the energy input and output of the agricultural waste collection, storage, and transportation system is needed. Through the theory of energy value analysis, combined with qualitative and quantitative comparative analysis methods and mathematical modeling, the resource utilization situation, environmental impact degree, and sustainability of the agricultural waste collection, storage, and transportation system are analyzed and compared from the perspective of two different agricultural waste collection and storage modes of agricultural waste collection and storage companies. The results show that the agricultural waste collection and storage system has a high degree of internal resource and environmental development, a high productivity, and a lower proportion of purchased renewable energy input compared to the manual and dispersed agricultural waste collection and storage model. The sustainable development capacity of the mechanical centralized model is obviously higher than that of the manual and dispersed agricultural waste collection and storage model.

Keywords Emergy analysis · Energy value · Agricultural waste power generation

6.1 Introduction

With the excessive consumption of conventional energy, the ecological environment has gradually deteriorated, and the development of clean renewable energy has attracted the attention of the government and the enterprises (Panwar et al. 2011). It is reported that the waste produced by Chinese agriculture, such as agricultural planting straw and livestock manure, has reached more than 5 billion tons every year. As green and renewable biomass resource, agricultural waste is often used to produce clean energy, which is an indispensable substitution of traditional fossil fuels. Biomass resources are expected to contribute between 20% and 40% of the world's primary energy consumption by the year 2050. It is significant to further explore the emergy application (power generation) of agricultural waste resources. Compared with traditional energy, crop straw is a renewable resource that can be completely degraded by microorganisms, and the regeneration period is short (Li et al. 2007; Wang et al. 2020a). Crop straw contains a large amount of cellulose and hemicellulose that can produce ethanol (Fang et al. 2010), agricultural fertilizers (Yin et al. 2018), farming feed (Seglah et al. 2019), papermaking (Espinosa et al. 2016), edible fungi (W. Lu et al. 2018), etc. The efficient use of crop straw can save the consumption of traditional energy, such as crude oil and natural gas (Croce et al. 2016). Crop straw has become one of the main biomass raw materials in China (Hong et al. 2016).

As an easy-scattered and low-density crop, straw has the defects of dispersion, seasonality, limited collection time, and perishable deterioration, as well as inconvenient transportation and storage, which seriously restrict its large-scale application (Jiang et al. 2021; Lu et al. 2020; Ren et al. 2019). In the process of straw recycling, the costs of straw collection, transportation, and storage system account for 35–50% of the total expenditure (Lu et al. 2020). Therefore, the efficient use of straw resource recycling depends on a modern and rational system of crop straw collection, transportation, and storage, such as good collection methods, an excellent transport network, and appropriate storage conditions (Liu et al. 2018; Wang et al. 2020b). The collection costs mainly include the straw purchase in the field, the profit earned by the straw recycling intermediary, straw baling, loading, and unloading in the field, etc. Transportation costs include labor, fuel, vehicle maintenance, insurance, equipment maintenance, etc. Storage expenditure refers to the construction of the storage station, site rental, site water, electricity, management fee, equipment maintenance, etc. (Shastri et al. 2011; Sun et al. 2017). In addition, any potential intermediate operation in the collection, transportation, and storage process means increased costs. The efficient use of biomass resources needs to be resolved urgently, and the key issue is how to establish a modern and rational system of crop straw collection, transportation, and storage (Mao et al. 2019).

In the field of straw recycling, experts and scholars have done a lot of research. Yufeng Sun et al. took the straw recycling company in Sheqi County, Nanyang City, as an example and combined field research and quantitative research methods to effectively solve the problems in the logistics of the collection, transportation, and storage of straw for power generation (Sun et al. 2017). The distributed storage and pretreatment resource supply model of crop straw established by the University of Illinois greatly reduced the cost and improved the efficiency of the supply system (Shastri et al. 2012). Tao Lin et al. established a Geographic Information System (GIS)-enabled biomass supply chain optimization model that reduced the straw supply cost (Lin et al. 2013). Gary et al. conducted a quantitative evaluation of available rice straw resources in three states in the Pacific Northwest based on the data of the US Department of Agriculture, and the regional distribution of available straw was also assessed to reduce the cost of straw collection and transportation (Banowetz et al. 2008). Xiaoxian Zhang et al. used the emergy analysis method to evaluate five types of straw-reuse technologies in the northeast of China and concluded that straw-briquetting and biogas production were the most beneficial technologies in terms of economy, environmental impact, and sustainability (Zhang and Ma 2015). Wenjing Liu et al. applied emergy analysis theory to comprehensively evaluate the corn-soybean rotation systems with and without the return of straw in the central region of the Loess Plateau (Liu et al. 2019b). Although experts and scholars have done a lot of work on straw recycling, the assessment of the appropriateness of the collection, transportation, and storage of straw from the field to the power plant still needs to be addressed.

Emergy analysis is a quantitative tool for evaluating the energy and resources needed to support a process or product. It can convert natural phenomena and human activities into a uniform metric, the solar emergy joule (sej) (Odum 1988). It is worth noting that there is a difference between emergy and energy. Emergy is mainly used to evaluate the value of things in terms of energy consumed in the past rather than existing energy. The formation process of green plants transforms solar energy from the lowest level to the higher-level green plant products, which are then transformed by consumers in the biological chain. In this process, energy decreases step by step, while emergy level, emergy value, and emergy conversion rate increase. Emergy theory can quantitatively analyze the true value of natural and human economic activities (Hau and Bakshi 2004).

Therefore, crop straw was used as an example to construct an agricultural waste power generation supply chain. The purpose of this research was to employ emergy analysis theory to construct an emergy indicator system for evaluating the straw collection, transportation, and storage system of the artificial and mechanical model of Laifa Straw Recycling Company. Some indexes, including emergy self-support ratio (ESR), environmental loading ratio (ELR), emergy yield ratio (EYR), and emergy sustainability index (ESI), were calculated to assess the economic and environmental benefits. Based on the established emergy indicator system for analyzing the relationship between the input and output of the two models, some suggestions were brought out for the long-term development of the straw recycling company.

6.2 Theoretical Foundation of Emergy Analysis of Agricultural Waste Power Generation Supply Chain

6.2.1 Emergy Concept

Emergy was proposed by American ecologist H.T. Odum in 1987 and defined as the sum of the energy consumed directly or indirectly in the process of product or service formation. It is worth noting that there is a difference between emergy and energy. Emergy is mainly used to evaluate the value of things in terms of energy consumed in the past rather than existing energy. The formation process of green plants transforms solar energy from the lowest level to the higher-level green plant products, which are then transformed by consumers in the biological chain. In this process, the energy decreases step by step, while emergy level, emergy value, and emergy conversion rate increase. H.T. Odum also referred to energy value as "embodied energy." In practical applications, solar emergy joules (sej) are always used as a unified standard to measure different forms and types of energy. In this way, different types of energy can be converted into a unified data unit for operation, thereby truly, accurately, and effectively measuring and comparing the actual value and real contribution corresponding to all different types of energy in the system (Sun et al. 2021).

Emergy transformity refers to the amount of another form of energy required to produce each unit of resource, product, or service. In practical research, it is generally referred to as solar transformity, which is equivalent to the amount of solar energy converted per unit of a certain substance or energy. The solar transformity plays a crucial role in energy value theory and is a rather important fundamental concept that reflects the energy quality level of a certain energy form.

6.2.2 Basic Steps of Emergy Analysis

Emergy analysis can be divided into six steps, including determining the boundaries and contents of the system, collecting and arranging data, drawing the emergy system diagram, constructing the emergy analysis table, establishing the emergy index system, and presenting the results and discussion.

6.2.3 Emergy Index Evaluation System

According to the different renewability properties of resource inputs in this paper, resource input was divided into renewable environmental resources (R) and nonrenewable environmental resources (N), which were used to calculate the emergy indexes. The purchased input F was separated into the nonrenewable proportion of purchased resources (F_N) and the renewable proportion of purchased resources (F_R). Y and T represent the total output emergy and the total emergy input of system, respectively (Guan et al. 2016). Then, the emergy-based indicators were calculated and used to assess performance in areas as diverse as resource utilization efficiency, environmental impact, and system sustainability. The calculation formulas are presented in Table 6.1.

According to the meanings corresponding to the indicators, all energy flows and energy conversion situations in the system are analyzed and evaluated to obtain the true attributes and interrelationships among energy flows, material flows, and currency flows in the system. Emergy self-support ratio (ESR) indicates the proportion of total emergy input from local natural resources. The higher the ratio is, the better autarkic ability of the system is, which is used to measure the dependence on natural environmental resources (Zhan et al. 2020). Emergy yield ratio (EYR) is the ratio of the output emergy of the system to the emergy of the purchased resources. A higher ratio indicates higher production efficiency, which reflects the efficiency degree of the system (Liu et al. 2019a). Environmental loading ratio (ELR) is the ratio of the input nonrenewable resource emergy value to the renewable resource emergy value.

Table 6.1 Emergy indicators and calculation formulas

Emergy indicator	Calculation formula	Significance	References
Emergy self-support ratio (ESR)	$(R + N)/T$	System dependence	Guan et al. (2016)
Emergy yield ratio (EYR)	$Y/(F_R + F_N)$	Efficiency degree of the system	Zhan et al. (2020)
Environmental loading ratio (ELR)	$(N + F_N)/(R + F_R)$	Environmental pressure	Liu et al. (2019a)
Emergy sustainability index (ESI)	EYR/ELR	Sustainability of the straw system	Yazdani et al. (2020)

The system's natural environmental pressure can be measured by ELR. A larger value of ELR means greater pressure of the system on the surrounding environment (Yazdani et al. 2020). Emergy sustainability index (ESI) measures the sustainability potential of the straw collection, transportation, and storage system. A higher value of ESI suggests greater sustainability of the straw collection, transportation, and storage system (Chen et al. 2006). The emergy indicator system was constructed from ESR, ELR, EYR, and ESI.

6.3 Emergy Analysis Model of Agricultural Waste Power Generation Supply Chain

6.3.1 *The Boundary of the Emergy System for the Artificial Model and the Mechanical Model*

Laifa Straw Recycling Company is a pintle-level subsidiary of an enterprise group, which purchased 0.83×10^8 kg corn straw and 1.17×10^8 kg wheat straw, with a total of 2.0×10^8 kg straw for power generation per year. It specializes in environmental protection, energy saving, and new energy industry investment, development, and utilization[17]. As shown in Figs. 6.1 and 6.2, the model of the straw collection, transportation, and storage system was divided into the artificial model and the mechanical model. For the artificial model, the straw was manually collected in the field and sent to the next storage place by tractors in bulk and then baled at the storage point, while for the mechanical model, the straw was baled by machines directly in the field and transported by tractors to the storage place. The boundary of the system was determined based on the ecological environment, social

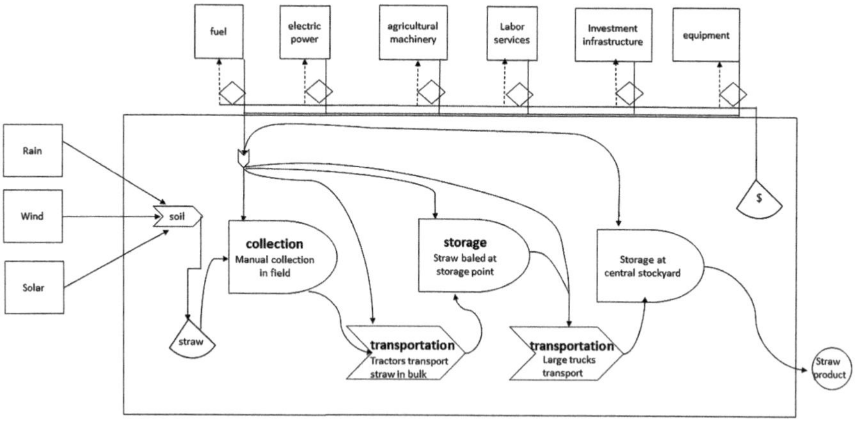

Fig. 6.1 Emergy system diagram of the artificial model

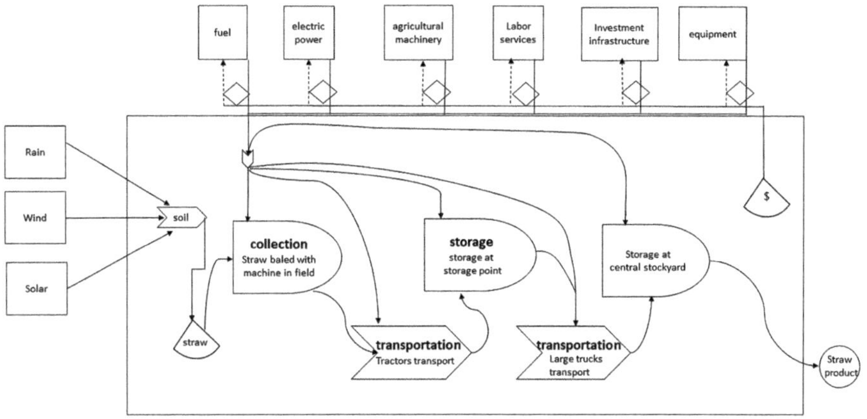

Fig. 6.2 Emergy system diagram of the mechanical model

economy, and other data of Sheqi County. The emergy flow, material flow, and currency flow of the straw system were analyzed by taking the collection, transportation, and storage system as the research subject.

6.3.2 Data Collection

The investigations of the straw recycling company in Sheqi County, China, comprised literature research, field research, and questionnaire surveys. By combining the natural environment and geographical conditions of the straw collection, transportation, and storage system in 2018, as well as the input and output data in the process of collection, transportation, and storage, and sorting out the relevant parameters such as material, labor,and capital input in the process of straw collection, transportation, and storage, the basic data needed for the emergy analysis of the straw collection, storage, and transportation parts were obtained and were listed in Tables 6.2, 6.3, 6.4, 6.5, 6.6, 6.7, and 6.8.

6.3.3 Establishment of Economic Mathematical Model and Emergy Analysis Model

The economic mathematical models in Table 6.9 were mainly based on our previous work. The specific economic input of the collection part, transportation part, and storage part of the two models were calculated using the economic mathematical model. The emergy input and output were calculated with the emergy currency

Table 6.2 Natural environmental data

Collection data	Numerical value	Unit	Provenance
Annual average total solar radiation	116.56	$kcal/cm^2$	Government website
Average annual precipitation	910.11	mm	Government website
Average altitude	130.00	m	Government website
Annual mean wind speed	3.20	m/s	Government website
Road twists and turns index in Sheqi County	1.30	/	Field research

Table 6.3 Emergy conversion coefficient

Project	Emergy conversion coefficient	Unit	Provenance
Corn straw	1.55 × 107	J/kg	Campbell and Tilley (2014)
Wheat straw	1.46 × 107	J/kg	Campbell and Tilley (2014)
Diesel oil	4.27 × 107	J/kg	Campbell and Tilley (2014)

Table 6.4 Solar emergy conversion rates

Project	Solar emergy conversion rate (sej J^{-1} or sej g^{-1})	Project	Solar emergy conversion rate (sej J^{-1})
Solar	1.00	Carbon dioxide	3.78 × 107
Wind	663.00	Straw	2.70×10^4
Rainwater chemical emergy	1.54×10^4	Electric power	1.59×10^5
Rainwater potential emergy	8.89×10^3	Fuel	6.60×10^4
Net loss of topsoil	7.40×10^4	Agricultural machinery power	7.50×10^7
Water	4.11×10^4	Labor	3.80×10^5

Table 6.5 Straw collection data

Project	Numerical value	Unit	Provenance
Annual purchase of straw	2×10^8	kg	Field research
Collection radius of central stockyard	15	km	Field research
Collection radius of straw storage point	5	km	Field research
Distance from storage point to central stockyard	10	km	Field research
Daily straw collection and loading per worker	800	kg	Field research
Daily wage per worker	17.88	$	Questionnaire survey
Baling machine working time per day	10	h	Questionnaire survey
Collection time of the artificial model	180	Day	Questionnaire survey
Collection time of the mechanical model	60	Day	Questionnaire survey

Table 6.6 Straw transportation data

Project	Farm tractor	Large trucks	Unit	Provenance
Full-load speed of equipment	20	35	km/h	Field research
Speed when equipment is empty	25	45	km/h	Field research
Fuel consumption for full-load transportation	0.28	0.40	kg/(kw*h)	Product instructions
Fuel consumption of no-load operation	0.24	0.30	kg/(kw*h)	Product instructions
Power load ratio (power/load)	9100	6100	kw/kg	Field research
Price of each equipment	8195	21,605	$	Market price
Equipment loading capacity (loose straw)	650	1500	kg	Field research
Equipment load (baled straw)	3000	6000	kg	Field research

Table 6.7 Storage and other data of straw

Project	Type	Numerical value	Unit	Provenance
Storage data	Rental price of straw storage point	119.20	$/(hm^2*year)	Field research
	Effective working days of power plant	330	Day	Field research
	Number of storage points	7	/	Field research
	Rental price of central yard	193.70	$/(hm^2*year)	Field research
	Density of straw bulk	40	kg/m^3	Reference
	Density of baled straw	200	kg/m^3	Reference
Baling machine	Power	45	kw	Field research
	Work efficiency	2500	kg/h	Field research
	Equipment price	8195	$	Field research
	Average fuel consumption per hour	11.50	kg/h	Field research
Grass grabbing machine	Power	22	kw	Field research
	Work efficiency	Loose straw:2900 Baled straw:5000	kg/h	Field research
	Equipment price	4589	$	Field research
	Average fuel consumption per hour	5.10	kg/h	Field research

Table 6.8 Calculation formulas for input factors

Input elements	Calculation formulas for emergy value
Solar emergy (*R*)	Average annual total solar radiation per unit area × Area of straw collection, transportation, and storage system × Solar emergy conversion rate
Wind emergy source (*R*)	Area of straw collection, transportation, and storage system × Atmospheric height × Air density × Wind speed × Solar emergy conversion rate
Rainwater potential emergy (*R*)	Average annual precipitation × Average altitude × Water density × Area of straw collection, transportation, and storage system × Gravitational acceleration × Emergy conversion rate
Rainwater chemical emergy (*R*)	Average annual precipitation × Water density × Area of straw collection, transportation, and storage system × Gravitational acceleration × Solar emergy conversion rate × Free emergy of water
Net loss of topsoil (*N*)	Soil erosion rate × Organic matter content per unit mass of soil × Organic matter emergy × Farmland area × Emergy conversion rate
Fuel oil purchase (F_N)	Fuel consumption × Emergy conversion coefficient × Emergy conversion rate
Straw purchase (F_R)	Total straw collected × Emergy conversion coefficient × Emergy conversion rate
Labor purchase (F_R)	Labor wage input ($) × Emergy currency ratio (4.94×10^{12} sej/$)

Note: The emergy value of water and electricity costs consumed by storage points in the system was ignored because it was less than 5% of the total emergy

ratio. Then, the specific values of ESR, EYR, ELR, and ESI were obtained, and the economic and environmental sustainability of the two collection, transportation, and storage models could be evaluated in the next discussion part.

Based on the data collected in the early stage, the main resources, products, and services invested in the system were calculated. Among them, the emergy input included renewable environmental resources, nonrenewable environmental resources, purchased nonrenewable resources, and purchased renewable resources. According to the solar emergy transformity corresponding to various resources, the material or monetary flows of different measurement units were converted into a unified energy value unit (sej). The emergy analysis model of the system was compiled to evaluate their role and position in the system.

6.4 Emergy Analysis of Agricultural Waste Power Generation Supply Chain

6.4.1 Economic Consumption of the Artificial Model and the Mechanical Model

The economic consumption calculated from the economic mathematical models of the artificial and mechanical models is listed in Tables 6.10 and 6.11. The key economic consumption of the two models is shown in Fig. 6.3.

Table 6.9 Main calculation formulas of the economic mathematical model

Formula	Significance
$A = \theta * S * F * \beta$	A was the annual total straw collection; β (0.5) stands for the comprehensive coefficient of power straw production; θ represents the grass-straw ratio (wheat: 1.03, corn: 1.37); the crop planting area was S (606 hm^2), and F was the crop yield per hectare (wheat: 450 kg/ mu, corn: 300 kg/mu)
$M1 = A * m/Z$	M1 stands for the total cost of workers; the daily wage and straw collected per worker were set to m and Z, respectively
$q1 = \left[\frac{B1}{V1} + \frac{B2}{V2}\right]\frac{\mu 1}{2}$	The fuel consumption per unit kilometer of the tractor was $q1$ kg/(km*h); the fuel consumption per unit power of a small tractor at full load was $B1$ kg/(kW * h), and the speed was $V1$ (km/h); the fuel consumption per unit power at no-load was $B2$ kg/(kW*h), and the speed was $V2$ (km/h); the power load ratio was $\mu 1$ kw/t
$C1 = \int_0^r 2\Pi \partial q r^2 dr = \frac{2}{3}\partial q r$	The unit fuel consumption of straw transportation from the field to the storage point was C1; the collection radius was r1 (5 km); the road twists and turns index was ∂, which was equal to 1.3 in this study
$C2 = A * C1$	C2 was the total fuel consumption from the field to the collection point
$L1 = \int_0^r 2\Pi \partial r^2 dr = \frac{2}{3}\partial r$	The average transport distance from field to the storage point was L1
$t1 = L1\ \partial\left(\frac{1}{V1} + \frac{1}{V2}\right) + 0.6$	The number of times for the vehicle to transport the straw from the field to the storage point and then back to the field was $t1$, and the total time of weighing, loading, unloading, and adjusting equipment was taken as 0.6 h
$\varepsilon 1 = \mathrm{Int}(t/t1)$	The number of times a vehicle could travel back and forth every day was $\varepsilon 1$, and the effective working time of agricultural equipment was t hours per day
$n1 = \mathrm{Int}\left(\frac{A}{\varepsilon_1 Q_1 d_1}\right) + 1$	The number of tractors required was $n1$, and the annual effective working days were $d1$ days
$M2 = n1 * p * d1 * m$	The total wage for tractor transportation workers was M2; $p1$ staff (one driver) and $p2$ staffs (one driver and one assistant) were required for tractor transportation and large truck transportation, respectively
$n2 = \mathrm{Int}\left(\frac{A}{t * \gamma 1 * d2}\right) + 1$	The number of baling machines required was $n2$, and the working efficiency of the baling machine was $\gamma 1$ ton/h
$Cd1 = \frac{Cd - 3\%Cd}{T} + 5\%Cd$	The annual cost of a single baling machine and the unit price of the baling machine were Cd1 and Cd, respectively

(continued)

Table 6.9 (continued)

Formula	Significance
CD = Cd1∗n2	The total annual cost of the baling machines was CD
OD=Od∗t∗d1∗n2	The fuel consumption of one baling machine per hour was Od, and the total fuel consumption of the baling machines was OD
$C_3 = A * q_2 * L_2$	The distance from the storage point to the central stockyard was L2, and the total fuel consumption from the storage to the central stockyard was C3
S1 = 1000A1/10,000$h\rho\tau$	The area of the central stockyard was S1; the storage height h of straw was 6 m; the storage density of straw and the land occupation factor were ρ(kg/m^3) and τ (0.8), respectively
M3 = S1m1	The unit rent cost of the central stockyard was m1 (yuan/mu), and the total rent cost of the central stockyard was M3
M4 = m∗5∗d3	The total management cost of workers in the central stockyard was M4
M5 = w1∗d3	The maintenance cost was w1 (447 $/day), and the total maintenance fee of the central stockyard was M5
A2 = A/N	A2 was the storage capacity of straw at each storage site, and N represented the required number of storage sites

6.4.2 Construction of the Emergy Analysis Table

Based on the collected data corresponding to the solar emergy transformity, the corresponding emergy values were calculated separately using the formula. These values were then classified according to resource categories, mainly including renewable natural resources, nonrenewable natural resources, purchased renewable resources, and purchased nonrenewable resources. After completing the calculations, the calculated results of the input and output were constructed as emergy analysis tables. The emergy value consumption of 2.0 × 108 kg straw in the artificial and mechanical models was calculated using the emergy currency ratio (4.94 × 1012sej/$)[27], and the data are listed in Tables 6.12 and 6.13.

6.4.3 Comparative Analysis of the Two Models

According to the Tables 6.12 and 6.13, the renewable environmental resources (R), nonrenewable environmental resources (N), total emergy value input of the system (T), output emergy values (Y), purchased renewable resources (F_R), and purchased nonrenewable resources (F_N) are all listed in Table 6.14. The emergy value input classification of the two systems are illustrated in Fig. 6.4.

As shown in Table 6.14 and Fig. 6.4, the emergy value input of the straw collection, transportation, and storage system was mainly from the purchase value. In the

Table 6.10 Economic calculation results of the artificial model

Project	Type	Numerical value	Unit
Collection	Wages of straw collection workers	4,470,000	\$
Transportation	Wages of tractor workers	611,496	\$
	Fuel consumption of tractors	93,060	kg
	Cost of tractor use	228,886.40	\$
	Wages of large truck drivers	177,012	\$
	Fuel consumption per large truck	110,380	kg
	Cost of large truck use	47,639.03	\$
Storage	Wages of baling machine workers at collection point	442,530	\$
	Cost of using baling machine at collection point	30,116.63	\$
	Fuel consumption of baling machine at collection point	948,750	kg
	Total rental fee of storage point	186,254.80	\$
	Management fee of storage site	82,605.6	\$
	Maintenance fee of storage site	344,190	\$
	Total rental fee of central stockyard	60,531.25	\$
	Management fee of central stockyard	29,502	\$
	Maintenance fee of central stockyard	147,510	\$
	Cost of using grass grabbing machine at central stockyard	8770	\$
	Fuel consumption of grass grabbing machine at central stockyard	218,790	kg
	Wages of grass grabbing machine workers at central stockyard	76,705	\$
Total	Fuel consumption of straw collection in the filed	0	/
	Fuel consumption of straw transportation	203,440	kg
	Fuel consumption of straw storage	1,167,540	kg
	Total fuel consumption	1,370,980	kg
	Cost of straw collection workers	4,470,000	\$
	Cost of straw transportation workers	788,508	\$
	Cost of straw storage workers	631,342.80	\$
	Total labor cost	5,889,851	\$
	Equipment and maintenance services fee	315,412	\$
	Cost of infrastructure inputs	738,486	\$

artificial model, the total emergy value input was 173.02 × 10^{18} sej. The purchased renewable resource (F_R) was 110.10 × 10^{18} sej, which accounted for 64% of the total emergy value input system. The second-highest emergy value input was the renewable environmental resource (*R*), which was 48.97 × 10^{18} sej and accounted for 28% of the emergy value input system. The emergy value input of nonrenewable environmental resource (*N*) and the purchased nonrenewable resource (F_N) were 4.88 × 10^{18} sej and 9.07 × 10^{18} sej, occupying 3% and 6%, respectively.

In the mechanical model, the total emergy value input was 148.11 × 10^{18} sej, which was lower than that of the artificial model. The purchased renewable resource (F_R) was 85.94 × 10^{18} sej, accounting for 58%, followed by the renewable

Table 6.11 Calculation results of the mechanical model

Project	Type	Numerical value	Unit
Collection	Wages of field baling workers	431,265.60	$
	Cost of baling machine use in the field	161,425.10	$
	Fuel consumption of baling machine in the field	924,600	kg
	Wages of grass grabbing machine workers	71,877.60	$
	Loading cost of using grass grabbing machine	45,199.03	$
	Fuel consumption of grass grabbing machine	205,020	kg
Transportation	Wages of tractor workers	133,027.20	$
	Fuel consumption of tractors	93,060	kg
	Cost of tractor use	149,378.50	$
	Wages of large truck workers	177,012	$
	Fuel consumption per large truck	110,380	kg
	Annual cost of large truck use	47,639.03	$
Storage	Total rental fee of collection point	37,255.96	$
	Labor management fee of storage site	82,605.6	$
	Maintenance fee of storage site	344,190	$
	Total rental fee of central stockyard	60,531.25	$
	Management fee of central stockyard	29,502	$
	Maintenance fee of central stockyard	147,510	$
	Cost of using grass grabbing machine at central stockyard	8770	$
	Fuel consumption of grass grabbing machine at central stockyard	218,790	kg
	Wages of grass grabbing workers at central stockyard	76,705	$
Total	Fuel consumption of straw collection	1,129,620	kg
	Fuel consumption of straw transportation in the field	203,440	kg
	Fuel consumption of straw storage	218,790	kg
	Total fuel consumption	1,551,850	kg
	Cost of straw collection workers	503,143.20	$
	Cost of straw transportation workers	310,039.20	$
	Cost of straw storage workers	188,812.80	$
	Total labor cost	1,001,995.20	$
	Equipment and maintenance service fees	412,411.59	$
	Cost of infrastructure inputs	589,487.21	$

environmental resource (R) (48.97×10^{18} sej), which accounted for 33%. The emergy value input of nonrenewable environmental resource (N) and the purchased nonrenewable resource (F_N) were 4.88×10^{18} sej and 9.32×10^{18} sej, respectively.

In the artificial model, the renewable emergy value ($R + F_R$) and nonrenewable resource ($N + F_N$) accounted for 92% and 8%, respectively, of the total emergy value input. For the mechanical model, the renewable emergy value ($R + F_R$) and nonrenewable resource ($N + F_N$) accounted for 91% and 9%, respectively. The total emergy value input of renewable resources ($R + F_R$) was significantly higher than

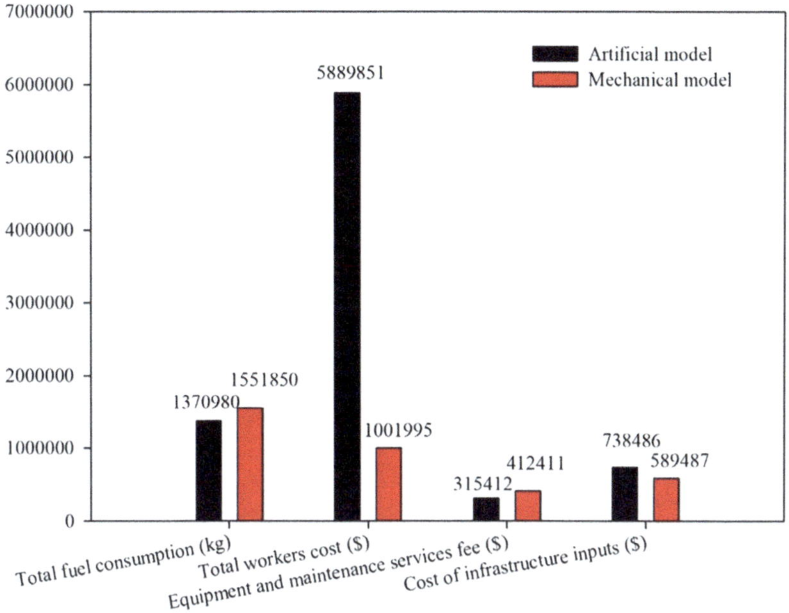

Fig. 6.3 Key economic consumption of the two models

that of nonrenewable resources ($N + F_N$) in the two models, so it was clear that the operation system of the company had low scientific and technological input, and the development degree was not high. The two models were all typical renewable resource utilization projects. According to Table 6.14, the emergy value output of the two models was the same value of 49.89×10^{18} sej, but the total emergy value inputs of the two models were greatly different. The main reasons, illustrated in Fig. 6.5, were differences in labor, fuel consumption, infrastructure investment, and equipment maintenance. The labor emergy value input of the artificial model was 29.1×10^{18} sej, which was obviously higher than that of the mechanical model (4.95×10^{18} sej). There were small differences between the two models in the emergy input of fuel consumption, infrastructure investment, and equipment maintenance. It is obvious that the artificial model required too much labor, which resulted in the high emergy input.

6.4.4 Comprehensive Emergy Indicators of the Two Models

The emergy indicator system was constructed from ESR, ELR, EYR, and ESI. The sustainability and operational efficiency of the straw collection, transportation, and storage system were analyzed. By using the emergy indicators in Fig. 6.6 and Table 6.15, the artificial model was compared with the mechanical model, and the development degree of the two models was evaluated.

Table 6.12 Emergy analysis of the artificial model

Classification	Project		Initial data	Unit of initial data	Solar emergy conversion rate	Emergy value ($\times 10^{18}$ sej)	References
R	Solar emergy		8.23×10^{17}	sej	1	0.82	Brown and Ulgiati (2004)
	Rainwater potential emergy		8.19×10^{14}	sej	8.89×10^{3}	7.28	Brown and Ulgiati (2016)
	Rainwater chemical emergy		3.18×10^{15}	sej	1.54×10^{4}	48.97	Brown and Ulgiati (2016)
	Wind emergy source		2.92×10^{15}	sej	663	1.94	Brown and Ulgiati (2016)
	Maximal value		/		/	48.97	
N	Net loss of topsoil		6.59×10^{13}	sej	7.40×10^{4}	4.88	Cheng et al. (2017)
	Subtotal		/		/	4.88	
F_R	Workers	Straw collection	4.47×10^{6}	$	4.94×10^{12}	22.08	Guan et al. (2016)
		Straw transportation	7.89×10^{5}	$	4.94×10^{12}	3.90	Guan et al. (2016)
		Straw storage	6.31×10^{5}	$	4.94×10^{12}	3.12	Guan et al. (2016)
	Corn straw		1.29×10^{15}	sej	2.70×10^{4}	34.83	Zhao et al. (2019)
	Wheat straw		1.71×10^{15}	sej	2.70×10^{4}	46.17	Zhao et al. (2019)
	Subtotal		/		/	110.10	
F_N	Fuel oil	Straw collection	0		6.60×10^{4}	0	
		Straw transportation	8.68×10^{12}	sej	6.60×10^{4}	0.57	Guan et al. (2016)
		Straw storage	4.98×10^{13}	sej	6.60×10^{4}	3.29	Guan et al. (2016)
	Investment in infrastructure		7.38×10^{5}	$	4.94×1012	3.65	Guan et al. (2016)
	Equipment maintenance		3.15×10^{5}	$	4.94×10^{12}	1.56	Guan et al. (2016)
	Subtotal		/		/	9.07	
Input *T*	Total		/		/	173.02	
Output *Y*	Straw products		1.01×10^{7}	$	4.94×10^{12}	49.89	Guan et al. (2016)

Table 6.13 Emergy analysis of mechanical model

Classification	Project		Initial data	Unit of initial data	Solar emergy conversion rate	Emergy value (× 10^{18} sej)	References
R	Solar emergy		8.23×10^{17}	sej	1	0.82	
	Rainwater potential emergy		8.19×10^{14}	sej	8.89×10^{3}	7.28	Brown and Ulgiati (2016)
	Rainwater chemical emergy		3.18×10^{15}	sej	1.54×10^{4}	48.97	Brown and Ulgiati (2016)
	Wind emergy source		2.92×10^{15}	sej	663	1.94	Brown and Ulgiati (2016)
	Maximal value		/		/	48.97	
N	Net loss of topsoil		6.59×10^{13}	sej	7.40×10^{4}	4.88	Cheng et al. (2017)
	Subtotal		/		/	4.88	
F_R	Workers	Straw collection	5.03×10^{5}	$	4.94×10^{12}	2.48	
		Straw transportation	3.10×10^{5}	$	4.94×10^{12}	1.53	
		Straw storage	1.89×10^{5}	$	4.94×10^{12}	0.94	
	Corn straw		1.29×10^{15}	sej	2.70×10^{4}	34.83	Zhao et al. (2019)
	Wheat straw		1.71×10^{15}	sej	2.70×10^{4}	46.17	Zhao et al. (2019)
	Subtotal		/		/	85.94	
F_N	Fuel oil	Straw collection	4.82×10^{13}	sej	6.60×10^{4}	3.18	
		Straw transportation	8.68×10^{12}	sej	6.60×10^{4}	0.57	
		Straw storage	9.33×10^{13}	sej	6.60×10^{4}	0.62	
	Investment in infrastructure		5.89×10^{5}	$	4.94×10^{12}	2.91	Guan et al. (2016)
	Equipment maintenance		4.12×10^{5}	$	4.94×10^{12}	2.04	Guan et al. (2016)
	Subtotal		/		/	9.32	
Input *T*	Total		/		/	148.11	
Output *Y*	Straw products		1.01×10^{7}	$	4.94×10^{12}	49.89	Guan et al. (2016)

Note: Several renewable environmental resources were caused by different phenomena resulting from the same climatic or geophysical effects. Therefore, only the chemical emergy of rainwater with the largest emergy value input was used to avoid double calculation of the emergy value

Table 6.14 Emergy values of the two models

Project	Artificial model emergy value input ($\times 10^{18}$ sej)	Mechanical model emergy value input ($\times 10^{18}$ sej)
R	48.97	48.97
N	4.88	4.88
F_R	110.10	85.94
F_N	9.07	9.32
T	173.02	148.11
Y	49.89	49.89

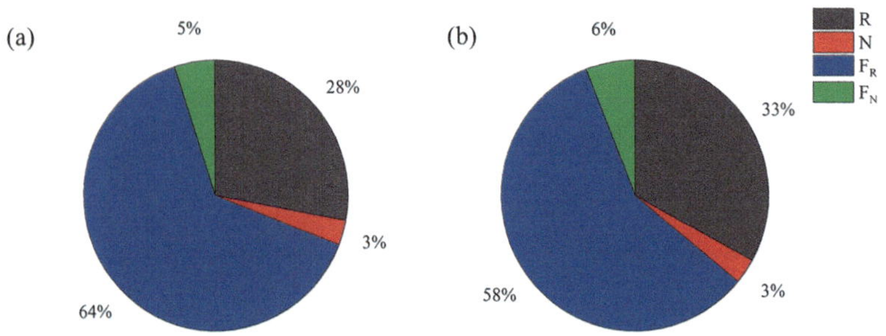

Fig. 6.4 Emergy value input classification: (**a**) artificial model and (**b**) mechanical model

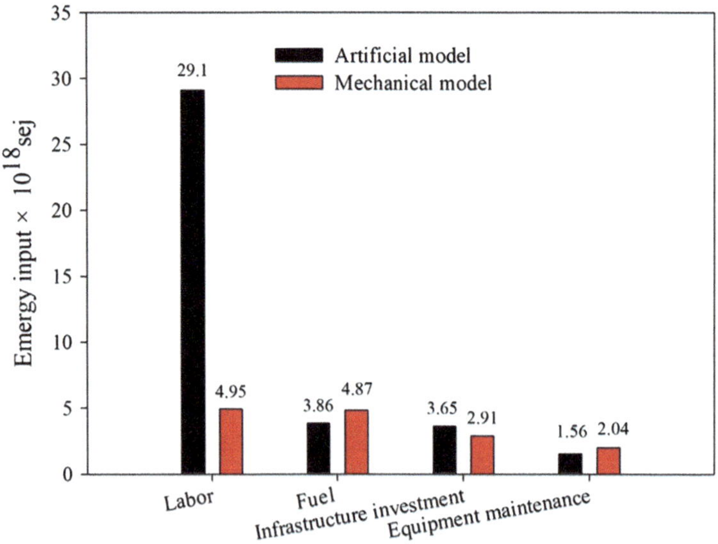

Fig. 6.5 Comparison of major emergy value input between the two models

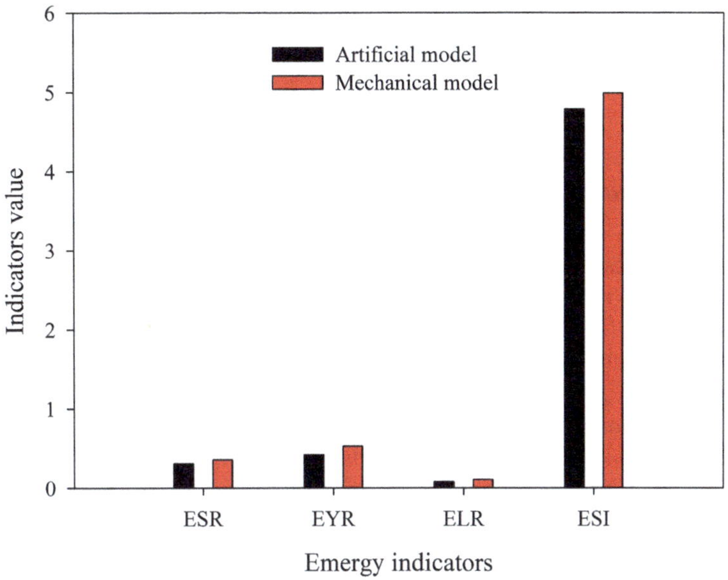

Fig. 6.6 Emergy indicators of the two models

Table 6.15 Comparison of the emergy index of the two models

Emergy value index	Artificial model	Mechanical model
ESR	0.31	0.36
EYR	0.42	0.53
ELR	0.08	0.11
ESI	4.79	4.99

The ESR of the artificial model and the mechanical model were 0.31 and 0.36, respectively. Both models were less dependent on the natural environment, but the ESR of the mechanical model was higher than that of the artificial model, indicating that the mechanical model system had a higher development degree of internal resources, and the demand for external resources in the system was relatively small. In the straw collection, transportation, and storage system, the EYR of the mechanical model was 0.53, which was higher than the artificial model (0.42), illustrating that the production efficiency of the mechanical model was higher than the artificial model. The main reason for the low production efficiency of the artificial model was the excessive labor input. The ELR of the artificial and mechanical models were 0.09 and 0.11, respectively. Both were lower than China's minimum value of 1.71[30]. It can be concluded that the renewable emergy value inputs of the two systems were high and they all showed low pressure on the environment, but the pressure on the environment of the artificial model was slightly lower than that of the mechanical model. ESI reflects the sustainable development level of the straw collection, transportation, and storage system. The ESI of the two models were 4.79 and 4.99, respectively. The values were between 1 and 10, showing that the two models were sustainable, but the sustainable development ability of the mechanical model was higher than that of the artificial model.

6.4.5 Comparison of Emergy Value of Two Models

As shown in Tables 6.11 and 6.13, the emergy value output of straw products was only 49.89×10^{18} sej, and the emergy value input of 2.0×10^8 kg straw was 148.11×10^{18}–173.02×10^{18} sej. In addition, the solar emergy contained in 2.0×10^8 kg straw itself was divided into two parts: One part was 34.80×10^{18}sej, which the 0.83×10^8 kg corn straw contained, and the other part was 46.11×10^{18}sej, which came from the solar emergy of 1.17×10^8 kg wheat straw. The total of the two parts was 80.91×10^{18} sej. Under the condition of emergy balance of the straw recycling company, the market price emergy of 2.0×10^8 kg straw should be equal to the total input solar emergy and higher than the solar emergy of the straw itself. But the solar emergy output of straw products was only 49.89×10^{18} sej, which was considerably lower than both the solar emergy input (148.11×10^{18}–173.02×10^{18} sej) and the solar emergy contained in the straw itself (80.91×10^{18} sej). The value of the straw itself was seriously underestimated, and the market price of straw was seriously low. This led to farmers' lack of understanding of straw resources. Farmers preferred to burn straw on the spot or discard it, resulting in waste of resources and environmental pollution.

According to the field investigation, the factory price of straw products per kg was 0.05 $ in the present study. The corn output was 12.16×10^8 kg, and the wheat output was 2.27×10^8 kg. The correlation emergy coefficient of crops and the formulas for calculating the emergy value of crops and straw are listed in Tables 6.16 and 6.17, respectively. The ratio of the market price emergy to the solar emergy is calculated and illustrated in Fig. 6.7.

As listed in Table 6.17, the emergy value of each kilogram commercial straw calculated from the local commercial price was only 0.25×10^{12} sej, which was about one-third of the emergy input of the company to recycle 1 kg commercial straw (0.74×10^{12}–0.86×10^{12} sej). As shown in Fig. 6.7, the ratio of the wheat market price emergy to the solar emergy of wheat itself was 1.07, slightly higher than 1.00, indicating that the market price emergy of wheat was reasonable. The ratio of the corn market price emergy to the solar emergy of corn itself was 0.76, less than 1.00, which meant that the market price emergy was unreasonable. Especially for the straw recycling, the ratio of the market price emergy to the solar emergy in the mechanical and artificial models were 0.34 and 0.29, respectively, both much lower than 1.00.This clearly shows that the value of straw itself is seriously underestimated, and the market price emergy of straw is too low.

Table 6.16 Correlation emergy coefficient of crops

Project	Corn	Wheat	Provenance
Emergy conversion coefficient (J/kg)	1.63×10^7	1.63×10^7	Lan (2002)
Solar emergy conversion rate (sej/J)	8.30×10^4	6.80×10^4	Croce et al. (2016)
Price per kilogram ($)	0.21	0.24	Field research

Table 6.17 Formulas for calculating the emergy value of 1 kg crop and 1 kg straw

Project	Formula	Numerical value ($\times 10^{12}$ sej)
Solar emergy of corn	Corn emergy conversion coefficient × Corn solar emergy conversion rate	1.35
Solar emergy of wheat	Wheat emergy conversion coefficient × Wheat solar emergy conversion rate	1.11
Market price emergy of corn	Unit price × Emergy currency ratio	1.03
Market price emergy of wheat	Unit price × Emergy currency ratio	1.18
Solar emergy of straw by artificial model	Total emergy value input $\times 2.0 \times 10^{-8}$	0.86
Solar emergy of straw by mechanical model	Total emergy value input $\times 2.0 \times 10^{-8}$	0.74
Market price emergy of straw	Total emergy value output $\times 2.0 \times 10^{-8}$	0.25

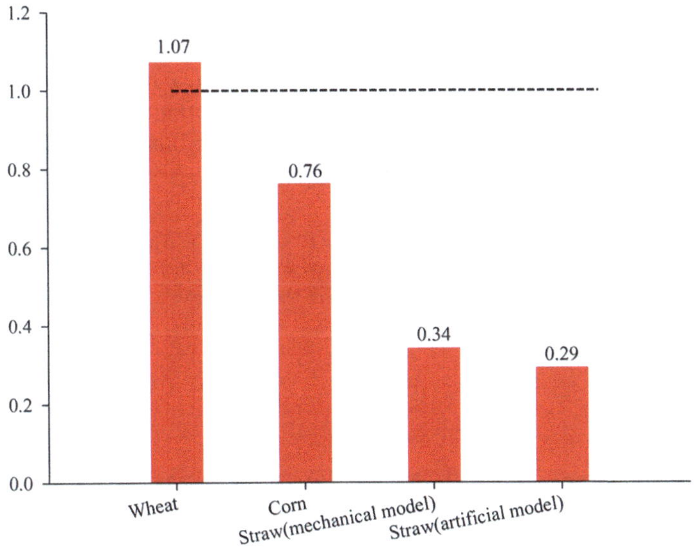

Fig. 6.7 Ratio of the market price emergy to the solar emergy value

6.5 Summary and Prospects

In this study, on the premise of emergy analysis theory, field survey, mathematical economic modeling, and quantitative analysis were employed to investigate the straw collection, storage, and transportation system of Laifa Straw Recycling Company in Sheqi County. The following conclusions were made from the present study:

According to different straw collection methods, the artificial model and the mechanical model were established and the emergy system diagrams of the two models were illustrated. The emergy value output of the straw products for the two models was 49.89×10^{18}sej, while the emergy input of the artificial model and the mechanical model was 173.02×10^{18}sej and 148.11×10^{18} sej, respectively. The main differences were due to labor wages, fuel consumption, infrastructure investment, and equipment maintenance. The emergy indicator system consisted of ESR, EYR, ELR, and ESI. ESR showed that the mechanical model had a higher development degree of internal resources; EYR indicated that the productivity of the mechanical model was higher than that of the artificial model; ELR illustrated that the renewable emergy value input of the two systems was high, and they exerted low pressure on the environment; ESI reflected that both models were sustainable, but the sustainable development ability of the mechanical model was higher than that of the artificial model. By analyzing the ratio of market price emergy to solar emergy, the ratio for wheat was 1.07, which implied the reasonable price of wheat, whereas for straw recycling, the ratios for the artificial and mechanical models were only 0.29 and 0.34, respectively. These results reveal that the value of straw resources is underestimated and that the market price emergy of straw is seriously low.

In conclusion, the emergy input and output of Laifa Straw Recycling Company are unbalanced. The company should adopt the mechanical model and optimize the structure of emergy value input to improve emergy efficiency. The government should also implement effective economic policies to promote the efficient use of straw resources and support straw recycling enterprises.

Acknowledgments This chapter has been reprinted from Y. Sun, Y. Wang, B. Yang, et al., *Emergy Evaluation of Straw Collection, Transportation and Storage System for Power Generation in China*. Energy, 2021, 231: 120792. Copyright (2021), with permission from Elsevier.

References

Banowetz GM, Boateng A, Steiner JJ, Griffith SM, Sethi V, El-Nashaar H (2008) Assessment of straw biomass feedstock resources in the Pacific Northwest. Biomass Bioenergy 32(7):629–634. https://doi.org/10.1016/j.biombioe.2007.12.014

Brown MT, Ulgiati S (2004) Energy quality, emergy, and transformity: H.T. Odum's contributions to quantifying and understanding systems. Ecol Model 178(1):201–213. https://doi.org/10.1016/j.ecolmodel.2004.03.002

Brown MT, Ulgiati S (2016) Emergy assessment of global renewable sources. Ecol Model 339:148–156. https://doi.org/10.1016/j.ecolmodel.2016.03.010

Campbell ET, Tilley DR (2014) The eco-price: how environmental emergy equates to currency. Ecosyst Serv 7:128–140. https://doi.org/10.1016/j.ecoser.2013.12.002

Chen GQ, Jiang MM, Chen B, Yang ZF, Lin C (2006) Emergy analysis of Chinese agriculture. Agric Ecosyst Environ 115(1):161–173. https://doi.org/10.1016/j.agee.2006.01.005

Cheng F, Lu H, Ren H, Zhou L, Zhang L, Li J et al (2017) Integrated emergy and economic evaluation of three typical rocky desertification control modes in karst areas of Guizhou Province, China. J Clean Prod 161:1104–1128. https://doi.org/10.1016/j.jclepro.2017.05.065

Croce S, Wei Q, D'Imporzano G, Dong R, Adani F (2016) Anaerobic digestion of straw and corn stover: the effect of biological process optimization and pre-treatment on total bio-methane yield and energy performance. Biotechnol Adv 34(8):1289–1304. https://doi.org/10.1016/j.biotechadv.2016.09.004

Espinosa E, Tarrés Q, Delgado-Aguilar M, González I, Mutjé P, Rodríguez A (2016) Suitability of wheat straw semichemical pulp for the fabrication of lignocellulosic nanofibres and their application to papermaking slurries. Cellulose 23(1):837–852. https://doi.org/10.1007/s10570-015-0807-8

Fang X, Shen Y, Zhao J, Bao X, Qu Y (2010) Status and prospect of lignocellulosic bioethanol production in China. Bioresour Technol 101(13):4814–4819. https://doi.org/10.1016/j.biortech.2009.11.050

Guan FC, Sha ZP, Zhang YY, Wang JF, Wang C (2016) Emergy assessment of three home courtyard agriculture production systems in Tibet Autonomous Region, China. J Zhejiang Univ Sci B 17(8):628–639. https://doi.org/10.1631/jzus.B1500154

Hau JL, Bakshi BR (2004) Promise and problems of emergy analysis. Ecol Model 178(1):215–225. https://doi.org/10.1016/j.ecolmodel.2003.12.016

Hong J, Ren L, Hong J, Xu C (2016) Environmental impact assessment of corn straw utilization in China. J Clean Prod 112:1700–1708. https://doi.org/10.1016/j.jclepro.2015.02.081

Jiang W, Yan T, Chen B (2021) Impact of media channels and social interactions on the adoption of straw return by Chinese farmers. Sci Total Environ 756:144078. https://doi.org/10.1016/j.scitotenv.2020.144078

Lan S (2002) Energy value analysis of ecological economic system. Chemical Industry Press, Frankfurt am Main

Li X, Wang S, Duan L, Hao J, Li C, Chen Y, Yang L (2007) Particulate and trace gas emissions from open burning of wheat straw and corn stover in China. Environ Sci Technol 41(17):6052–6058. https://doi.org/10.1021/es0705137

Lin T, Rodríguez LF, Shastri YN, Hansen AC, Ting K (2013) GIS-enabled biomass-ethanol supply chain optimization: model development and Miscanthus application. Biofuels Bioprod Biorefin 7(3):314–333. https://doi.org/10.1002/bbb.1394

Liu H, Ou X, Yuan J, Yan X (2018) Experience of producing natural gas from corn straw in China. Resour Conserv Recycl 135:216–224. https://doi.org/10.1016/j.resconrec.2017.10.005

Liu Z, Wang Y, Geng Y, Li R, Dong H, Xue B et al (2019a) Toward sustainable crop production in China: an emergy-based evaluation. J Clean Prod 206:11–26. https://doi.org/10.1016/j.jclepro.2018.09.183

Liu W, Wang J, Sun L, Wang T, Li C, Chen B (2019b) Sustainability evaluation of soybean-corn rotation systems in the Loess Plateau Region of Shaanxi, China. J Clean Prod 210:1229–1237. https://doi.org/10.1016/j.jclepro.2018.11.121

Lu W, Yu S, Ma Y, Huang H (2018) Integrated economic and environmental analysis of agricultural straw reuse in edible fungi industry. PeerJ 6:e4624. https://doi.org/10.7717/peerj.4624

Lu H, Hu L, Zheng W, Yao S, Qian L (2020) Impact of household land endowment and environmental cognition on the willingness to implement straw incorporation in China. J Clean Prod 262:121479. https://doi.org/10.1016/j.jclepro.2020.121479

Mao J, Jiang X, Zhang X (2019) Research on smart surveillance system of internet of things of straw recycling process based on optimizing genetic algorithm. J Intell Fuzzy Syst 37:1–7. https://doi.org/10.3233/JIFS-179306

Odum HT (1988) Self-organization, transformity, and information. Science 242(4882):1132–1139. https://doi.org/10.1126/science.242.4882.1132

Panwar NL, Kaushik SC, Kothari S (2011) Role of renewable energy sources in environmental protection: a review. Renew Sust Energ Rev 15(3):1513–1524. https://doi.org/10.1016/j.rser.2010.11.037

Ren J, Yu P, Xu X (2019) Straw utilization in China—status and recommendations. Sustainability 11(6):1762. Retrieved from https://www.mdpi.com/2071-1050/11/6/1762

Seglah PA, Wang Y, Wang H, Bi Y (2019) Estimation and efficient utilization of straw resources in Ghana. Sustainability 11(15):4172. Retrieved from https://www.mdpi.com/2071-1050/11/15/4172

Shastri Y, Hansen A, Rodríguez L, Ting KC (2011) Development and application of BioFeed model for optimization of herbaceous biomass feedstock production. Biomass Bioenergy 35(7):2961–2974. https://doi.org/10.1016/j.biombioe.2011.03.035

Shastri YN, Rodriguez LF, Hansen AC, Ting KC (2012) Impact of distributed storage and pre-processing on Miscanthus production and provision systems. 6(1):21–31. https://doi.org/10.1002/bbb.326

Sun Y, Cai W, Chen B, Guo X, Hu J, Jiao Y (2017) Economic analysis of fuel collection, storage, and transportation in straw power generation in China. Energy 132:194–203. https://doi.org/10.1016/j.energy.2017.05.077

Sun YF, Wang YP, Yang B, Zheng ZP, Wang C, Chen B et al (2021) Emergy evaluation of straw collection, transportation and storage system for power generation in China. Energy 231. https://doi.org/10.1016/j.energy.2021.120792

Wang Q, Wang L, Li X, Xin J, Liu Z, Sun Y et al (2020a) Emission characteristics of size distribution, chemical composition and light absorption of particles from field-scale crop residue burning in Northeast China. Sci Total Environ 710:136304. https://doi.org/10.1016/j.scitotenv.2019.136304

Wang Z, Wang Z, Tahir N, Wang H, Li J, Xu G (2020b) Study of synergetic development in straw power supply chain: straw price and government subsidy as incentive. Energy Policy 146:111788. https://doi.org/10.1016/j.enpol.2020.111788

Yazdani S, Salimipour E, Moghaddam MS (2020) A comparison between a natural gas power plant and a municipal solid waste incineration power plant based on an emergy analysis. J Clean Prod 274:123158. https://doi.org/10.1016/j.jclepro.2020.123158

Yin H, Zhao W, Li T, Cheng X, Liu Q (2018) Balancing straw returning and chemical fertilizers in China: role of straw nutrient resources. Renew Sust Energ Rev 81:2695–2702. https://doi.org/10.1016/j.rser.2017.06.076

Zhan C, Zhao R, Hu S (2020) Emergy-based sustainability assessment of forest ecosystem with the aid of mountain eco-hydrological model in Huanjiang County, China. J Clean Prod 251:14. Article No. 119638. https://doi.org/10.1016/j.jclepro.2019.119638

Zhang X, Ma F (2015) Emergy evaluation of different straw reuse technologies in Northeast China. 7(9):11360–11377. Retrieved from https://www.mdpi.com/2071-1050/7/9/11360

Zhao H, Zhai X, Guo L, Liu K, Huang D, Yang Y et al (2019) Assessing the efficiency and sustainability of wheat production systems in different climate zones in China using emergy analysis. J Clean Prod 235:724–732. https://doi.org/10.1016/j.jclepro.2019.06.251

Chapter 7
Research on the Revenue Coordination of Agricultural Waste Power Generation Supply Chain

Zhenfeng Wang, Tianjiao Hou, and Zhanwu Wang

Contents

Abstract Agricultural waste has the characteristics of natural carbon neutrality and stable output guarantee, which provides an effective way to cope with the current climate change and energy crisis. However, in actual production, the proportion of agricultural waste used for power generation is low. In order to improve the current situation of insufficient fuel supply in power plants and the reduction of income caused by the decline of national compensation, this chapter takes the straw direct combustion power generation supply chain as the research object. First, the economic costs of each stage are considered, including the collection, transportation, storage, and power generation processes. Then, combined with the carbon emissions of coal-fired power plants and straw power plants in the whole life cycle, the

Z. Wang · T. Hou · Z. Wang (✉)
College of Mechanical and Electrical Engineering, Henan Agricultural University, Zhengzhou, Henan Province, China
e-mail: wzw6997@henau.edu.cn

Z. Wang et al. (eds.), *Agricultural Waste Power Generation Supply Chain*, Industrial Ecology and Environmental Management 7,
https://doi.org/10.1007/978-3-032-20087-7_7

market carbon price is taken into account. The Stackelberg master-slave game model under decentralized decision-making mode and the overall revenue model of supply chain under centralized decision-making mode are constructed, respectively. By comparing and analyzing the optimal solution of the model, it is found that in the case of centralized decision-making, the amount of straw in the supply chain increases significantly, and the total revenue is improved. Then, based on the supply of centralized decision-making, the revenue-sharing contract is used for coordination. Through the combination design of contract parameters and acquisition price, the total revenue of centralized decision-making is redistributed. Finally, the feasibility and effectiveness of the model are verified by numerical calculation. The results show that the revenue-sharing contract can bring more straw raw materials to the power plant and improve the total revenue of the supply chain. At the same time, the reasonable combination of straw purchase price and revenue-sharing coefficient can also improve the revenue of each participating member in the supply chain to achieve a win-win situation.

Keywords Carbon trading price · Power generation supply chain · Stackelberg game · Revenue-sharing contract

7.1 Introduction

With the rapid development of the economy, the problems of energy shortage and environmental pollution have aroused widespread concern. In 2020, China's power generation sector emitted about 5.4 billion tons of carbon dioxide, accounting for 47% of the total emissions from the national energy system (IEA 2021). The higher emissions are due to the fact that China's existing power generation structure is dominated by coal combustion (Xiao et al. 2021). The gradual depletion of fossil fuels (Huang et al. 2022) and the proposal of the dual carbon goal have promoted the further low-carbon transformation of the power industry. Biomass can achieve negative carbon emissions by combining carbon capture and storage technologies (Emenike et al. 2020). European developed countries, represented by France, the United Kingdom, and Germany, plan to withdraw from all coal-fired power generation in 2022, 2025, and 2050, respectively. The transformation of coal-fired power plants into biomass power plants is one of the important means (Wang et al. 2020b). In China, biomass power generation technologies mainly include direct combustion, gasification power generation, biomass and coal co-firing power generation, and biogas power generation (Zhao and Yan 2012). Biomass used as fuel mainly includes agricultural and forestry production residues (Fan et al. 2022), which are widely distributed and recyclable. Since the distribution of forest biomass is more concentrated than that of agricultural biomass, its production and supply system is more developed (Zhang et al. 2016b). Therefore, we chose agricultural biomass as the research object in this study.

As a fuel, straw can replace some coal resources and reduce the environmental burden, and on the other hand, it can promote rural employment and increase farmers' income. There have been numerous studies describing the properties and potential of straw for power generation (Le et al. 2021; Logeswaran et al. 2020; Kashif et al. 2020; Zhang et al. 2016a). Since the official implementation of the Renewable Energy Law in 2006, China has successively issued a series of support policies, including preferential access to the grid, price subsidies, and tax incentives. The straw power generation industry has developed rapidly. However, due to the higher operating costs and insufficient supply of raw materials, most straw power generation enterprises are in a state of financial deficit (Wu et al. 2021a). China produces about 1 billion tons of crop straw every year (Mou et al. 2021). The shortage of raw materials is not due to the shortage of resources, but the lack of management and coordination of the supply chain. The specific performance is that the low price of raw materials makes few farmers willing to sell straw, and the shortage of raw materials for power plants makes it impossible to fully cover the cost. So, power plants cannot raise prices to purchase more raw materials; this vicious circle hinders the sustainable development of the supply chain. In addition, the 2021 Biomass Power Generation Project Construction Work Plan divides agricultural and forestry biomass power generation projects into two categories: competitive configuration and noncompetitive configuration. The electricity price of noncompetitive configuration projects will continue to be 0.75 CNY/kWh, and the electricity price of competitive configuration projects is lower than 0.75 CNY/kWh. Electricity price is directly related to the income of a straw power generation project, and it is the most direct manifestation of whether the project is profitable (Wu et al. 2017). In the short term, the gradual withdrawal of state subsidies will also bring certain economic pressure to the straw power generation industry. Besides, China's carbon emissions trading was launched on July 16, 2021, and the power generation industry is the first industry to be included. Under the new policy and market background, how to ensure the supply of straw and further increase the total revenue of the supply chain has become the key to the smooth operation of the straw power generation industry.

From an economic point of view, we summarize the factors that affect the revenue of the straw power generation supply chain into two aspects: cost and supply. Compared with coal-fired power generation, straw power generation has higher initial construction costs and later operating costs (Nagel et al. 2000). The supply chain acts as a bridge connecting the origin and the sale place. One of the main goals of its management is to minimize costs (Mafakheri and Nasiri 2014). Therefore, relevant scholars analyze it from the perspective of supply chain economic cost. The straw power generation supply chain includes collection, transportation, processing, and storage (Tan et al. 2017). One study pointed out that straw collection cost, transportation cost, and loading cost are the most important components, accounting for more than 75% of the total cost (Wu et al. 2021b). The straw recycling radius is the most sensitive factor affecting logistics cost (Xu and Chen 2020). Since individual farmers determine the use of straw, and the characteristics of straw itself are large in quantity, scattered in distribution, and harvested seasonally, it is difficult for large-scale collection and management (Cao et al. 2016). High land rental and management costs will lead to increased storage costs (Sun et al. 2017). The power generation cost is related to the in-plant price, equipment utilization rate, equipment maintenance cost, production electricity consumption, and total investment cost (Zhang et al. 2013). Different straw supply methods can also have a significant impact on the cost of the supply chain. Zhang et al. (2016b) proposed that village committees set up biomass raw material cooperatives to be responsible for collection, transportation, and temporary storage. Cao et al. (2016) proposed a straw collection and transportation mode based on the transfer station and the village as the minimum collection unit. Wu et al. (2021b) found that the Farmer-Broker-Centralized Storage Site-Factory mode is a low-cost supply mode suitable for large-scale utilization of straw. Wen (2015) pointed out that the hybrid acquisition model can significantly reduce costs compared with the middleman acquisition model. Sun et al. (2017) analyzed the economics of mechanical centralized mode and artificial discrete mode. In this study, we follow the practice of Wang et al. (2020a) and Wen and Zhou (2018), choosing the supply mode of Farmers-Brokers-Power Plants for research. This model is also the most common supply method in China at present.

Adequate resources and well-functioning markets can ensure continuous biomass supply and are important prerequisites for the development of sustainable bioenergy systems (Iakovou et al. 2010). China implements the agricultural household responsibility system, and the straw holders are thousands of scattered farmers. For a long time, straw has been used as fuelwood, feed, fertilizer, etc., and farmers are not very willing to sell straw (Wen and Zhang 2015). Because they are busy sowing the next crop, farmers usually do not have enough time to collect straw (Wu et al. 2021b). To encourage farmers to cooperate with intermediaries in straw supply activities, it is necessary to provide farmers with economic incentives as a short-term solution (Wang and Watanabe 2016). Some researchers have analyzed the mechanism and impact of government subsidies. The results of Wang et al. (2020a) showed that a government subsidy of 53 CNY/t could provide synergistic benefits to the existing straw supply chain system. Wen and Zhou (2018) found that subsidies may work better when farmers are more sensitive to currency. Lin and He (2017) combined emission reduction and subsidies to examine the cost-effectiveness of subsidies for biomass power generation in China. He et al. (2022) studied government regulations for straw bioenergy supply chains, including penalties and subsidy policies. Sun et al. (2019) empirically tested the impact of public policies on crop straw utilization. On the one hand, government subsidies for straw trading activities can motivate supply chain members (Singh 2015), increase trading volume (Wen and Zhou 2018), and reduce production costs (Wang et al. 2014). On the other hand, subsidies also bring a heavy burden to the government (Lin and He 2017). We are concerned that government subsidies are gradually withdrawn. To cope with the capital problem under the new situation, the straw power generation industry needs to further rely on the market to adjust supply. Biomass power generation has obvious environmental advantages. The specific performance is that the use of biomass does not lead to the accumulation of CO_2 in the atmosphere, because the amount of carbon emitted during the transformation process is roughly the same as the amount of carbon absorbed during plant growth (Singh 2015). Zhang et al. (2018) gave the formula for calculating the carbon content of biomass raw material waste. Compared with coal-fired power generation, rice straw power generation can reduce about 1.79 kg $CO_{2\text{-eq/kWh}}$ (Shafie et al. 2014). Studies have further pointed out that the greenhouse gas emissions reduced by straw can gain additional benefits from Certified Emission Reduction (Suramaythangkoor and Gheewala 2011). Wang et al. (2014) also consider the potential revenue of certified greenhouse gas emission reductions as part of the total revenue of biomass power generation companies, and the calculation method is the annual certified emission reduction multiplied by the carbon price in the international carbon trading market. As the national carbon trading market is still in its infancy, few studies have integrated the market carbon price into the straw power generation supply chain. To measure the environmental benefits of straw-based electricity generation, we took this part of the revenue into account in the model-building process.

This study mainly discusses the coordination between stakeholders in the supply chain, and many researchers use the method of game theory to analyze it. Nasiri and Zaccour (2009) first proposed a sequential game model considering distributors, facility developers, and participating farmers to analyze the biomass power generation process in 2009. On this basis, Wen and Zhang (2015) used the concept of the sequential game and coordination to design the straw acquisition model and found that the more straw directly obtained by the power plant from farmers, the higher the profit. Wu et al. (2021a) pointed out that the biomass supply and the profits of supply chain members under the cooperative game are higher than those under the noncooperative game. Wang et al. (2016) used Stackelberg game theory to model the supply chain of straw power generation in China and pointed out that under incentive conditions, the amount of straw and the profits of stakeholders are expected to increase. In addition to game theory, supply chain contracts have also been used to solve coordination problems among members. Revenue-sharing contracts (RSC) are one of the most commonly used types of contracts (Canbulut et al. 2021). They first appeared in the video rental industry. Since RSC can coordinate the interests of supply chain members and alleviate the double marginal effect (Bart et al. 2021), they are now widely used in fresh agricultural products, port services, medical equipment supply, and other fields. In the field of the biomass supply chain, Fan et al. (2019) designed the contract of "protected price + subsidy" between farmers and manufacturers, and the contract of "repurchase + revenue sharing" between intermediaries and manufacturers, which solved the problem of insufficient supply of raw materials. In another article, Fan et al. (2022) established a combined contract model based on surplus compensation, shortage penalty, and revenue sharing, analyzed the impact of government subsidies on member decision-making, and discussed changes in member profits and supply chain profits.

Through the literature review, we find that research on the straw power generation supply chain mainly focuses on economic cost accounting, acquisition mode selection, and government subsidies. Few studies explore the coordination of the straw power generation supply chain from the perspective of contracts. Under the background of the new policy and carbon market, more attention should be paid to the reasonable distribution of the benefits of each member. Inspired by the RSC of the "supplier-retailer" structure in the traditional supply chain and the idea of a sequential game, the purpose of this study is to establish a decision-making model suitable for the supply chain of straw power generation in China to achieve supply chain coordination from the perspective of contract. Through the adjustment of contract parameters, the participating members are guided to make decisions in accordance with maximizing the overall revenue of the supply chain and achieving reasonable distribution of the benefits to each member. Although many scholars have used the solution method of game theory in combination with RSC (Song and Gao 2018; El Ouardighi 2014; Mao et al. 2022), there are few studies on this aspect in the biomass supply chain. In addition, we also introduced market carbon price parameters into the model to quantitatively calculate the environmental benefits, hoping to make up for the negative impact of the decline of state subsidies on project profitability to a certain extent.

7.2 Model

7.2.1 Settings and Assumptions

According to different roles, the participating members of China's straw power generation supply chain can be defined as farmers, brokers, and power plants. Among them, farmers are producers of straw raw materials and are responsible for straw collection. Brokers are the collective name for straw brokers and enterprises engaged in straw transportation and storage, which benefit from the price difference in the process of straw acquisition. Power plants are consumers of straw raw materials. The structure diagram of the crop straw power generation supply chain can be described as the one-way flow of straw raw materials, as shown in Fig. 7.1.

To reduce the complexity of the model, according to the actual situation in China, we have made the following settings:

1. In China, the State Grid is responsible for establishing channels for power transmission and regulating the balance between power generation and power consumption to ensure the reliability of the power supply. According to the

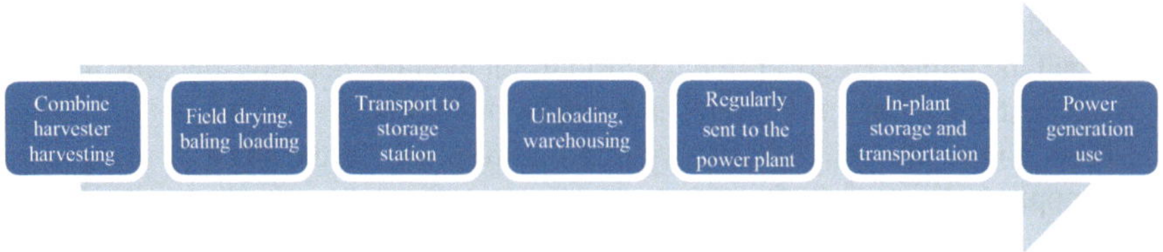

Fig. 7.1 Structure diagram of straw power generation supply chain

Administrative Measures for the Full Guaranteed Acquisition of Renewable Energy Power Generation, the grid enterprises shall purchase the full amount of biomass energy on-grid electricity. Therefore, the power plant only needs to bear the cost of generating electricity.
2. The 2021 Biomass Power Generation Project Construction Work Plan pointed out that the total central subsidy funds for biomass power generation in 2021 is 2.5 billion yuan, of which 2 billion yuan is used to arrange noncompetitive allocation projects, and 0.5 billion yuan is used to arrange competitive allocation projects. Considering that the current subsidy amount is still dominated by noncompetitive configuration projects, we continue to use the purchase price of 0.75 CNY/kWh for calculation.
3. Taking into account the environmental benefits of straw power generation, the carbon price in the carbon trading market is introduced as a measure.
4. As a by-product of crops, straw does not need to bear the production cost.
5. In the transportation process, the vehicles are fully loaded at the beginning and emptied on the return; the distance ratio between the two is 1:1.
6. Due to the obvious seasonal characteristics of straw, for the convenience of statistics, the following models take an annual measurement cycle.

7.2.2 *Cost of Straw Power Generation Supply Chain*

Biomass supply chains for energy production typically consist of four steps: collection and pretreatment, transport, storage, and energy conversion. Therefore, we divide the cost of the straw power generation supply chain into four parts: collection cost, transportation cost, storage cost, and power generation cost (Wang et al. 2023).

7.2.2.1 Collection Cost

After the crops are harvested, the straw scattered in the field is dried by natural links to make its moisture content reach the corresponding standard, and then it is bundled and collected by a baler and loaded by straw grabber. C_c (CNY/t) represents the total cost of the collection link. C_{har} (CNY/t), C_{bal} (CNY/t), and C_{loa} (CNY/t) represent the mechanical harvesting cost, baling cost, and loading cost of unit straw, respectively. The mechanical harvesting cost is 20 (CNY/t) (Sun et al. 2018). $Q(t)$ represents the amount of straw provided by farmers.

$$C_c = C_{har} + C_{bal} + C_{loa} \tag{7.1}$$

Based on Sun et al. (2018), we can obtain:

$$C_{bal} = \frac{n_{bal} \times c_{bal}}{Q} = \frac{c_{bal}^{1} + c_{bal}^{2} + c_{bal}^{3} + c_{bal}^{4}}{Q_{bal} \times t_{bal}}$$
$$= \frac{p_{bal}^{1} \times t_{bal} \times n_{per} + q_{bal}^{2} \times t_{bal} \times P_{oil} + \left(p_{bal}^{3} - p_{bal(t)}^{3}\right) / t_{bal}^{3} + 5\% p_{bal}^{3}}{Q_{bal} \times t_{bal}} \tag{7.2}$$

In Eq. (7.2), n_{bal} represents the number of machines required for straw baling, and c_{bal} (CNY) represents the annual operating cost of each machine. Q_{bal} (t/d) represents the amount of straw processed by each machine per day, t_{bal} (d) represents the annual working time. $c_{bal}^{1}\left(\mathrm{CNY}/\mathrm{y}\right)$, $c_{bal}^{2}\left(\mathrm{CNY}/\mathrm{y}\right)$, $c_{bal}^{3}\left(\mathrm{CNY}/\mathrm{y}\right)$, and $c_{bal}^{4}\left(\mathrm{CNY}/\mathrm{y}\right)$ represent the annual labor cost, fuel cost, depreciation cost, and maintenance cost of each baling machine, respectively. $p_{bal}^{1}\left(\mathrm{CNY}/\mathrm{d}\right)$ represents the daily salary of each operator, and n_{per} represents the number of operators. $q_{bal}^{2}\left(\mathrm{L}/\mathrm{d}\right)$ represents the daily fuel consumption of each machine, and P_{oil} (CNY/L) represents the fuel price. $p_{bal}^{3}\left(\mathrm{CNY}\right)$ represents the purchase price of each piece of mechanical equipment, t_{bal}^{3} represents the average service life of the equipment, and the general value is $t_{bal}^{3} = 10\left(y\right)$. $p_{bal(t)}^{3}\left(\mathrm{CNY}\right)$ represents the residual value of each piece of equipment after scrapping, which is 5% of the purchase price of the equipment. Maintenance costs mainly include routine maintenance and equipment repair costs, which are taken as 5% of the purchase price of the equipment.

The loading cost is treated in the same way as the mechanical baling cost. One baling machine corresponds to one operator, and the working time of the machine is equal to that of the operator. The daily working time is calculated as 8 h. However, loading and unloading require one person to operate the grass-grabbing equipment, in addition to two people for loading and finishing.

7.2.2.2 Off-Site Transportation Cost

According to the research by Zong (2017) and Sun et al. (2017), the transportation process of straw raw materials is divided into two stages. The first stage involves transportation from the field to the nearby storage station, and a small agricultural tractor is used for transportation. The second stage involves transportation from the storage station to the power plant, and a large truck is utilized for transportation. Assuming that the storage stations are evenly distributed and that the collection range is a circle with its center, the radiation radius is denoted by R, and the average distance of the broker to collect straw raw materials is 2/3 R (Wen 2015). The detailed transportation cost calculations are as follows:

1. Labor cost

Involving operations such as stacking, loading, and unloading of straw, based on the average wage, $C_{tra}^{1}\left(CNY/t\right)$ represents the labor cost.

2. Fuel cost

q_{oil} (kg/tkm) represents the ton-kilometer fuel consumption of the cargo.

$$q_{oil} = \frac{\left(q_f^2 \times t_f + q_n^2 \times t_n\right) \times N_{tra}}{2m_{tra} \times l}$$

$$v_f = l / t_f$$

$$v_n = l / t_n$$

Then:

$$q_{oil} = \frac{\left(q_f^2 / v_f + q_n^2 / v_n\right) \times N_{tra}}{2m_{tra}} \tag{7.3}$$

R_{tra} (km) represents the transportation distance of Stage I.

$$R_{tra} = \frac{2}{3} R \times \beta \tag{7.4}$$

L_{tra} (km) represents the transportation distance of Stage II:

$$L_{tra} = l_{tra} \times \beta \tag{7.5}$$

$C_{tra}^2 \left(\text{CNY} / \text{t}\right)$ represents the fuel cost per unit weight of straw.

$$C_{tra}^2 = \frac{\left[R_{tra} \times q_{oil}^1 + L_{tra} \times q_{oil}^2\right] \times P_{oil}}{\rho_{oil}} \tag{7.6}$$

In the formula, $q_f^2 \left(\text{kg} / \text{kWh}\right)$ and $q_n^2 \left(\text{kg} / \text{kWh}\right)$ represent the fuel consumption at full load and empty load, respectively. t_f (h) and t_n (h) represent their corresponding transportation times. v_f (km/h) and v_n (km/h) represent the average vehicle speeds, respectively. N_{tra} (kW) represents the rated power, and m_{tra} (t) represents the vehicle load. l (km) represents the transport distance. β represents the road tortuosity factor, which is the complexity of road conditions, and generally takes the value of $\sqrt{2}$. l_{tra} (km) represents the average transportation distance in the second stage. ρ_{oil} (kg/L) represents the fuel density. $q_{oil}^1 \left(\text{kg} / \text{tkm}\right)$ and $q_{oil}^2 \left(\text{kg} / \text{tkm}\right)$ represent the fuel consumption per ton-kilometer of cargo transportation in Stage I and Stage II, respectively.

3. Additional cost

Additional costs mainly include depreciation of the fixed cost of transport vehicles, equipment maintenance cost, interest, taxes, etc. The total cost of this part is estimated as 10% of the total labor cost and fuel cost.

C_T (CNY) represent the total cost of straw transportation outside the factory.

$$C_T = 1.1Q\left(C_{tra}^1 + C_{tra}^2\right) \tag{7.7}$$

7.2.2.3 Storage Cost

Considering the seasonal output and scattered geographical distribution of crop straw, it needs to be stored in moderation. According to the research conclusions that "Power plants generally build 4–10 purchase stations" (Zhang et al. 2013). The model assumes that there are *n* storage stations and one power plant warehouse. The total storage cost is divided into two parts: one is the annual fixed cost under the total storage capacity, and the other is the variable cost of bundled storage.

1. Fixed cost

   $C_{sto}^1\left(\text{CNY}\right)$ represents the fixed cost, which includes warehouse rental cost, fixed labor wage, and fixed lighting cost. ρ_{sto} (kg/m^3) represents the storage density of straw. h (m) represents the height, and τ represents the land occupation coefficient. S_{sto} (hm^2) represents the total area of the storage station. $P_{sto}^1\left(\text{CNY / hm}^2\right)$ represents the annual rental cost per unit of space. $c_{sto}^1\left(\text{CNY / hm}^2\right)$ represents the other fixed costs except for the rental cost.

$$C_{sto}^1 = S_{sto} \times \left(P_{sto}^1 + c_{sto}^1\right) = \frac{1000Q}{10{,}000\rho_{sto}h\tau} \times \left(P_{sto}^1 + c_{sto}^1\right) \tag{7.8}$$

2. Variable cost

   $C_{sto}^2\left(\text{CNY / t}\right)$ represents the variable costs incurred during warehouse operation, including utilities, equipment maintenance, and employee overtime. C_s (CNY) represents the total storage cost.

$$C_s = C_{sto}^1 + QC_{sto}^2 \tag{7.9}$$

7.2.2.4 Power Generation Cost

C_P (CNY/kWh) represents the conversion cost per unit of straw power. C_F (CNY) represents the annual fixed cost of the straw power plant, including depreciation, personnel remuneration, management fees, financial expenses, and other expenses that are not directly related to the increase or decrease of power generation, which can be calculated according to the average life method.

7.2.3 *Decentralized Decision Model of Straw Power Supply Chain*

The basic dynamic models commonly used in supply chain research include two categories: the decentralized decision-making model and the centralized decision-making model (Song and Gao 2018). In the decentralized decision-making model, farmers, brokers, and power plants all conform to the assumption of a rational economic man, with the ultimate goal of maximizing their benefits. They also know the relevant information about each other, and there is no obstruction or information blocking.

7.2.3.1 Income Model of Farmers

Assuming that the collection cost is borne by the farmers, P_1 (CNY/t) represents the purchase price per unit of straw given by brokers to farmers. G_f represents the farmers' income.

$$G_f = (P_1 - C_c)Q \tag{7.10}$$

According to supply theory in economics, the straw quantity Q provided by farmers is assumed to be a piecewise function, as shown in Fig. 7.2. Within the unit collection cost, the farmer's supply is 0. When the purchase price of the broker is greater than the farmer's unit collection cost, it is assumed that the amount of straw Q increases with the increase in purchase price P_1. The *Supplementary Notice on Matters Concerning the Several Opinions on Promoting the Healthy Development of Non-Water Renewable Energy Power Generation*, issued in 2020, points out that the reasonable utilization hours in the whole life cycle of biomass power generation projects are 82,500 h, and that, after 15 years from the date of grid connection, biomass power generation projects will no longer enjoy central financial subsidy funds,

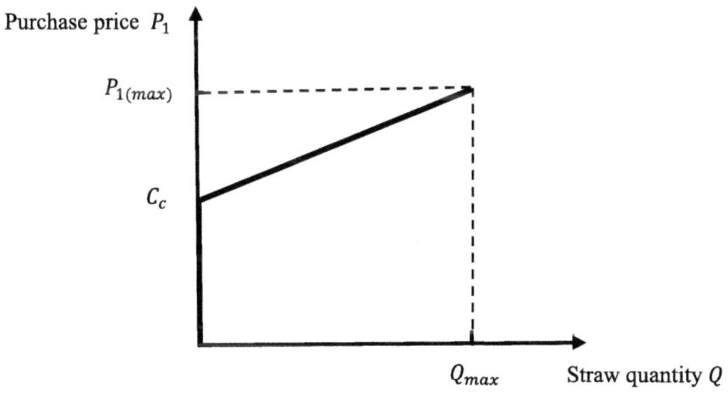

Fig. 7.2 Relationship between straw quantity and purchase price

regardless of whether the project reaches the full life cycle subsidy power. High construction and operating costs result in low commercial profits for biomass power plants, and most companies have to rely on government subsidies. With the gradual decline of the electricity price subsidy, if electricity is sold at the local benchmark electricity price for thermal power, under the constraint of high cost, the company will lose money. Therefore, the average annual subsidy hours $T = 5500$ h are selected for calculation. I (kW) represents the installed capacity of the biomass power plant. Then, the upper limit of straw purchased by the power plant each year can be obtained as follows:

$$Q_{\max} = \frac{TIk}{1-\gamma} \tag{7.11}$$

$$Q = \begin{cases} 0 & 0 \le P_1 \le C_c \\ a(P_1 - C_c) & C_c < P_1 < P_{1(\max)} \end{cases} \tag{7.12}$$

In Eq. (7.12), $a > 0$ is the slope, which is a constant coefficient reflecting the elasticity of the amount of straw sold by farmers to the purchase price. It is jointly affected by farmers' awareness and straw yield within the collection range. $P_{1(max)}$ represents the highest purchase price that brokers can offer when the amount of straw purchased by the power plant reaches the upper limit.

7.2.3.2 Income Model of Brokers

The cost input of the brokers roughly includes transportation and storage costs. P_2 (CNY/t) represents the unit straw purchase price given by the power plant to the brokers. Affected by factors such as planting structure, weather conditions, storage, and transportation, we compromise the quality of straw. It is assumed that the loss coefficient is γ, where $\gamma \in (0, 1)$. G_b represents the income of the brokers.

$$\begin{aligned} G_b &= (1-\gamma)P_2Q - P_1Q - C_T - C_s \\ &= Q\left[(1-\gamma)P_2 - P_1 - 1.1\times\left(C_{tra}^1 + C_{tra}^2\right) - \left(\frac{P_{sto}^1 + c_{sto}^1}{10\rho_{sto}h\tau} + C_{sto}^2\right)\right] \end{aligned} \tag{7.13}$$

7.2.3.3 Income Model of Power Plants

In addition to the fuel purchase cost, straw power plants also need to bear certain in-plant costs. The benefits of straw power plants can be summarized into two parts, namely, power generation benefits and carbon emission reduction benefits. The carbon emission values per unit of electricity generated by coal-fired power plants and straw power plants in the whole life cycle are calculated, respectively. The carbon emission reduction per unit y (t/kWh) is obtained by subtracting the two, and the carbon price is set as x (CNY/t). The *Opinions on Deepening the Reform of the Ecological Protection Compensation System*, issued on September 12, 2021, pointed out that voluntary greenhouse gas emission reduction projects in forestry, renewable energy, methane utilization, and other fields are included in the national carbon emission trading market. China's carbon emission trading was opened on July 16, 2021.

According to the daily transaction data published by the Shanghai Environment Energy Exchange, the K-line chart from July 16, 2021, to July 15, 2022, is drawn, as shown in Fig. 7.3.

As shown in Fig. 7.3, the price of China's carbon trading market is about 40–60 (CNY/t), and 50 (CNY/t) is selected for calculation. P_3 represents the unit on-grid electricity price, which is a fixed value, that is, $P_3 = 0.75$ (CNY/kWh). k (t/kWh) represents the power generation conversion coefficient. G_p represents the income of the biomass power plant.

$$G_p = Q(1-\gamma)\left[k^{-1}.(P_3 + xy - C_P) - P_2\right] - C_F \tag{7.14}$$

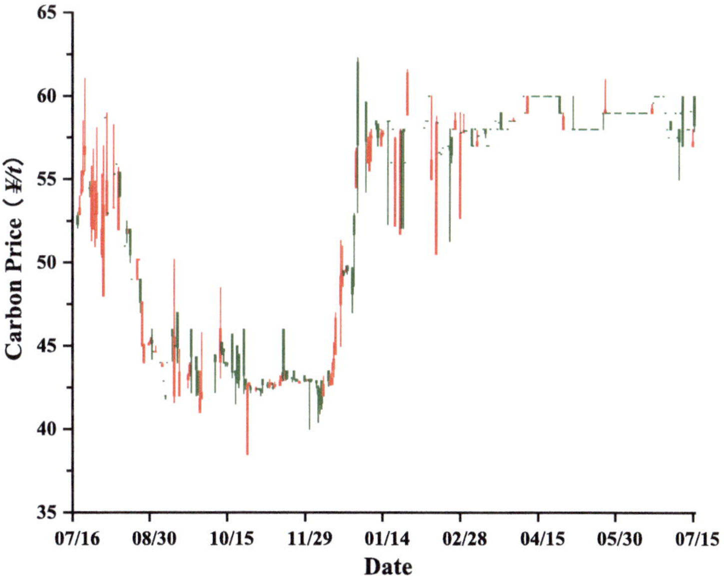

Fig. 7.3 K-line chart of daily transaction data in the national carbon market

7.2.3.4 Model Solving

The decentralized decision model is solved according to Stackelberg game theory. The power plant is regarded as the primary leader, the brokers are taken as the secondary leaders, the farmers are the followers, and the crop straw power supply chain is a two-stage dynamic game process with complete information. According to the reverse induction method, the farmers make supply decisions based on previous experience in the first stage. The brokers comprehensively calculate the purchase price paid to farmers based on their costs and the amount of straw provided by farmers. Then, the power plants determine the optimal price paid to brokers based on the farmers' supply quantity and brokers' acquisition price. In the second stage, farmers and brokers make the final decision based on the feedback of the decision-making price of the power plant, that is, to obtain the refined Nash equilibrium solution of the subgame.

For the convenience of calculation, the unit cost of straw storage and transportation is abbreviated as A.

$$A = 1.1 \times \left(C_{tra}^{1} + C_{tra}^{2}\right) + \left(\frac{P_{sto}^{1} + c_{sto}^{1}}{10\rho_{sto}h\tau} + C_{sto}^{2}\right) \tag{7.15}$$

In reality, the amount of straw given by farmers is:

$$Q = a\left(P_1 - C_c\right) \quad C_c < P_1 < P_{1(\max)} \tag{7.16}$$

When $Q = Q_{max}$, we can obtain $P_{1(\max)} = \dfrac{TIk}{a(1-\gamma)} + C_c$. At this time, the brokers consider the farmers' supply information. Thus, Eq. (7.16) is substituted into the profit function of the brokers to obtain G_b.

$$G_b = a\left(P_1 - C_c\right)\left[\left(1-\gamma\right)P_2 - P_1 - A\right] \quad C_c < P_1 < P_{1(\max)} \tag{7.17}$$

The first-order and second-order partial derivatives of Eq. (7.17) with respect to P_1 are obtained as follows:

$$\frac{\partial G_b}{\partial P_1} = a\left[\left(1-\gamma\right)P_2 - 2P_1 - A + C_c\right]$$

$$\frac{\partial^2 G_b}{\partial P_1^2} = -2a < 0$$

Since the second-order derivative is less than zero, G_b has a local maximum obtained by setting $\partial G_b/\partial P_1 = 0$. Thus, we have:

$$P_1 = \frac{1}{2}\left[(1-\gamma)P_2 - A + C_c\right] \qquad P_{2(\min)} < P_2 < P_{2(\max)} \tag{7.18}$$

When $P_1 = C_c$, $P_{2(\min)} = \dfrac{C_c + A}{1-\gamma}$; when $P_1 = P_{1(max)}$, $P_{2(\max)} = \dfrac{2TIk}{a(1-\gamma)^2} + \dfrac{C_c + A}{1-\gamma}$.

The power plant considers the supply information of farmers and the price information of brokers. Therefore, Eqs. (7.16) and (7.18) are substituted into the profit function of the power plant to obtain G_p.

$$G_p = \frac{a}{2}(1-\gamma)\left[(1-\gamma)P_2 - A - C_c\right]\left[k^{-1}.(P_3 + xy - C_P) - P_2\right] - C_F \qquad P_{2(\min)} < P_2 < P_{2(\max)} \tag{7.19}$$

The first-order and second-order partial derivatives of Eq. (7.19) with respect to P_2 are obtained as follows:

$$\frac{\partial G_p}{\partial P_2} = \frac{a}{2}(1-\gamma)\left\{(1-\gamma)\left[k^{-1}.(P_3 + xy - C_P) - 2P_2\right] + A + C_c\right\}$$

$$\frac{\partial^2 G_p}{\partial P_2^2} = -a(1-\gamma)^2 < 0$$

To further maximize power plant revenue, by setting $\partial G_p/\partial P_2 = 0$, the optimal pricing of the straw power plant is obtained as:

$$P_2^* = \frac{1}{2}k^{-1}.(P_3 + xy - C_P) + \frac{A + C_c}{2(1-\gamma)} \qquad P_{2(\min)} < P_2^* < P_{2(\max)} \tag{7.20}$$

Substituting Eq. (7.20) into Eqs. (7.18) and (7.16) in turn gives the subgame refined Nash equilibrium solutions of brokers and farmers as:

$$P_1^* = \frac{1}{4}\left[k^{-1}(1-\gamma)(P_3 + xy - C_P) - A + 3C_c\right] \qquad C_c < P_1^* < P_{1(\max)} \tag{7.21}$$

$$Q^* = \frac{a}{4}\left[k^{-1}(1-\gamma)(P_3 + xy - C_P) - A - C_c\right] \qquad Q^* < Q_{\max} \tag{7.22}$$

Substituting Eq. (7.21) into the initial constraint condition $P_1^* > C_c$ gives:

$$k^{-1}(1-\gamma)(P_3+xy-C_P)-A-C_c>0 \tag{7.23}$$

Substituting Eq. (7.22) into the initial constraint condition $Q^* < Q_{max}$ gives:

$$k^{-1}(1-\gamma)(P_3+xy-C_P)-A-C_c<\frac{4Q_{\max}}{a} \tag{7.24}$$

Combining Eqs. (7.23) and (7.24) gives the effective intervals of Eqs. (7.20), (7.21), and (7.22) as:

$$0<k^{-1}(1-\gamma)(P_3+xy-C_P)-A-C_c<\frac{4Q_{\max}}{a} \tag{7.25}$$

When the straw power plant does not reach the upper limit of straw collection, it is denoted as constraint condition 1, which satisfies Eq. (7.25). When the straw power plant reaches the upper limit of straw collection, it is denoted as constraint 2, satisfying $k^{-1}(1-\gamma)(P_3+xy-C_P)-A-C_c=\frac{4Q_{\max}}{a}$. To sum up, the game equilibrium solutions of farmers, brokers, and power plants are obtained as:

$$\begin{bmatrix} Q^* \\ P_1^* \\ P_2^* \end{bmatrix} = \begin{cases} \begin{bmatrix} \frac{a}{4}\left[k^{-1}(1-\gamma)(P_3+xy-C_P)-A-C_c\right] \\ \frac{1}{4}\left[k^{-1}(1-\gamma)(P_3+xy-C_P)-A+3C_c\right] \\ \frac{1}{2}k^{-1}.(P_3+xy-C_P)+\frac{A+C_c}{2(1-\gamma)} \end{bmatrix} & \text{constraint 1} \\ \begin{bmatrix} \frac{TIk}{1-\gamma} \\ \frac{TIk}{a(1-\gamma)}+C_c \\ \frac{2TIk}{a(1-\gamma)^2}+\frac{C_c+A}{1-\gamma} \end{bmatrix} & \text{constraint 2} \end{cases} \tag{7.26}$$

The equilibrium decision is substituted into the three income functions to obtain:

$$\begin{bmatrix} G_f^* \\ G_b^* \\ G_p^* \end{bmatrix} = \begin{cases} \begin{bmatrix} \frac{a}{16}\left[k^{-1}.(1-\gamma)(P_3+xy-C_P)-A-C_c\right]^2 \\ \frac{a}{16}\left[k^{-1}.(1-\gamma)(P_3+xy-C_P)-A-C_c\right]^2 \\ \frac{a}{8}\left[k^{-1}.(1-\gamma)(P_3+xy-C_P)-A-C_c\right]^2 - C_F \end{bmatrix} & \text{constraint 1} \\ \begin{bmatrix} \frac{(TIk)^2}{a(1-\gamma)^2} \\ \frac{(TIk)^2}{a(1-\gamma)^2} \\ TIk\left[k^{-1}.(P_3+xy-C_P)-\frac{2TIk}{a(1-\gamma)^2}-\frac{C_c+A}{1-\gamma}\right]-C_F \end{bmatrix} & \text{constraint 2} \end{cases} \tag{7.27}$$

7.2.4 *Centralized Decision Model of Straw Power Supply Chain*

In the centralized decision-making model, farmers, brokers, and power plants are interconnected as a system. The straw power plant is the core enterprise and terminal point of the supply chain, and the straw acquisition mode it adopts determines the structure of the supply chain. Therefore, the model sets the power plant as the sole decision-maker and ensures the maximum benefit of the whole system by determining the optimal straw purchase amount. G_t represents the overall revenue of the supply chain.

$$G_t = Q\left[k^{-1}.(1-\gamma)(P_3+xy-C_P)-A-C_c\right]-C_F \tag{7.28}$$

It can be seen that the income of unit straw is related to the cost of storage and transportation of raw materials, the variable cost of power generation, and the degree of loss. In the case of fixed biomass electricity prices, participating in the carbon trading market can provide additional income for power plants. The conversion coefficient of power generation reflects the efficiency of converting biomass energy into electrical energy, which is related to technical means. Affected by fixed costs such as plant construction and equipment purchase, the fixed cost allocated to unit straw decreases with the increase in straw amount. Therefore, power plants consider purchasing as much straw as possible to meet power generation requirements. The amount of straw purchased under the condition of maximizing income can be obtained as:

$$Q_{\max} = \frac{TIk}{1-\gamma} \tag{7.29}$$

The total revenue is:

$$G_t^* = TIk\left[k^{-1}.\left(P_3 + xy - C_P\right) - \frac{A + C_c}{1-\gamma}\right] - C_F \tag{7.30}$$

Combined with the model in Sect. 7.2.3, the relevant variables are compared and analyzed. The following conclusions can be drawn:

1. Under the condition of constraint 1, $Q_{max} > Q^*$, $G_t^* > G_f^* + G_b^* + G_p^*$, that is, the quantity of straw supplied and the total revenue of the overall decision will be larger than the decentralized decision-making, demonstrating that the decentralized decision-making has not reached the optimal state of the supply chain.
2. Under the critical value for the decentralized decision-making model, when constraint 2 is met, the straw supply reaches the maximum value ($Q^* = Q_{max}$). The combined income of farmers, brokers, and power plants equals the total income under centralized decision-making $\left(G_f^* + G_b^* + G_p^* = G_t^*\right)$. However, there is no guarantee that the three distributed incomes are optimal at the same time.
3. Although the centralized decision-making model has greatly improved the amount of straw and the total revenue of the supply chain, the model aims at maximizing the overall profit of the supply chain and does not pay attention to how to distribute profits among farmers, brokers, and power plants. The model requires a single decision- maker to control the operation process of the supply chain and master all the information, which is difficult to achieve in the real environment. Therefore, we introduce the RSC into the straw power generation supply chain, hoping that, while the straw supply is maximized, the income of each participant can also be improved.

7.2.5 Coordination Model with Revenue-Sharing Contract

In the supply chain of crop straw power generation, straw power plants are continuously dependent on raw materials. Relevant research indicates that China's straw power generation industry has faced a shortage of raw materials. Some local governments have proposed price incentives for farmers or brokers engaged in straw trading activities (Wen and Zhou 2018). Therefore, power plants need to take into account the interests of farmers and brokers while ensuring their profitability, and for this purpose, they need to provide appropriate incentives to prevent supply interruptions caused by the damage to the interests of farmers and brokers. This paper selects RSC to coordinate the supply chain of straw power generation, while the total profit is redistributed based on the supply quantity at the time of centralized decision-making. The straw supply quantity under RSC can be obtained according to Eq. (7.29):

$$Q_{\max} = \frac{TIk}{1-\gamma}$$

To encourage the three parties to accept the contract, it is necessary to ensure that the profits of all members are Pareto-improved. The premise for the realization of the contract is that the profits of farmers, brokers, and power plants should be at least not lower than the profits under the condition that the contract has not been implemented before. Specifically, the power plant and the brokers purchase crop straw at certain prices. P_2^- represents the purchase price by the power plant, and P_1^- represents the purchase price by brokers. After completing the power generation process, the power plant distributes part of its revenue to the brokers, recorded as ω_1, while the brokers redistribute part of the revenue to farmers, denoted by ω_2. Thus, the revenue functions of the three participants can be obtained as:

1. Revenue function of the power plant:

$$G_p^- = Q(1-\gamma)\left\{k^{-1}\left[(1-\omega_1)(P_3+xy)-C_P\right]-P_2^-\right\}-C_F \tag{7.31}$$

2. Revenue function of brokers:

$$G_b^- = Q\left\{(1-\gamma)(1-\omega_2)\left[P_2^- + \omega_1 k^{-1}(P_3+xy)\right]-P_1^- - A\right\} \tag{7.32}$$

3. Revenue function of farmers:

$$G_f^- = Q\left\{\omega_2(1-\gamma)\left[P_2^- + \omega_1 k^{-1}(P_3+xy)\right]+P_1^- - C_c\right\} \tag{7.33}$$

7.3 Results and Discussion

7.3.1 Numerical Calculation

To more directly reflect the impact of supply quantity, purchase price, and contract parameters in the straw power generation supply chain on the final income of farmers, brokers, and power plants, This section verifies the effectiveness of RSC for supply chain revenue coordination. The data simulation of the model built above is carried out below. Some parameters are listed in Tables 7.1, 7.2, 7.3, 7.4, 7.5, and 7.6.

According to Tables 7.5 and 7.6, the carbon emissions of coal-fired power generation over the whole life cycle is 1091.78 (g/kWh), while that of straw power

Table 7.1 Parameter settings in the collection stage

Parameter	Abb	Values		Unit	Data source
		Mobile baler	Grass grabber		
Amount of straw treated per machine per day	Q_{bal}	20	40	t/d	Sun et al. (2018)
Annual working time	t_{bal}	60	60	d	Sun et al. (2021)
Daily wage per operator	p_{bal}^{1}	200	200	CNY/d	Sun et al. (2018)
Number of operators	n_{per}	1	3	–	Field research
Daily fuel consumption per machine	q_{bal}^{2}	56	16	L/d	Sun et al. (2018)
Diesel price	P_{oil}	6.49	6.49	CNY/L	Annual price in (2021)
Purchase price per device	p_{bal}^{3}	46,500	31,500	CNY	Sun et al. (2018)

Table 7.2 Parameter settings in the transportation stage

Parameter	Abb	Values		Unit	Data source
		Miniature tractor	Transformer		
Labor cost	C_{tra}^{1}	35.16	35.16	CNY/t	Zong et al. (2017)
Full-load fuel consumption	q_{f}^{2}	0.28	0.4	kg/kWh	Sun et al. (2017)
No-load fuel consumption	q_{n}^{2}	0.24	0.3	kg/kWh	Sun et al. (2017)
Power-load ratio (power/load)	N_{tra}/m_{tra}	9.1	6.1	kW/t	Sun et al. (2017)
Vehicle load	m_{tra}	3	6	t	Sun et al. (2017)
Average speed at full load	v_{f}	20	35	km/h	Sun et al. (2017)
Average speed at no load	v_{n}	25	45	km/h	Sun et al. (2017)
Diesel density	ρ_{oil}	0.84	0.84	kg/L	Wang et al. (2020a, b)
Collection range radiation radius	R_{tra}	20	–	km	Field Research
Second-stage transport distance	l_{tra}	–	10	km	Sun et al. (2017)

generation is 38.54 (g/kWh), accounting for only 3.53%. It can be seen that straw power generation provides obvious environmental benefits. It can bring 0.053 (CNY/kWh) environmental benefits to the straw power supply chain considering the market carbon trading price of 50 (CNY/t).

Considering that the most common installed capacity of biomass power generation systems in China is between 20 and 30 MW (Lin and He 2017), taking the

Table 7.3 Parameter settings in the storage stage

Parameter	Abb	Values	Unit	Data source
Straw storage density	ρ_{sto}	200	kg/m^3	Sun et al. (2017)
Height	h	6	m	Sun et al. (2021)
Area coefficient	τ	0.8	–	Sun et al. (2021)
Annual lease cost per unit site	P_{sto}^1	12,000	CNY/(hm^2 • y)	Sun et al. (2017)
Fixed cost other than rental cost	c_{sto}^1	35,400	CNY/hm^2	Chang (2014)
Variable cost	C_{sto}^2	32.42	CNY/t	Chang (2014)

Table 7.4 Parameter settings in the power generation stage

Parameter	Abb	Values	Unit	Data source
Power generation conversion coefficient	k	1.4×10^{-3}	t/kWh	Wei et al. (2016)
Straw loss coefficient	γ	0.1	–	Wen (2015)
Conversion cost of unit straw electricity	C_P	0.12	CNY/kWh	Field research
Annual depreciation of straw power plant fixed cost	C_F	1.2×10^7	CNY	Field research
Elasticity of straw quantity to purchase price	a	3450	–	Wen (2015)

Table 7.5 Full life-cycle carbon emissions of coal-fired power generation (g/kWh)

Output material	Coal mining and pretreatment (coal washing)	Transport and storage	Coal-fired power generation	Total output	Data source
CO_2	275.096	14.849	801.77	1091.715	Wang (2017)
CO	–	0.061	–	0.061	Wang (2017)

30 MW power plant as an example, the equilibrium incomes of farmers, brokers, and power plants under decentralized decision-making are obtained according to Eq. (7.27) as:

$$\begin{bmatrix} G_f^* \\ G_b^* \\ G_P^* \end{bmatrix} = \begin{bmatrix} 15,242,017.314 \\ 15,242,017.314 \\ 18,484,034.629 \end{bmatrix} \tag{7.34}$$

Under the RSC, the three benefits can be expressed as:

Table 7.6 Full life cycle carbon emissions of straw power generation (10^{-3} g/kg)

Output material	Field pretreatment	Field transportation	Logistics transport	Storage and transportation	Power plant pretreatment	Total output	Data source
CO_2	4942.401	1315.268	8852.768	1308.310	10,919.363	27,338.11	Wang (2017)
CO	103.853	27.637	50.733	7.468	0	189.691	Wang (2017)

$$\begin{bmatrix} G_p^- \\ G_b^- \\ G_f^- \end{bmatrix} = \begin{bmatrix} 100{,}695{,}000 - 231{,}000 \times \left(573.571\omega_1 + P_2^-\right) \\ 231{,}000 \times \left[\left(1-\omega_2\right)\left(573.571\omega_1 + P_2^-\right) - 10P_1^- / 9\right] - 25{,}641{,}000 \\ 231{,}000 \times \left[\omega_2 . \left(573.571\omega_1 + P_2^-\right) + 10P_1^- / 9\right] - 18{,}813{,}666.667 \end{bmatrix} \tag{7.35}$$

Equation (7.35) can be divided into the following two parts:

$$\begin{bmatrix} G_p^- \\ \mathrm{G}_{\mathrm{b}(1)}^- \end{bmatrix} = \begin{bmatrix} 100{,}695{,}000 - 231{,}000 \times \left(573.571\omega_1 + P_2^-\right) \\ 231{,}000 \times \left(573.571\omega_1 + P_2^-\right) - 25{,}641{,}000 \end{bmatrix} \tag{7.36}$$

a

$$\begin{bmatrix} \mathrm{G}_{\mathrm{b}(2)}^- \\ \mathrm{G}_{\mathrm{f}}^- \end{bmatrix} = \begin{bmatrix} 231{,}000 \times \left[\left(1-\omega_2\right)\left(573.571\omega_1 + P_2^-\right) - 10P_1^- / 9\right] - 25{,}641{,}000 \\ 231{,}000 \times \left[\omega_2 . \left(573.571\omega_1 + P_2^-\right) + 10P_1^- / 9\right] - 18{,}813{,}666.667 \end{bmatrix} \tag{7.37}$$

Combining $G_p^- \geq G_p^*$, $\mathrm{G}_{\mathrm{b}(1)}^- \geq G_b^*$, and $\mathrm{G}_{\mathrm{b}(2)}^- + \mathrm{G}_{\mathrm{f}}^- \geq G_b^* + \mathrm{G}_{\mathrm{f}}^*$, the effective interval 1 of Eq. (7.35) can be obtained as follows:

$$324.410 \leq 573.571\omega_1 + P_2^- \leq 355.891 \tag{7.38}$$

Let $B = 573.571\omega_1 + P_2^-$. Combining $\mathrm{G}_{\mathrm{b}(2)}^- \geq G_b^*$ and $\mathrm{G}_{\mathrm{f}}^- \geq \mathrm{G}_{\mathrm{f}}^*$, the effective interval 2 of Eq. (7.35) is obtained as:

$$147.428 \leq \omega_2 B + 10P_1^- / 9 \leq B - 176.983 \tag{7.39}$$

So far, the effective solution set of Eq. (7.35) has been obtained. Some point sets whose income values are greater than those in Eq. (7.34) are selected, and Origin software is used to draw a graph (Fig. 7.4).

From Fig. 7.4, with the increase of $573.571\omega_1 + P_2^-$, the magnitude of the increase in power plant revenue is reduced. If $573.571\omega_1 + P_2^-$ remains unchanged, with the increase of $\omega_2 B + 10P_1^- / 9$, the magnitude of the increase in farmers' revenue is increased, and the magnitude of the increase in broker's revenue is decreased, but it will not affect the revenue of power plants.

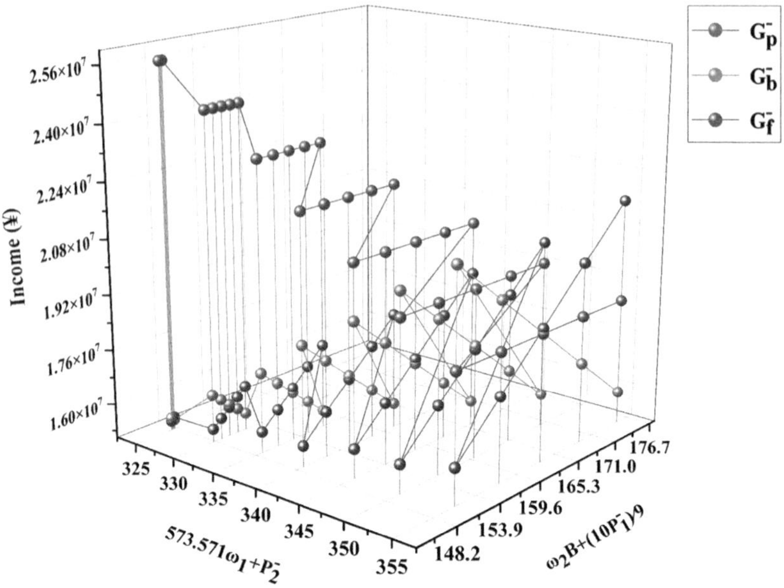

Fig. 7.4 Partial income of farmers, brokers, and power plants under RSC

7.3.2 Sensitivity Analysis

Considering the small fluctuation range of parameters such as straw loss coefficient, power generation conversion coefficient, economic cost, and carbon emission reduction benefits, it is difficult to measure the elasticity "a" of straw quantity on the purchase price, and it is not easy to determine the parameter value due to the combined influence of regional straw yield and farmers' awareness of selling. Therefore, based on the data in Sect. 7.3.1, we perform a sensitivity analysis on the parameter "a" under constraint 1 and constraint 2. The results are shown in Figs. 7.5 and 7.6. Referring to the calculation formula of the sensitivity coefficient in the literature (Chao et al. 2015), the sensitivity coefficient of each parameter is obtained, and the results are shown in Table 7.7.

As can be seen from Table 7.7, in the case of constraint 1, the sensitivity coefficients of Q, G_t, G_p, and G_b are all greater than zero and have a positive correlation with "a." Taking G_t as an example, a 1% reduction in "a" will reduce G_t by about 1.16% (mean value). As shown in Fig. 7.5, when other parameters are kept unchanged, with the increase of "a," Q, G_t, G_p, and G_b first increase and then remain unchanged. The reason for this trend is that the purchase price is fixed. As the elasticity of the amount of straw sold by farmers with respect to the purchase price increases, farmers are willing to provide more straw, which also brings more economic benefits to brokers and power plants. The turning point occurs because the amount of straw purchased by the power plant reaches its maximum limit at this time, and after that, with the increase of "a," the power plant will not continue to purchase.

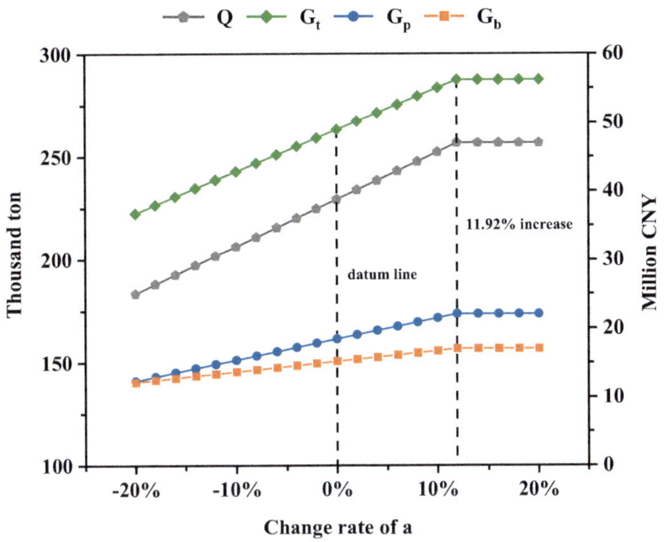

Fig. 7.5 Sensitivity test for parameter "a" under constraint 1

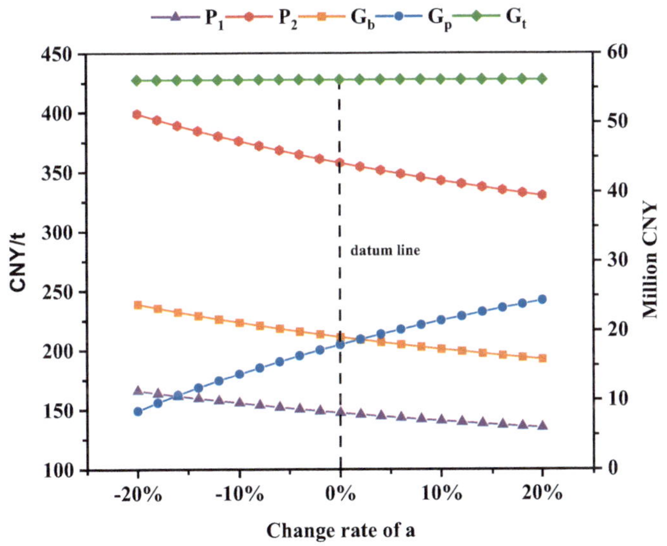

Fig. 7.6 Sensitivity test for parameter "a" under constraint 2

Table 7.7 Parameter sensitivity coefficient changes

$\Delta a/a$	$Q^1(a)$	$G_b^1(a)$	$G_p^1(a)$	$G_t^1(a)$	$P_1^2(a)$	$P_2^2(a)$	$G_b^2(a)$	$G_p^2(a)$	$G_t^2(a)$
−20%	1.00	1.00	1.65	1.25	−0.63	−0.58	−1.25	2.64	0
−16%	1.00	1.00	1.65	1.25	−0.60	−0.55	−1.19	2.52	0
−12%	1.00	1.00	1.65	1.25	−0.57	−0.53	−1.13	2.40	0
−8%	1.00	1.00	1.65	1.25	−0.55	−0.50	−1.08	2.30	0
−4%	1.00	1.00	1.64	1.25	−0.52	−0.48	−1.03	2.20	0
4%	1.00	1.00	1.66	1.24	−0.48	−0.44	−0.97	2.03	0
8%	1.00	1.00	1.65	1.24	−0.47	−0.43	−0.93	1.96	0
12%	0.99	1.00	1.65	1.24	−0.45	−0.41	−0.89	1.88	0
16%	0.75	0.75	1.23	0.93	−0.43	−0.40	−0.86	1.82	0
20%	0.60	0.60	0.98	0.74	−0.42	−0.39	−0.83	1.76	0

Note: Superscripts 1 and 2 represent constraint 1 and constraint 2, respectively

In the case of constraint 2, the sensitivity coefficients of P_1, P_2, and G_b are all less than zero and have a negative correlation with "a." The sensitivity coefficient of G_p is greater than zero and has a positive correlation with "a." Taking P_1 as an example, a 1% increase in "a" reduces P_1 by about 0.51% (mean value). The sensitivity coefficient of G_t is equal to zero, indicating that changes in "a" do not affect the total return. In Fig. 7.6, G_t is a horizontal line. This is because the amount of straw remains constant at this time. As the elasticity of the amount of straw sold by farmers with respect to the purchase price increases, farmers are more willing to provide straw. Even if power plants and brokers lower the purchase price, power plants can eventually get enough straw. Since the total income is fixed, an increase in the income of the power plant will cause the income of farmers and brokers to decrease. This phenomenon also shows that the reasonable location of straw power plants is very important. In areas with high straw yield and low-income levels of farmers, farmers are more willing to sell straw, that is, the value of "a" is larger, which is more beneficial to the sustainable development of power plants.

7.3.3 Comparison and Analysis

To demonstrate that the RSC can effectively coordinate the crop straw power supply chain and enable all participants to achieve a win-win situation, a comparative analysis of the three models was performed in Chap. 2 to verify the correctness of the results. In Sect. 7.2.3.4, the decentralized decision model has two constraints, so it is calculated separately. The results are shown in Table 7.8.

The following inferences can be drawn from Table 7.8.

Inference 1 Under the condition of constraint 1, that is, when the straw supply has not reached the upper limit of power plant collection, the straw quantity under centralized decision-making is significantly greater than that under decentralized

Table 7.8 Equilibrium solutions for different models

	Decentralized decision model under constraint 1	Decentralized decision model under constraint 2	Centralized decision model	RSC coordination model
Straw amount Q(t)	229,314.11	256,666.67	256,666.67	256,666.67
Farmers' unit income (CNY/t)	139.77	147.70	–	$147.428 \le \omega_2 B + 10P_1^- / 9 \le B - 176.983$
Brokers' unit income (CNY/t)	340.15	357.77	–	$324.410 \le 573.571\omega_1 + P_2^- \le 355.891$
Power plant income G_P(CNY)	18,484,034.63	18,050,317.23	–	$18,484,034.63 \le G_P \le 25,756,298.71$
Brokers' income G_b(CNY)	15,242,017.31	19,095,008.05	–	$15,242,017.31 \le G_b \le 22,514,281.38$
Farmers' income G_f(CNY)	15,242,017.31	19,095,008.05	–	$15,242,017.31 \le G_f \le 22,514,281.38$
Total supply chain revenue G_t (CNY)	48,968,069.25	56,240,333.33	56,240,333.33	56,240,333.33

decision-making, while the total benefit of the supply chain is also higher than that under decentralized decision.

Proof 1 The optimal value of the straw quantity under centralized decision-making is 256,666.67 tons, which is 11.928% greater than 229,314.11 tons obtained under decentralized decision-making. The total revenue of the supply chain under centralized decision-making is 56,240,333.33 (CNY), which is 14.851% higher than that under decentralized decision-making of 48,968,069.25 (CNY).

Inference 2 Under the condition of constraint 2, although the straw quantities under decentralized and centralized decision-making are equal to the total supply chain income, compared with the optimal solution under constraint 1, the incomes of brokers and farmers increase, while the income of the power plant decreases. As expected, even if the final solution of the model has reached the maximum value, the benefit of all participating members of the supply chain has not been increased by decentralized decision-making, indicating that the total benefit has not been reasonably distributed at this time.

Proof 2 Under constraint 2, the balanced solution of power plant revenue is 1,850,317.23 (CNY), reduced by 2.346% compared with 18,484,034.63 (CNY) under constraint 1.

Inference 3 After introducing the RSC, the supply quantity of straw into the supply chain and the final total revenue increase. The incomes of power plants, brokers, and farmers also increase to varying degrees with changes in the acquisition price and the proportion of contract sharing. From the viewpoint of brokers, the yield per unit of straw is a solution set. Since the gain brought by the increase in straw quantity makes up for the loss caused by the decrease in the purchase price, the yield can increase even if the price is lower than the previous purchase price (constraint 1 state). From the perspective of farmers, brokers transfer part of their income to farmers, which makes the income per unit of straw higher than the previous purchase price (constraint 1 state), while the final income increases with the increase in straw quantity.

Proof 3 When the purchase price paid by the power plant to brokers, plus a certain proportion of its earnings, is maintained within the numerical interval [324.410, 355.891], the final earnings of the brokers can be greater than 15,242,017.31 yuan, that is, the optimal solution obtained under constraint 1. The larger the value is, the greater the final earnings of the brokers. The maximum limit is to ensure that the revenue of the power plant is not damaged. Similarly, when the brokers' purchase price plus a certain proportion of their income remains within the numerical interval [147.428, B-176.983], the income obtained by farmers will also be greater than the optimal solution under constraint 1. It can be concluded that determining the specific value of RSC and purchase price depends on the position of farmers, brokers, and power plants in the supply chain and their negotiating ability.

7.3.4 Suggestions

1. The introduction of the RSC can realize the coordination of the supply chain. Therefore, power plants need to change the previous concept of completely relying on farmers and brokers to provide raw materials and actively cooperate with brokers and farmers to give them appropriate incentives to ensure an adequate supply of raw materials.
2. The gradual withdrawal of state subsidies has brought a greater impact on the straw power generation industry. Specifically, the unit power generation cost invested by the power plant is about 0.66 CNY/kWh, while the current benchmark price of coal power is about 0.38 CNY/kWh. To stabilize the income of the straw power generation industry, it is recommended that the government maintain a certain subsidy intensity.
3. Compared with coal-fired power generation, straw power generation can consume a large amounts of agricultural waste and bring better environmental benefits. As can be seen from Fig. 7.3, the carbon price in 2022 increased significantly compared with 2021, and power plants can obtain some benefits by participating in market carbon trading. However, the current *Measures for the Administration of Carbon Emissions Trading (Trial)* stipulates that key emission entities can use the state-certified voluntary emission reductions to offset the payment of carbon emission quotas every year, and the offset ratio shall not exceed 5% of the carbon emission quotas that should be paid off. In light of the reality of the country's subsidy and slope decline, it is recommended to appropriately expand the proportion of this part.

7.4 Conclusion

This paper takes the three-level crop straw supply chain composed of farmers, brokers, and power plants as the research object. While quantifying the costs of straw collection, storage, transportation, and power generation, we took into account the market carbon price to discuss the issue of revenue coordination in the straw power generation supply chain. We established three decision-making models and found that the straw supply and total supply chain revenue under decentralized decision-making are lower than those under centralized decision-making, but centralized decision-making does not pay attention to the distribution of member benefits, and it is difficult to achieve in reality. The contract decision-making model is further designed, and the analysis results are verified by a numerical case. The results show that the RSC can bring more straw raw materials to straw power plants and improve the total revenue of the supply chain. The reasonable combination of the straw purchase price and revenue-sharing coefficient also increases the benefits of each participant to varying degrees.

There are several limitations to this study. First, we regard the straw quantity as a linear function of the purchase price. However, in practice, the straw supply is affected by many factors, and the relationship between the straw quantity and the purchase price may also be presented in the form of a curve. Therefore, in further work, we will consider the impact of different supply functions on supply chain returns. Second, due to regional differences and differences in crop straw types, there are differences in straw acquisition modes. Future research can be carried out on the coordination of supply chains under different acquisition modes. Third, it is also a point worth exploring to study the benefits of straw power generation from the perspective of economic cost.

Acknowledgments This chapter has been reprinted from Z. Wang, T. Hou, and Z. Wang, *Revenue Coordination of Straw Power Generation Supply Chain Under Carbon Trading Environment*, Waste and Biomass Valorization, 2023, 14(5):1769–1785. Copyright (2023), with permission from Springer.

References

Bart N, Chernonog T, Avinadav T (2021) Revenue-sharing contracts in supply chains: a comprehensive literature review. Int J Prod Res 59(21):6633–6658
Canbulut G, Kose E, Arik OA (2021) Handling of revenue sharing contracts within the scope of game theory. Soft Comput 25(1):321–329
Cao J, Pang B, Mo XP, Xu FY (2016) A new model that using transfer stations for straw collection and transportation in the rural areas of China: a case of Jinghai, Tianjin. Renew Energy 99:911–918
Chang CY (2014) Research on biomass-fired power generation fuel supply chain inventory system optimization. Dissertation. Nanjing University of Aeronautics and Astronautics [in Chinese]
Chao R, Guo JQ, Feng XX, Gu GL (2015) Uncertainty and sensitivity analysis of parameters calculated by inference formula method. Pearl River 36(04):29–31
El Ouardighi F (2014) Supply quality management with optimal wholesale price and revenue sharing contracts: a two-stage game approach. Int J Prod Econ 156:260–268
Emenike O, Michailos S, Finney KN, Hughes KJ, Ingham D, Pourkashanian M (2020) Initial techno-economic screening of BECCS technologies in power generation for a range of biomass feedstock. Sustain Energy Technol Assess 40:100743
Fan K, Li XN, Wang L, Wang M (2019) Two-stage supply chain contract coordination of solid biomass fuel involving multiple suppliers. Comput Ind Eng 135:1167–1174
Fan K, Mao WJ, Qu H, Li XN, Wang M (2022) Study on government subsidy in a two-level supply chain of direct-fired biomass power generation based on contract coordination. J Ind Manag Optim 19:2436–2451
He N, Jiang ZZ, Huang S, Li KY (2022) Evolutionary game analysis for government regulations in a straw-based bioenergy supply chain. Int J Prod Res. https://doi.org/10.1080/00207543.2022.2030067
Huang M, Zhu L, Zhang WB, Zhu LJ, Ma ZQ, Chen DY (2022) Insight into the synergistic reaction mechanism of biomass pseudo components and low-density polyethylene for the production of light aromatics through co-catalytic fast pyrolysis over hierarchical HZSM-5. Fuel 324:124699
Iakovou E, Karagiannidis A, Vlachos D, Toka A, Malamakis A (2010) Waste biomass-to-energy supply chain management: a critical synthesis. Waste Manag 30(10):1860–1870
IEA (2021) An energy sector roadmap to carbon neutrality in China. IEA, Paris

Kashif M, Awan MB, Nawaz S, Amjad M, Talib B, Farooq M, Nizami AS, Rehan M (2020) Untapped renewable energy potential of crop residues in Pakistan: challenges and future directions. J Environ Manag 256:109924
Le HA, Thu DH, Khoi NQ (2021) Rice straw-base power generation: a potential and economic cost-benefit analysis for a small power plant (10 MWe) in Vietnam. J Mater Cycles Waste Manag 23(6):2232–2241
Lin BQ, He JX (2017) Is biomass power a good choice for governments in China? Renew Sust Energ Rev 73:1218–1230
Logeswaran J, Shamsuddin A, Silitonga AS, Mahlia TMI (2020) Prospect of using rice straw for power generation: a review. Environ Sci Pollut Res Int 27(21):25956–25969
Mafakheri F, Nasiri F (2014) Modeling of biomass-to-energy supply chain operations: applications, challenges and research directions. Energy Policy 67:116–126
Mao HY, Guo YX, Zhang YY, Zhou SL, Liu CH (2022) Low-carbon technology service mode with revenue-sharing contract considering advance funding risk. Environ Sci Pollut Res Int 29(45):68842–68856
Mou AHW, Li BH, Zhou CYG, Zheng DYJ, Dong ERJ, Cao FJX (2021) Estimating winter wheat straw amount and spatial distribution in Qihe County, China, using GF-1 satellite images. J Renewable Sustainable Energy 13(1):013102
Nagel J (2000) Determination of an economic energy supply structure based on biomass using a mixed-integer linear optimization model. Ecol Eng 16(1):91–102
Nasiri F, Zaccour G (2009) An exploratory game-theoretic analysis of biomass electricity generation supply chain. Energy Policy 37(11):4514–4522
Shafie SM, Masjuki HH, Mahlia TMI (2014) Life cycle assessment of rice straw-based power generation in Malaysia. Energy 70:401–410
Singh J (2015) Overview of electric power potential of surplus agricultural biomass from economic, social, environmental and technical perspective—a case study of Punjab. Renew Sust Energ Rev 42:286–297
Song HH, Gao XX (2018) Green supply chain game model and analysis under revenue-sharing contract. J Clean Prod 170:183–192
Sun YF, Cai WC, Chen B, Guo XY, Hu JJ, Jiao YZ (2017) Economic analysis of fuel collection, storage, and transportation in straw power generation in China. Energy 132:194–203
Sun N, Wang YJ, Gao CY, Wang HY, Qin C, Bi YY (2018) Cost analysis of straw collection, storage and transportation—a case study of Henan Province. Chin J Agric Resour Reg Plann 39(05):91–96+230
Sun DQ, Ge Y, Zhou YH (2019) Punishing and rewarding: how do policy measures affect crop straw use by farmers? An empirical analysis of Jiangsu Province of China. Energy Policy 134:110882
Sun YF, Wang YP, Yang B, Zheng ZP, Wang C, Chen B, Li SL, Ying JL, Liu XP, Chen L, Mu WL (2021) Emergy evaluation of straw collection, transportation and storage system for power generation in China. Energy 231:120792
Suramaythangkoor T, Gheewala SH (2011) Implementability of rice straw utilization and greenhouse gas emission reductions for heat and power in Thailand. Waste Biomass Valoriz 2(2):133–147
Tan QL, Wang TR, Zhang YM, Miao XY, Zhu J (2017) Nonlinear multi-objective optimization model for a biomass direct-fired power generation supply chain using a case study in China. Energy 139:1066–1079
Wang PG (2017) Emission analysis of air and PM pollution on straw power plant supply chain based on life cycle assessment. Trans Chin Soc Agric Eng 33(14):229–237
Wang LL, Watanabe T (2016) Factors affecting farmers' risk perceptions regarding biomass supply: a case study of the national bioenergy industry in Northeast China. J Clean Prod 139:517–526
Wang XW, Cai YP, Dai C (2014) Evaluating China's biomass power production investment based on a policy benefit real options model. Energy 73:751–761

Wang ZW, Wang ZF, Tahir N, Wang H, Li J, Xu GY (2020a) Study of synergetic development in straw power supply chain: straw price and government subsidy as incentive. Energy Policy 146:111788
Wang ZW, Wang ZF, Xu GY, Ren JZ, Wang H, Li J (2020b) Sustainability assessment of straw direct combustion power generation in China: from the environmental and economic perspectives of straw substitute to coal. J Clean Prod 273:122890
Wang Z, Hou T, Wang Z (2023) Revenue coordination of straw power generation supply chain under carbon trading environment. Waste Biomass Valoriz 14(5):1769–1785. https://doi.org/10.1007/s12649-022-01960-1
Wei YM, Zhang XP, Luo KY, Liu Y (2016) External performance of biomass power generation in China based on life cycle analysis. J Biobased Mater Bioenergy 10(5):385–396
Wen W (2015) Game analysis on feedstock supply chain of straw power industry in China. Dissertation. Nanjing University of Aeronautics and Astronautics
Wen W, Zhang Q (2015) A design of straw acquisition mode for China's straw power plant based on supply chain coordination. Renew Energy 76:369–374
Wen W, Zhou P (2018) Impacts of regional governmental incentives on the straw power industry in China: a game-theoretic analysis. J Clean Prod 203:1095–1105
Wu YN, Li LWY, Xu RH, Chen KF, Hu Y, Lin XS (2017) Risk assessment in straw-based power generation public-private partnership projects in China: a fuzzy synthetic evaluation analysis. J Clean Prod 161:977–990
Wu JJ, Zhang J, Yi WM, Cai HZ, Li Y, Su ZP (2021a) A game-theoretic analysis of incentive effects for agribiomass power generation supply chain in China. Energies 14(3):546
Wu JJ, Zhang J, Yi WM, Cai HZ, Su ZP, Li Y (2021b) Economic analysis of different straw supply modes in China. Energy 237:121594
Xiao J, Li GH, Xie L, Wang SY, Yu LA (2021) Decarbonizing China's power sector by 2030 with consideration of technological progress and cross-regional power transmission. Energy Policy 150:112150
Xu XL, Chen YJ (2020) A comprehensive model to analyze straw recycling logistics costs for sustainable development: evidence from biomass power generation. Environ Prog Sustain Energy 39(4):e13394
Zhang Q, Zhou DQ, Zhou P, Ding H (2013) Cost analysis of straw-based power generation in Jiangsu Province, China. Appl Energy 102:785–793
Zhang MM, Zhou DQ, Ding H, Jin JL (2016a) Biomass power generation investment in China: a real options evaluation. Sustainability 8(6):563
Zhang XP, Luo KY, Tan QL (2016b) A feedstock supply model integrating the official organization for China's biomass generation plants. Energy Policy 97:276–290
Zhang CY, Ho SH, Chen WH, Xie YP, Liu ZQ, Chang JS (2018) Torrefaction performance and energy usage of biomass wastes and their correlations with torrefaction severity index. Appl Energy 220:598–604
Zhao ZY, Yan H (2012) Assessment of the biomass power generation industry in China. Renew Energy 37(1):53–60
Zong XY (2017) Study on transportation cost model and acquisition path of straw direct combustion power generation. Dissertation. North China Electric Power University [in Chinese]

Chapter 8
Sustainability Evaluation of Agricultural Waste Power Generation Supply Chain

Zhanwu Wang, Zhenfeng Wang, Guangyin Xu, Heng Wang, and Jin Li

Contents

Abstract Biomass power generation can not only reduce dependence on coal but also convert agricultural waste into energy and alleviate environmental pollution. In order to address the major challenges such as the sustainability of the straw power generation (SPG) and to provide a scientific basis from environmental and economic perspectives, a sustainability assessment model was established to evaluate

Z. Wang · Z. Wang · G. Xu (✉) · H. Wang · J. Li
College of Mechanical and Electrical Engineering, Henan Agricultural University, Zhengzhou, Henan Province, China
e-mail: xgy4175@henau.edu.cn

Z. Wang et al. (eds.), *Agricultural Waste Power Generation Supply Chain*, Industrial Ecology and Environmental Management 7,
https://doi.org/10.1007/978-3-032-20087-7_8

the performance of SPG. Life cycle assessment (LCA), life cycle cost (LCC), and the Best-Worst Method (BWM) were integrated into the model. Four SPG scenarios were studied. The results showed that agricultural waste alternative coal-fired power generation had weak economic sustainability, with the ratio of benefit to cost of SPG ranging from 0.8752 to 1.1866. While for the environmental dimension, a strong sustainability was obtained, with the greenhouse gas (GHG) abatement potential of −0.8903 t CO^2-eq unit agricultural waste. The findings obtained from the four SPG scenarios suggested that direct combustion SPG had great sustainability potential in China, and several factors such as agricultural waste supply chain model, pollutant treatment technology, transport vehicles, and agricultural waste purchase cost were the main factors which lead to unfavorable economic performance. We suggest that SPG schemes, transport models, and financial subsidies should be carefully designed to improve the sustainability of agricultural waste power generation.

Keywords Sustainability · Life cycle assessment · Best-Worst Method · Agricultural waste power supply chain

8.1 Introduction

The gap between energy supply and demand is becoming increasingly prominent worldwide. Moreover, environmental and energy security issues are becoming more and more serious. In China, 70–80% of the electricity comes from thermal power generation, and the main fuel is coal. With the development of renewable energy, the proportion of coal-based power generation has decreased year by year. In 2016, it decreased by 33.476 million tons of coal compared with 2015. China is one of the largest agricultural countries with abundant biomass resources. As a renewable and clean resource, biomass can be substituted for traditional fossil fuels. The utilization of crop agricultural waste was 720 million tons in 2015, of which 390 million tons (43.2%), 170 million tons (18.8%), 20 million tons (2.7%), 40 million tons (4%), and 100 million tons (11.4%) of total agricultural waste biomass were utilized as fertilizer, feedstuff, raw material, agricultural waste base stock, and energy, respectively. China's 13th Five-Year Plan for renewable energy development clearly indicates that the proportion of renewable energy consumption in China will be 20% in 2030. Thus, the utilization of agricultural waste for energy production has tremendous development space.

Biomass can replace coal for power generation and alleviate environmental pollution (Wen and Zhang 2015). The developed countries in Europe, represented by France, Britain, and Germany, have set clear targets and road maps for coal-fired power reduction, with plans to phase out all coal-fired power generation by 2022, 2025, and 2050, respectively, and to convert coal-fired power plants into biomass plants. At present, the United Kingdom is implementing the project of transforming coal-fired power stations into biomass power stations, in which the installed capacity and power generation capacity of biomass power generation increased by 12% and 27%, respectively, in 2015. The coal consumption of Denmark decreased by nearly 60% from 2000 to 2015, while the biomass energy consumption in agriculture and forestry increased by 100% in the same period, and the coal-fired power plants in Denmark have been gradually transformed into biomass power plants. China is also gradually reducing the proportion of coal-fired power generation and encouraging the use of biomass to replace coal-fired power generation. Increasing biomass power generations have been constructed with a total installed capacity of 21.16 million KW by 2019. However, compared with developed countries, biomass power generation started late and developed slowly in China (Wang et al. 2020). Moreover, there is a lack of clear understanding of biomass replacing coal-fired power generation, which also faces the following problems: I Are the environmental and economic benefits of biomass direct combustion power generation better than coal-fired power generation? II There are many schemes of biomass direct combustion power generation, but which one is the best? III How about the sustainability of biomass direct combustion power generation?

It is widely known that the open burning of agricultural residue emits a large amount of pollutants. Sun et al. (2016) showed that about 2707.34 Tg of CO_2 was emitted by agriculture residue in open burning from 1996 to 2013. Wang et al. (2018b) showed that the average annual biomass burning emissions in China from 2000 to 2012 led to emissions of 880.66 Mt. CO_2, 96.59 Mt. CO_2-eq for CH_4, and 16.81 Mt. CO_2-eq for N_2O. Agricultural waste power generation can provide clean energy and solve the pollution problem caused by burning agricultural waste directly in the field (Wen and Zhang 2015). Agricultural waste power generation can not only reduce the use of fossil fuels but also increase the value of agricultural waste (Wang et al. 2017b). There has been a lot of research on biomass power generation. Zhao and Yan (2012) used SWOT analysis method to assess the biomass power generation industry in China. Tang (2011) developed a real options model for evaluating the biomass power generation investment in China. Wang et al. (2015) studied direct combustion power generation of biomass. Shafie et al. (2013) studied the economic feasibility of rice agricultural waste co-firing at coal power plants. Rountree (2019) pointed out that renewable-based electricity was yet to become a competitive alternative to fossil fuel-based electricity. The straw power generation (SPG) industry is still lacking advanced development in comparison to other types of renewable energy, such as wind and solar energy, mostly due to the high collection cost of biomass (Cao et al. 2016). An efficient supply chain system can reduce the transport costs, ensure the sustainable supply of agricultural waste, and promote the development of agricultural waste power generation enterprises. However, due to many special characteristics such as small scale and scattered planting in China, it is difficult to implement the agricultural waste supply chain system. The transport costs can be reduced by optimizing the supply chain system (Ebadian et al. 2013). Ng and Maravelias (2017) analyzed the impact of agricultural waste transportation volume and distance. Golecha and Gan (2016) studied the relationship between biomass transport costs and fuel productivity. Obviously, reducing the transport cost of agricultural waste is an urgent problem to be solved.

Constructing a sustainable agricultural waste power generation system is in line with China's development strategy. A set of criteria to assess sustainability is needed before the construction of the agricultural waste power plants. Nishiguchi and Tabata (2016) conducted economic and environmental assessments of wood biomass energy. Kylili et al. (2016) conducted environmental assessment of transportation and processing mechanisms of biomass granulation. Wang et al. 2017b developed an uncertain comprehensive multi-attribute evaluation method to evaluate the biomass polygeneration system. However, reports on detailed assessment methods for agricultural waste power generation projects are still missing. To our knowledge, in most studies about agricultural waste utilization, the authors have only considered one indicator. Song et al. (2017) studied the economics of agricultural waste pellet fuel. Wang et al. (2017a) studied life cycle environmental impacts of agricultural waste briquette fuel utilization. Giorio et al. (2019) measured the main emissions from the energy conversion of pruning residues in vineyards and conducted an environmental assessment. Wang et al. (2018a) evaluated agricultural waste utilization from the view of regional energy, environment, and socioeconomic benefits. However, there are few studies on SPG.

Sustainability refers to a process or state that can be maintained for a long time; environmental and economic performances are the main factors that affect the sustainability. In the sustainability study of biomass energy utilization, mature evaluation methods exist for different evaluation indicators. Life cycle assessment (LCA) is the main method for environmental assessment. Giuntoli et al. (2013) adopted the LCA method to study the environmental impacts of power generation from wheat agricultural waste bales and pellets. Wang et al. (2017b) studied the environmental impact of agricultural waste briquette fuel by LCA. Life cycle cost (LCC) is one of the main methods for evaluating economic indicators. Ren et al. (2015) used the LCC method to optimize the cost of the biomass supply chain. Sun et al. (2017) carried out an economic evaluation of agricultural waste storage and transportation system. Best-Worst Method (BWM) is often utilized to determine the weight of indicators. Wang et al. (2019) used BWM to rank the risk factors and study the energy performance contracting system. It can be seen that these methods have been widely applied in different fields.

Given that there is no detailed sustainability study on direct combustion SPG in China, we integrated the LCA, LCC, and BWM methods to develop a sustainability assessment model of agricultural waste power generation from environmental and economic perspectives. The objective is to evaluate the sustainability of agricultural waste direct combustion power generation in China and to determine the main factors affecting agricultural waste power generation.

8.2 Construction of Sustainable Development Evaluation Model of Agricultural Waste Power Generation Supply Chain

In this section, a methodological framework for sustainability assessment of direct combustion SPG systems is presented. It can be summarized into four steps: (1) BWM for determining the weights of the criteria, (2) life cycle environmental assessment, (3) life cycle economic assessment, and (4) sustainability assessment model. The methodological framework of this study is presented in Fig. 8.1.

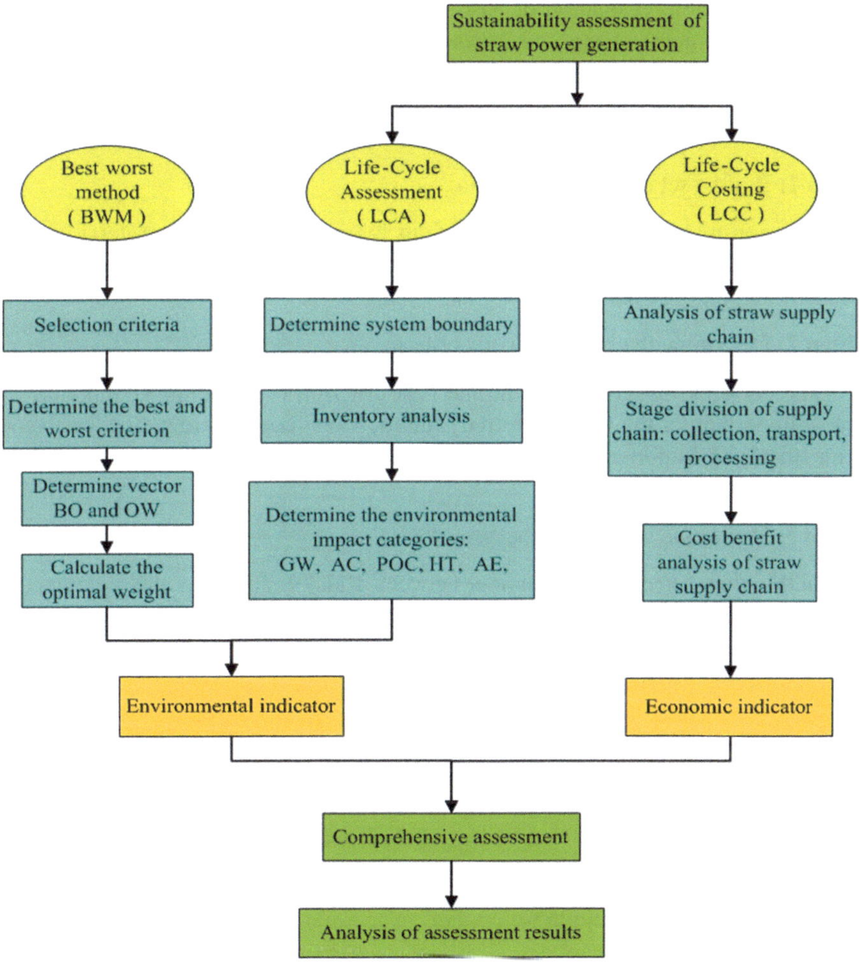

Fig. 8.1 Methodological framework of this study

8.2.1 Best-Worst Method

In decision-making problems, the accuracy of the weights of the criteria has significant impacts on the decision results, which makes the determination of weights more crucial. Rezaei (2015) proposed a new method, the so-called Best-Worst Method (BWM), to determine the weight. Compared with other methods (e.g., the Analytic Hierarchy Process), BWM has two main advantages: (1) BWM needs less pairwise comparisons and is easy for operations, and (2) BWM can effectively determine the weights of the criteria according to the opinions and preferences of decision-makers with higher consistency (Liao et al. 2019). The BWM method has been widely used in various fields, for example, supplier selection (Rezaei et al. 2016), sustainable evaluation of gasoline supply chain (Ahmad et al. 2017), and selection of biomass energy conversion technology (van de Kaa et al. 2017). Ren et al. (2018) studied the multi-stakeholder system combined the Best-Worst Method to select the ballast water treatment technology. The specific steps of BWM are as follows (Rezaei 2015):

Step 1: Build a set of decision criteria

Decision-makers usually make decisions by selecting the main criteria in complex decision-making problems. Suppose the decision-makers select n criteria, which can be expressed as $(k_1, k_2, \ldots, k_n)$.

Step 2: Determine the best and worst criteria.

The decision-makers determine the best criterion (e.g., the most superior, the most important, or the most preferred) and the worst criterion (e.g., the least superior, the least important, or the least preferred), and marks them as k_B and k_W, respectively.

Step 3: Determine vector BO (best-to-others).

In this step, selection of a number between 1 and 9 represents the preference of the best criterion over all the other criteria. The best-to-others vector can be obtained as BO = $(a_{B1}, a_{B2} \ldots, a_{Bn})$, where a_{Bj} indicates the preference of the best criterion k_B over criterion k_j. It is clear that $a_{BB} = 1$.

Step 4: Determine vector OW (others-to-worst).

Similar to the above step, selection of number between 1 and 9 represents the preference of all the other criteria over the worst criterion. The others-to-worst vector can be obtained as OW = $(a_{1W}, a_{2W}, \ldots, a_{nW})^T$, where a_{jW} indicates the preference of the criterion k_j over the worst criterion k_W. It is clear that $a_{WW} = 1$.

Step 5: Calculate the optimal weight $(\omega_1^*, \omega_2^*, \ldots, \omega_n^*)$.

For each pair of ω_B/ω_j and ω_j/ω_W, we set $\omega_B/\omega_j = a_{Bj}$ and $\omega_j/\omega_W = a_{jW}$. To satisfy these conditions for all j, we can find a result ξ that satisfies equation (1).

$$\xi = \min_{j} \max \left\{ \left| \frac{\omega_B}{\omega_j} - a_{Bj} \right|, \left| \frac{\omega_j}{\omega_W} - a_{jW} \right| \right\} \tag{8.1}$$

Considering the nonnegativity of the weights, Eq. (8.1) can be converted into Eq. (8.2).

$$\begin{gathered} \min_{\mathrm{j}} \max \left\{ \left| \frac{\omega_B}{\omega_j} - a_{Bj} \right|, \left| \frac{\omega_j}{\omega_W} - a_{jW} \right| \right\} \\ s.t. \\ \sum_j \omega_j = 1 \\ \omega_j \geq 0, j = 1, 2, \ldots, n \end{gathered} \tag{8.2}$$

Equation (8.2) can be converted into the following problem, which is more conducive to solving the weights.

$$\begin{gathered} \min \tfrac{3}{4} \\ \mathrm{B} \qquad \text{s.t.} \\ \left| \frac{\omega_B}{\omega_j} - a_{Bj} \right| \leq \tfrac{3}{4} \\ \left| \frac{\omega_j}{\omega_W} - a_{jW} \right| \leq \tfrac{3}{4} \\ \sum_j \omega_j = 1 \\ \omega_j \geq 0, j = 1, 2, \cdots, n \end{gathered} \tag{8.3}$$

By solving Eq. (8.3), the optimal weights $\omega_1^*, \omega_2^*, \ldots, \omega_n^*$ can be obtained, as well as the value of ξ^*.

Step 6: Consistency check.

The consistency ratio (CR) is effective index that reflects the degree of consistency in the preference relation. It can be calculated by Eq. (8.4).

$$CR = \frac{\xi^*}{CI} \tag{8.4}$$

where CI is the consistency index (Rezaei 2015). The value of CI can be found in Table 8.1. CR stands for consistency ratio, CR $\in [0, 1]$. CR $\rightarrow 0$ indicates greater consistency, and CR $\rightarrow 1$ indicates less consistency (Rezaei 2015; Wu et al. 2019).

Table 8.1 Consistency index (CI) table (Rezaei 2015)

a_{BW}	1	2	3	4	5	6	7	8	9
CI (max ξ)	0.00	0.44	1.00	1.63	2.30	3.00	3.73	4.47	5.23

8.2.2 *Life Cycle Environmental Assessment*

Life cycle assessment (LCA) has been widely employed to evaluate the environmental impact of a product's life from raw material acquisition, production, use, and disposal. Ren et al. (2014) studied the life cycle energy efficiency of biomass fuels. The LCA method has four main steps: (1) target and system boundary, (2) life cycle inventory, (3) impact assessment, and (4) life cycle interpretation.

As the SPG system is complex, the following assumptions are made in the modeling process:

1. The environmental impact of infrastructure is not considered.
2. The supply of agricultural waste is sufficient, and the agricultural waste supply chain is coordinated.
3. During the agricultural waste collection process, part of the agricultural waste is reserved for returning to the field.

8.2.2.1 Target and System Boundary

In the present study, corn agricultural waste was used for power generation as an example to illustrate the developed model. The goal was to assess the environmental impact of agricultural waste power generation and to carry out sustainability assessment by calculating the environmental indicators. The life cycle of straw can be divided into three parts: planting stage, transportation, and burning. However, the differences in pretreatment (baling, briquetting, and granulation), storage, and transportation modes cause diversity in straw supply. The boundary division of the straw power generation system is shown in Fig. 8.2.

8.2.2.2 Life Cycle Inventory Analysis

The second step of the LCA is inventory analysis.

1. Stage of planting

Straw is the by-product of agricultural cultivation and is also called agricultural waste. The pollutant emission coefficients are shown in Table 8.2.

The amount of pollutant emissions during the planting stage can be calculated by using Eq. (8.5).

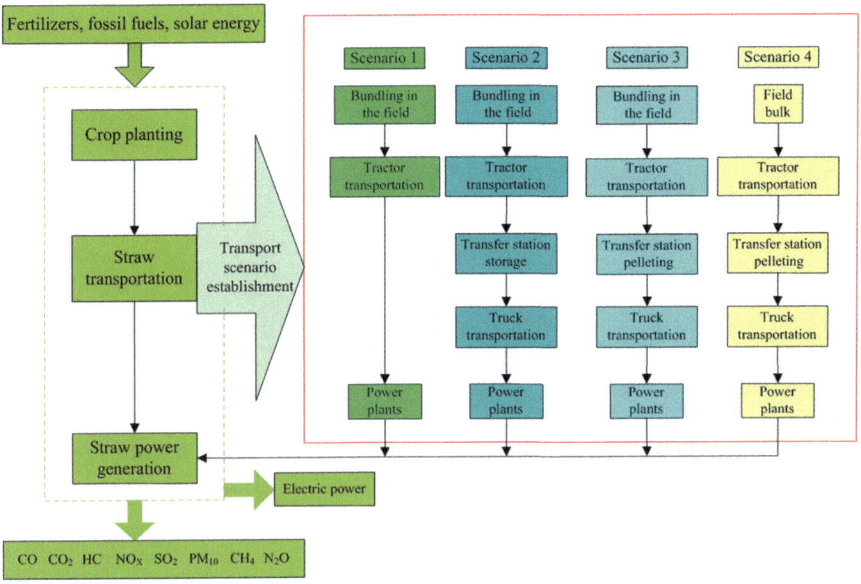

Fig. 8.2 System boundary of straw power generation system

Table 8.2 Pollutant emission coefficients of corn agricultural waste at planting stage (g/t)

	CO_2	CO	NO_X	HC	SO_2	PM_{10}	CH_4	N_2O	Reference
Coefficient	84,860	46.31	101.35	/	87.43	9.22	186.12	1.44	Xia et al. (2012)

$$Q_i^P = Q_{\text{straw}} \times \partial_i \tag{8.5}$$

where Q_i^P represents the i-th pollutant emissions in the straw planting stage, Q_{straw} represents the amount of straw collected, and ∂_i is the emission coefficient of the i-th pollutant.

2. Stage of transport

Straw transport consists of the processes of baling, transportation, and granulation. Transport vehicles emit pollutants when burning diesel. Different transport vehicles are used at different times. The emission coefficients of diesel combustion are shown in Table 8.3.

The pollutant emissions in the transport stage can be calculated by using Eq. (8.6).

$$Q_{ji} = K_j \times \varphi_{ji} \tag{8.6}$$

where Q_{ji} represents the emission of the i-th pollutant from the j-th vehicle in the transportation stage, K_j is the amount of diesel used in the j-th vehicle, and φ_{ji} represents the emission coefficient of the i-th pollutant from the j-th vehicle.

Table 8.3 Emission coefficients of diesel combustion (g/L)

	CO_2	CO	NO_X	HC	SO_2	PM_{10}	CH_4	N_2O	Reference
Baling machine	2616.637	14.995	34.988	4.598	0.588	5.199	0.149	0.068	(Wang et al. 2017a)
Tractor	2664.456	17.054	44.400	4.440	0.576	1.486	0.151	0.074	(Hu et al. 2014)
Truck	2616.637	14.995	34.988	4.598	0.588	0.199	0.149	0.068	(Wang et al. 2017a)

3. Stage of straw burning

Straw is a kind of cleaner energy. The emission of CO_2 from combustion is equal to that from straw planting. Therefore, the emission of CO_2 from straw combustion was not calculated herein. The other pollutant emission coefficients are shown in Table 8.4.

The amount of pollutant emissions from the power generation stage can be calculated by using Eq. (8.7).

$$Q_i^b = Q_{\text{straw}} \times \vartheta_i \tag{8.7}$$

where Q_i^b represents emissions of pollutants in the straw burning stage. Q_{straw} is the amount of straw burned, and ϑ_i is the emission coefficient of the i-th pollutant from straw burning stage.

8.2.2.3 Calculation of Environmental Indicator

The discharge of pollutants can create various environmental problems. For example, CO_2, CH_4, NO_X, and N_2O emissions cause global warming (GW). The present work is focused on well-known existing environmental pollution categories, such as global warming (GW), acidification (AC), photochemical oxidation (PO), human toxicity (HT), and aerosol (AE) (Ren 2018; Valente et al. 2018).

The environmental indicator is the ratio of the environmental impact value of using fossil fuels to the use of biomass fuels (Xia et al. 2012). It can be calculated in five steps as follows.

Step 1: Characterization calculation

$$EP(j) = \sum_{i=1}^{n} Q(j)_i \times EF(j)_i \tag{8.8}$$

where $EP(j)$ represents the characterization result of the j-th environmental impact category, $Q(j)_i$ is the emission amount of the i-th pollutant, and $EF(j)_i$ is the equivalent coefficient of the i-th pollutant.

Step 2: Data standardization

Table 8.4 Emission coefficients of pollutants from straw burning

	CO_2	CO	NO_X	HC	SO_2	PM_{10}	CH_4	N_2O	References
Value (g/kg)	/	0.198	0.217	/	0.034	0.197	0.033	0.960	Hong et al. (2016), Hu et al. (2014)

$$N_j = EP(j)/S(j) \tag{8.9}$$

where N_j is the standardized result of the *j*-th environmental impact category, *EP*(*j*) is the characterization result, and *S*(*j*) is the standardized reference value.

Step 3: Determination of weight

During the process of environmental impact assessment, weight represents the degree of different impact types on the total target. The Best-Worst Method was chosen to calculate the weights, as described in Sect. 8.2.1.

Step 4: Calculating environmental impact potential

$$EIP = \omega_j \times N_j \tag{8.10}$$

where *EIP* represents the environmental impact potential, ω_j is the weight of the *j*-th impact category, and N_j is the standardized result of the *j*-th environmental impact category.

Step 5: Calculating the environmental indicator

By calculating the environmental potential of different fuels, the environmental indicator can be calculated using Eq. (8.11).

$$E_1 = EIP_{\text{coal}} / EIP_{\text{straw}} \tag{8.11}$$

where E_1 is the results of the environmental indicator, EIP_{coal} is the environmental impact potential of coal-fired power generation, and EIP_{straw} is the environmental impact potential of straw-fired power generation.

8.2.3 *Life Cycle Economic Assessment*

8.2.3.1 Collection Cost

There are two main pathways to collect straw in China: manual collection and mechanical collection; thus, cost can also be calculated through two different methods.

1. Manual collection cost

It was found that the density of manually collected straw was 40 kg/m^3. The cost of manual collection can be calculated using Eq. (8.12).

$$C_{\text{collp}} = \left(P_1 / q_{pe} + p_0\right) \times Q_{\text{straw}} \tag{8.12}$$

where C_{collp} represents the cost of manual collection, P_1 is the daily wage of workers, q_{pe} is the amount of straw that each worker can collect per day, p_0 is the purchase price of straw, and Q_{straw} is the amount of straw collected.

2. Mechanical collection cost

This chapter investigates the 9YFK-3043A square baling machine produced by Huaxi-Yutian enterprise, which can shape the straw into square bales with a length of 1.2 m, a width of 0.3 m, a height of 0.43 m, and a weight of 20 kg. It can collect 32 tons of straw per day with a fuel consumption of 7.5 L/ha. The cost of mechanical collection can be calculated using Eq. (8.13).

$$C_{\text{collm}} = \left(\left(P_2 + C_{de1} + C_{m1} + C_{f1}\right) / q_{b1} + p_0\right) \times Q_{\text{straw}} \tag{8.13}$$

where C_{collm} represents the cost of mechanical collection, P_2 is the daily salary of the mechanical driver, C_{de1} is the depreciation cost of the baling machine, C_{m1} is the maintenance cost, C_{f1} is the fuel cost, q_{b1} is the baling quantity, p_0 is the purchase price of straw, and Q_{straw} is the amount of straw collected.

8.2.3.2 Primary Transport Cost

The straw must be transported to the storage place after balling in the field. Small tractors are the main means of transportation from the field to the transfer station. The transportation cost can be calculated using Eq. (8.14).

$$C_{tr} = C_p + C_{f2} + C_{de2} + C_{m2} + C_u \tag{8.14}$$

$$C_{f2} = q_s \times Q_{\text{straw}} \times L_1 / \rho \times P_3 \tag{8.15}$$

where C_{tr} is the transportation cost from the field to the transfer station, C_p is the labor cost, C_{f2} is the fuel cost, L_1 is the distance between the field and the transfer station, q_s is the unit fuel consumption of straw transported by the tractor, ρ is the density of diesel, P_3 is the price of diesel, C_{de2} is the depreciation cost of the transport tool, C_{m2} is the maintenance cost, and C_u is the handling cost.

8.2.3.3 Storage Cost

Straw cannot be converted in a short time after collection. To ensure the quality of straw, it is necessary to establish a storage center for short-term storage. The storage cost can be calculated using Eq. (8.16).

$$C_{st} = \text{A} \times P_4 + P_5 \times Q_{\text{straw}} + C_s \tag{8.16}$$

$$A = Q_{\text{straw}} / h_0 \varnothing_0 \rho_1 \times 10^{-4} \tag{8.17}$$

where C_{st} is the storage cost, A is the effective storage area, P_4 is the rental price of land, P_5 is the management cost of unit straw, h_0 is the straw storage height, $\varnothing_0$ is the straw occupancy coefficient, ρ_1 is the straw density, and C_s is the stacking cost of straw.

8.2.3.4 Processing Cost

The processing of straw briquetting mainly includes crushing, drying, transportation, compression, cooling, and quality inspection. The processing cost can be calculated using Eq. (8.18).

$$C_{pr} = \left(C_{pe2} + C_g + C_{ele} + C_{de3} + C_{m3}\right) \times Q_{\text{straw}} \tag{8.18}$$

$$C_{ele} = P_6 \times E \tag{8.19}$$

where C_{pr} is the processing cost of straw, C_{pe2} is the labor cost of the unit production, C_g is the operational cost of the feeding machine, C_{de3} is the depreciation cost of production, C_{m3} is the maintenance cost of production, C_{ele} is the electricity cost, E is the power consumption of the unit production, and P_6 is the electricity price for agricultural production.

8.2.3.5 Secondary Transport Cost

In this stage, straw is transferred from the storage area to the power plant. The method of transport used in this stage is the large logistics truck. The calculation method of transport costs refers to the primary transport in Sect. 8.2.3.2.

8.2.3.6 Burning Cost

As the end user of straw, power plants must pay a certain cost for their operational processes. The cost of the biomass power plant mainly consists of operation, power generation, and pollutant treatment costs. The operation cost mainly includes human resource, management, and financial costs. Power generation costs include straw processing, equipment depreciation, and maintenance costs. The operating cost of the power plant can be calculated using Eq. (8.20).

$$C_{op} = C_1 + C_2 + C_3 + C_4 + C_5 \tag{8.20}$$

where C_{op} represents the total cost of straw power generation, C_1 is the operation cost, C_2 is the depreciation cost, C_3 is the maintenance cost, C_4 is the straw pretreatment cost, and C_5 represents the pollutant treatment cost.

8.2.3.7 Calculation of Economic Indicator

The economic indicator refers to the ratio of the benefit of straw power generation to the cost. The benefit mainly derives from the sale of electricity. The price of electricity purchased by the Chinese government is 0.75 Yuan/KWh. The cost of power generation mainly includes the raw material, collection, transportation, processing, and burning costs. The economic indicator can be calculated using Eq. (8.21).

$$E_2 = W_1 / C_{\mathrm{straw}} \tag{8.21}$$

$$W_1 = q \times p \tag{8.22}$$

$$C_{\mathrm{straw}} = C_{\mathrm{coll}} + C_{tr} + C_{st} + C_{pr} + C_{\mathrm{op}} \tag{8.23}$$

where E_2 is the economic indicator of the straw power generation, W_1 is the benefit of straw power generation, C_{straw} is the total cost of power generation, q is the quantity of power generation, and p is the government electricity purchase price.

8.2.4 Sustainability Assessment Modeling

By analyzing the environmental and economic data, relevant environmental and economic indicators can be determined. In order to understand the intuitive sustainability of SPG, the environmental indicator as the X-axis and the economic indicator as the Y-axis were used to construct the sustainability discriminant graph, as shown in Fig. 8.3.

There are four regions in the sustainability discriminant graph. Region I represents sustainability from both environmental and economic aspects. Region II represents only economic sustainability, whereas region III represents unsustainability in both environmental and economic aspects, and region IV shows sustainability only in the environmental aspect.

8.3 Case Analysis of Agricultural Waste Power Generation Supply Chain

8.3.1 System Parameters

In order to validate the developed model, the Henan Y SPG (power generation scale: 30 MW) was studied. The collection radius of straw was 40 km. The sustainability assessment of straw power generation was studied based on a system with a production of 10,000 kWh of electricity. The power generation efficiency of Y straw power

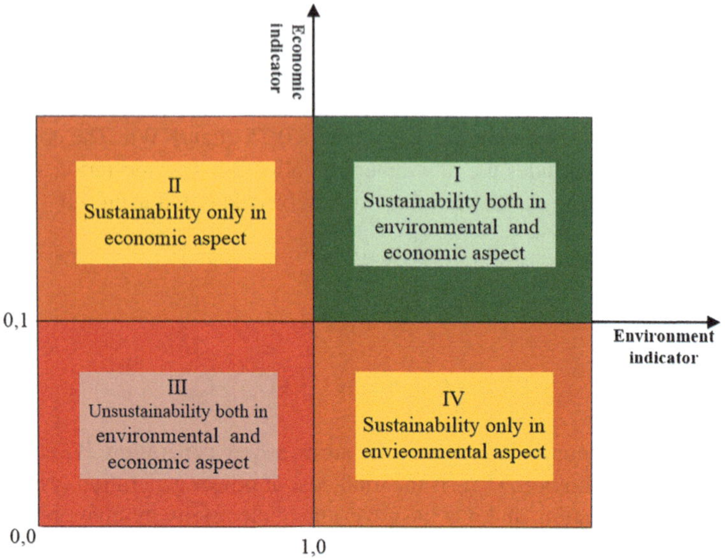

Fig. 8.3 Sustainability discriminant graph of straw substituting for coal in power generation

Table 8.5 Parameters of agricultural machinery

Parameter	Baler	Tractor	Truck	Unit	Data source
Price	98,000	50,000	150,000	Yuan	Field research
Fuel consumption in bulk transport	/	0.136	/	L/(tkm)	Field research
Fuel consumption in baling transport	/	0.101	0.078	L/(tkm)	Field research
Straw transportation fuel consumption	/	/	0.053	L/(tkm)	Field research
Depreciation cost	0.839	0.428	1.284	Yuan/t	Field research
Maintenance cost	2.552	0.237	0.710	Yuan/t	Field research

plant was 1.4 kg/kWh. Because of external environmental and physical changes, the loss rate of straw during storage and processing was assumed to be 5% and 16.67%, respectively. For the production of 10,000 kWh of electricity, approximately 18 tons of straw should be collected. The parameters of agricultural machinery are described in Table 8.5, whereas the parameters of the straw power generation at different stages are summarized in Table 8.6.

8.3.2 *Pollutant Emissions from the Straw Life Cycle*

The straw supply chain uses different transport vehicles at different stages. Equations (8.5)–(8.7) can be used to calculate pollutant emissions at different stages. The results are summarized in Table 8.7.

The results of pollutant emissions of SPG in different scenarios are summarized in Table 8.8.

Table 8.6 Parameters of the straw power generation system

Stage	Parameter	Value	Unit	Data source
Collection stage	p_0	25	Yuan/t	Field research
	P_1	100	Yuan	Field research
	q_{pe}	1	t/day	Field research
	q_{b1}	32	t	Field research
	C_{f1}	240	Yuan	Field research
	P_2	400	Yuan/day	Field research
Transport stage	C_p	200	Yuan/day	Field research
	ρ	0.84	kg/L	Field research
	P_3	6.93	Yuan/L	Henan oil price
	L_1	10	km	Field research
	L_2	30	km	Field research
	C_u	10	Yuan/t	Field research
Storage stage	P_4	800	Yuan/mu	Field research
	P_4	1300	Yuan/mu	Field research
	h_0	5	Meter	Field research
	$\varnothing_0$	0.8		Sun et al. (2017)
	P_5	120	Yuan/day	Field research
	C_s	10	Yuan/t	Field research
Processing stage	C_{pe2}	36.63	Yuan/t	Field research
	C_g	1.42	Yuan/t	Field research
	C_{de3}	4.5	Yuan/t	Field research
	C_{m3}	5	Yuan/t	Field research
	E	68	kWh/T	Hu et al. (2014)
	P_6	0.6	Yuan/kWh	Hu et al. (2014)
Burning stage	C_1	0.08331	Yuan/kWh	Field research
	C_2	0.10347	Yuan/kWh	Field research
	C_3	0.04994	Yuan/kWh	Field research
	C_4	0.07	Yuan/kWh	Field research
	C_5	0.044	Yuan/kWh	Field research

8.3.3 Pollutant Emissions from Coal Power Generation

The coal power generation system mainly includes three stages: mining, transportation, and burning. Pollutant emissions from coal burning mainly include SO_2, NO_X, CO_2, CO, HC, N_2O, CH_4, and PM_{10}. The emission coefficients are described in Table 8.9.

Pollutant emission amounts from different stages of the coal power generation system can be calculated by using Eq. (8.24).

$$Q_{ji}^2 = Q_{\text{coal}} \times \varphi_{ji}^2 \tag{8.24}$$

Table 8.7 Emission amounts of pollutants (g/10,000kWh)

	Stage	CO_2	CO	NO_X	HC	SO_2	PM_{10}	CH_4	N_2O
Scenario 1	Planting	1,527,480	833.58	1824.30	/	1573.74	165.96	3350.16	25.92
	Collection	58874.33	337.39	787.23	103.46	13.23	116.98	3.35	1.53
	Transport	193759.20	1240.17	3228.77	322.88	41.89	108.06	10.98	5.38
	Burning	/	2772	3038	/	476	2758	462	/
Scenario 2	Planting	1,527,480	833.58	1824.30	/	1573.74	165.96	3350.16	25.92
	Collection	58874.33	337.39	787.23	103.46	13.23	116.98	3.35	1.53
	P transport	48439.81	310.04	807.19	80.72	10.47	27.02	2.75	1.35
	S transport	104702.10	600.01	1400.01	183.98	23.53	7.96	5.96	2.72
	Burning	/	2772	3038	/	476	2758	462	/
Scenario 3	Planting	1,527,480	833.58	1824.30	/	1573.74	165.96	3350.16	25.92
	Collection	58874.33	337.39	787.23	103.46	13.23	116.98	3.35	1.53
	P transport	48439.81	310.04	807.19	80.72	10.47	27.02	2.75	1.35
	S transport	71143.74	407.70	951.29	125.02	15.99	5.41	4.05	1.85
	Burning	/	2772	3038	/	476	2758	462	/
Scenario 4	Planting	1,527,480	833.58	1824.30	/	1573.74	165.96	3350.16	25.92
	P transport	65225.88	417.48	1086.91	108.69	14.10	36.38	3.70	1.81
	S transport	71143.74	407.70	951.29	125.02	15.99	5.41	4.05	1.85
	Burning	/	2772	3038	/	476	2758	462	/

Table 8.8 Total amount of pollutants discharged under different scenarios (g/10,000 kWh)

	Scenario 1	Scenario 2	Scenario 3	Scenario 4
CO_2	1780113.53	1739496.24	1705937.88	1647063.55
CO	5183.13	4853.02	4660.71	4323.32
NO_X	8878.30	7856.73	7408.01	6620.78
HC	426.33	368.15	309.19	205.74
SO_2	2104.86	2096.97	2089.43	2076.20
PM_{10}	3148.98	3075.90	3073.35	2956.39
CH_4	3826.49	3824.22	3822.31	3818.96
N_2O	32.83	31.52	30.65	29.12

Table 8.9 Emission coefficients of pollutants from coal burning

	CO_2	CO	NO_X	HC	SO_2	PM_{10}	CH_4	N_2O
Mining stage (kg/t)	500	0.330	0.370	0.160	0.021	5	/	/
Transport stage (g/L)	2616.637	14.995	34.988	4.598	0.588	0.199	0.149	0.0680
Burning stage (g/MJ)	105.087	0.012	0.270	/	3.249	0.012	/	0.0003

Table 8.10 Pollutant emissions of coal power generation system (unit: g)

	CO_2	CO	NO_X	HC	SO_2	PM_{10}	CH_4	N_2O
Mining	861,500	568.590	637.510	275.680	36.183	8615	0	0
Transport	318549.4	1825.491	4259.439	559.761	71.583	24.226	18.139	8.278
Burning	3,786,067	1428.851	32443.32	0	390112.3	1440.858	0	36.021
Total	4966116.4	3822.932	37340.27	835.441	390,220	10080.08	18.139	44.299

where Q_{ji}^2 is the pollutant emission amount, Q_{coal} is the quantity of coal burned, and φ_{ji}^2 is the emission coefficient of pollutants.

The pollutant emission amounts in the coal power generation system are summarized in Table 8.10.

8.3.4 Calculation of Sustainability Assessment Indicators

Through the calculation of pollutant emissions during different stages of the SPG system, the obtained data can be further processed to calculate environmental impact indicators. The impact categories and equivalent factors of pollutants are described in Table 8.11.

The Best-Worst Method is employed to calculate the weight of environmental pollutant impact categories. Through the guidance of relevant experts, we chose

Table 8.11 Environmental impact categories and equivalent factors of pollutants

Impact category	Standardized benchmarks	Unit	Weight	Pollutants	Equivalent factor
Global warming (GW)	8700	kg CO_2-eq	ω_1^*	CO_2	1
				CH_4	21
				NO_X	310
				N_2O	310
Photochemical oxidation (PO)	34.72	kg C_2H_4-eq	ω_2^*	HC	0.416
				CH_4	0.007
Human toxicity (HT)	197.21	kg 1,4-DCB	ω_3^*	CO	0.012
				NO_X	0.780
Aerosol (AE)	18	Kg	ω_4^*	PM_{10}	1
Acidification (AC)	36	Kg SO_2-eq	ω_5^*	SO_2	1
				NO_X	0.700

Table 8.12 Assessment indicators for different scenarios of straw power generation

Scenarios	Environmental indicators	Economic indicators
Scenario 1	9.1101	1.1068
Scenario 2	9.8802	1.1866
Scenario 3	10.2911	1.1632
Scenario 4	10.8306	0.8752

a_{BW} = 5 and set B0 = $(1, 2, 1, 5, 3)$ and OW = $(5, 3, 5, 1, 2)^T$. According to Eq. (8.3), the optimal weight can be calculated as $\omega_1^* = 0.34, \omega_2^* = 0.189, \omega_3^* = 0.283, \omega_4^* = 0.067, \omega_5^* = 0.121$, and $\xi^* = 0.2$. Using the consistency check by Eq. (8.4), we obtain CR = 0.087, which is in agreement with the consistency check, and the results are authenticated.

The environmental and economic indicators of different scenarios can be calculated using Eqs. (8.11) and (8.21). The results are described in Table 8.12.

8.3.5 Assessment Results of Sustainability

According to the established sustainability discriminant graph, it is easy to get the sustainability assessment results of different straw power generation scenarios, as shown in Fig. 8.4.

It is very clear that the straw power generation has shown better sustainability in scenario 1, scenario 2, and scenario 3, both in economic and environmental aspects. Scenario 3 has a better environmental sustainability evaluation indicator, and scenario 2 has the best economic indicator. Among these four scenarios, scenario 4 had the best environmental sustainability evaluation indicator; however, the economic

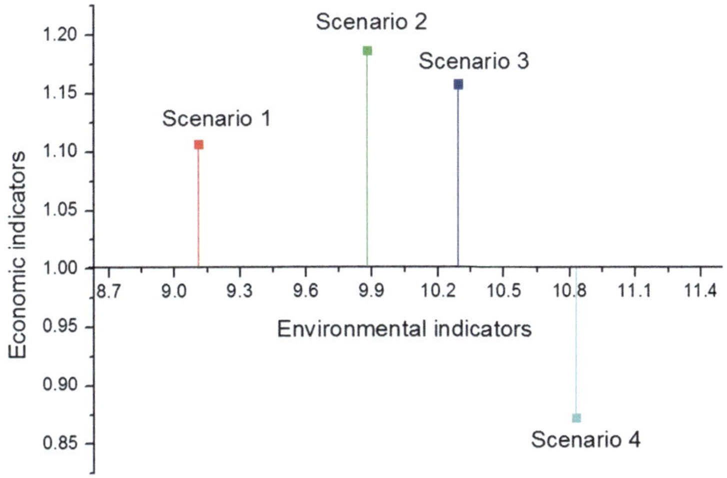

Fig. 8.4 Sustainability assessment results of different power generation scenarios

indicator is less than 1, demonstrating that it cannot create profits for enterprises. Thus, this straw power generation scheme was not considered by enterprise decision-makers.

8.4 Sustainability Evaluation Results of Agricultural Waste Power Generation Supply Chain

8.4.1 Analysis of Research Results

Many previous studies used LCA method to assess the environmental impact of biomass utilization. Ji et al. (2018) showed that greenhouse gas (GHG) abatement potential was –0.9 t CO_2-eq t^{-1} straw. In our research, the GHG abatement potential was –0.8903 t CO_2-eq t^{-1} straw. Wang (2017) studied different biomass supply models and pointed out that CO_2 emissions from straw bulk transport were 4186.619 10^{-3} g/kg and from baling transport were 5635.834 10^{-3} g/kg. In this study, the CO_2 emissions from straw bulk transport and baling transport were 3952.43 10^{-3} g/kg and 5816.7834 10^{-3} g/kg, respectively. It is necessary to understand the emission status of pollutants during different stages of the straw life cycle. The pollutants discharged during different stages can be analyzed, and the proportions of pollutant emissions are shown in Fig. 8.5 (the average values of pollutant emissions in four scenarios were used).

From Fig. 8.5, it can be observed that CO_2, SO_2, CH_4, and N_2O are mainly derived during the planting and transport stages, accounting for 89%, 75%, 88%, and 84%, respectively. The pollutant HC originates during transport stage. The pollutants

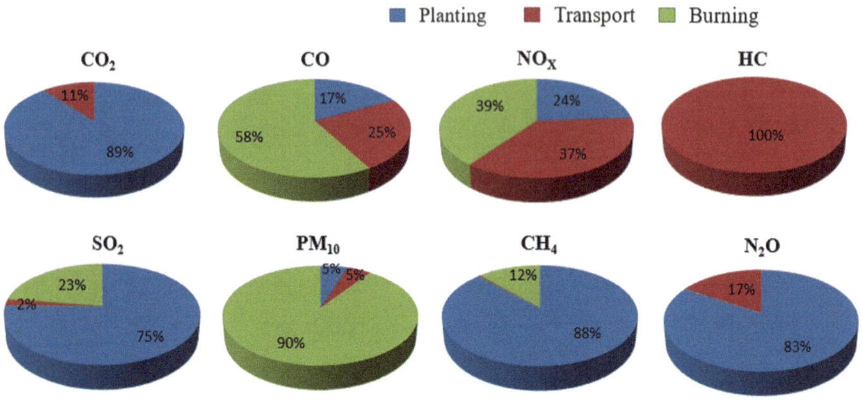

Fig. 8.5 Emission proportion of different pollutants in different stages

NO_X, CO, and PM_{10} produced mainly during the burning stage, accounting for 39%, 58%, and 90%, respectively.

It has been observed that changes in straw transportation modes show considerable influence on pollutant emissions. Liu et al. (2017) studied different logistics models for SPG and found that the transport model can significantly affect emission reduction goals. Proskurina et al. (2017) pointed out that biomass pretreatment can increase density and exhibit good flow ability in logistics processes. This is conducive to improving transport efficiency and reducing pollutant emissions during transport. Nunes et al. (2014) showed that straw granulation has a significant impact on transport. Accordingly, straw processing technology provides a great contribution to reducing pollutant emissions. In this study, the pollutant emissions from the transport stage in different scenarios are shown in Fig. 8.6.

From Fig. 8.6, it is clear that emission of pollutants changed significantly with the change in the transport mode of straw. By comparing the four scenarios, we observed that reducing the use of machinery without considering economic benefits can effectively reduce pollutant emissions. However, this does not meet the goals of the enterprise. By considering production efficiency, the pollutant emissions of scenario 3 were the lowest due to the use of straw briquetting technology, which confirms that straw briquetting technology has a remarkable impact on pollutant emissions.

A comparison of CO_2 emissions from straw direct combustion power generation and coal power generation is shown in Fig. 8.7. The CO_2 emissions from straw power generation were far below those from coal power generation, which suggests that straw power generation has great potential to reduce pollutant emissions.

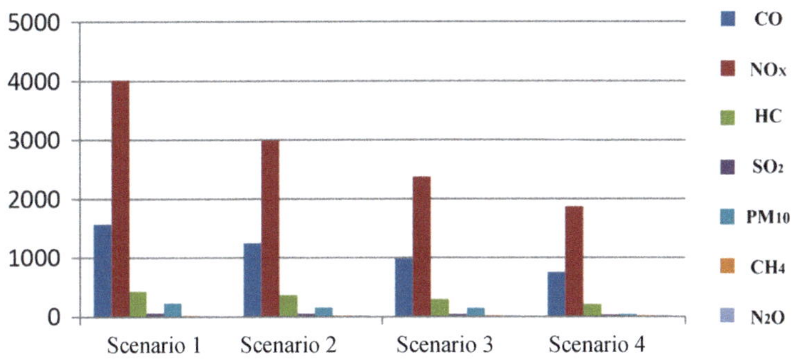

Fig. 8.6 Pollutant emissions from the transport stage in different scenarios (unit: g)

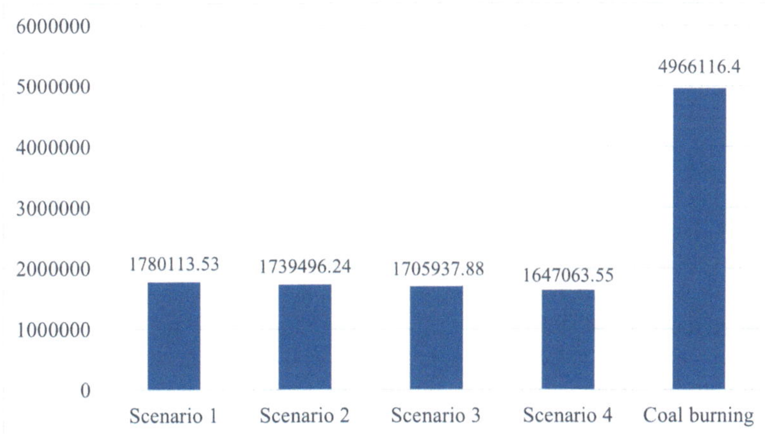

Fig. 8.7 CO_2 emissions in different scenarios (unit: g)

8.4.2 *Sensitivity Analysis of Influencing Factors*

Straw power generation has good development potential based on sustainability assessment. However, SPG is a complex system in which environmental and economic indicators can be influenced by many factors (e.g., straw purchase price, amount of straw collection, power plant scale, types of transport vehicles, straw compression rate, operational status, and treatment technology of pollutants). In order to understand the influence of these factors on the sustainability of straw power generation, sensitivity analysis needs to be carried out.

8.4.2.1 Impact of the Scale of Power Plants

Power plants need to expand their scale to produce more electricity in order to fulfill the consumption demands. Expansion of the power plant scale leads to increases in the straw collection radius. It is assumed that the straw collection radius increases

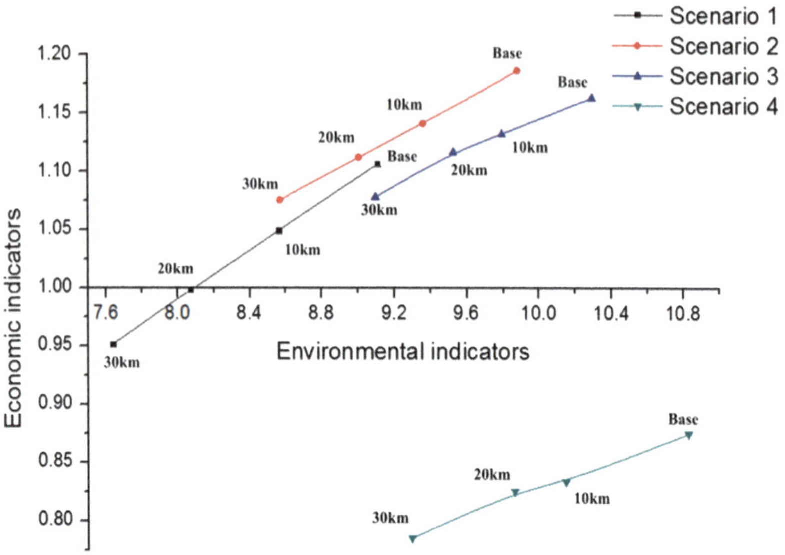

Fig. 8.8 Impact of the scale of straw power plants on sustainability assessment

by 10 km, 20 km, and 30 km. The variation trend of the sustainability assessment results is shown in Fig. 8.8.

Figure 8.8 showed that economic and environmental indicators decrease with an increase in the scale of straw power plants. When the collection radius increased by 20 km, the economic indicator of scenario 1 was 0.998, indicating that enterprises gains no profit. A comparison of scenario 2 and scenario 3 showed a difference of 0.01 between the economic indicators, while the difference in environmental indicators was 0.503. Power plants should select the scheme of scenario 3 to generate electricity. When the straw collection radius increased by 30 km, the economic indicator of scenario 3 exceeded that of scenario 2, which suggested that scenario 3 was the best choice for power plants.

8.4.2.2 Impact of Transport Vehicles

During the field research process, we figured out that straw is transported mainly by tractors. There are many shortcomings associated with the use of tractors, as it consumes more fuel and have a small tonnage. Y. Sun et al. (2017) conducted an economic analysis of straw collection and pointed out that fuel prices affect transportation costs. Tauro et al. (2018) analyzed the impact of trucks with different tonnage on logistics costs. Tsalidis et al. (2014) pointed out that the transportation stage of biomass had the highest contribution to pollutant discharge. El Hanandeh (2015) showed that longer transportation distances added environmental burdens. To analyze the impact of transportation activities on sustainability, this study

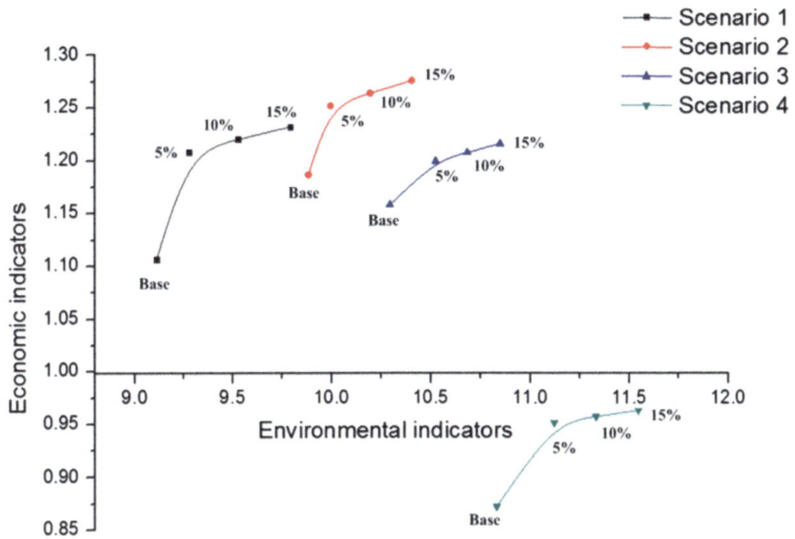

Fig. 8.9 Impact of transport vehicle changes on sustainability assessment

assumes that straw power plants should adopt a new transport vehicle, which has a large load capacity with lower fuel consumption, with reductions of 5%, 10%, and 15%. The effects of the above assumptions on the sustainability assessment are shown in Fig. 8.9.

Figure 8.9 shows that the sustainability of straw power generation can be enhanced by selecting efficient transport vehicles. The selection of low fuel consumption vehicles has shown a considerable impact in scenario 1. When fuel consumption is decreased by 5%, the economic indicators is increased by 9.15%. A comparison of scenario 2 and scenario 3 revealed that scenario 2 was more sensitive to changes in transport vehicles and had better economic indicators. Simultaneously, the use of efficient transport vehicles can improve the economic indicators of scenario 4. According to the above analysis, we suggest that straw power plants optimize transport vehicles to improve sustainability.

8.4.2.3 Impact of Pollutant Treatment Technology

China is actively developing renewable energy to reduce its dependence on fossil fuels and to minimize the pollutant emissions. Straw power generation can not only deal with a large amount of straw but also alleviate greenhouse gas emissions (Ji et al. 2018). Giuntoli et al. (2013) showed that straw-fired power plants can realize high greenhouse gas savings. Karkania et al. (2012) studied agricultural waste utilization and pointed out that the concentration of pollutants was low after combustion treatment. Pollutant emissions from straw power generation are strictly monitored by environmental protection departments. Power plants need to adopt new

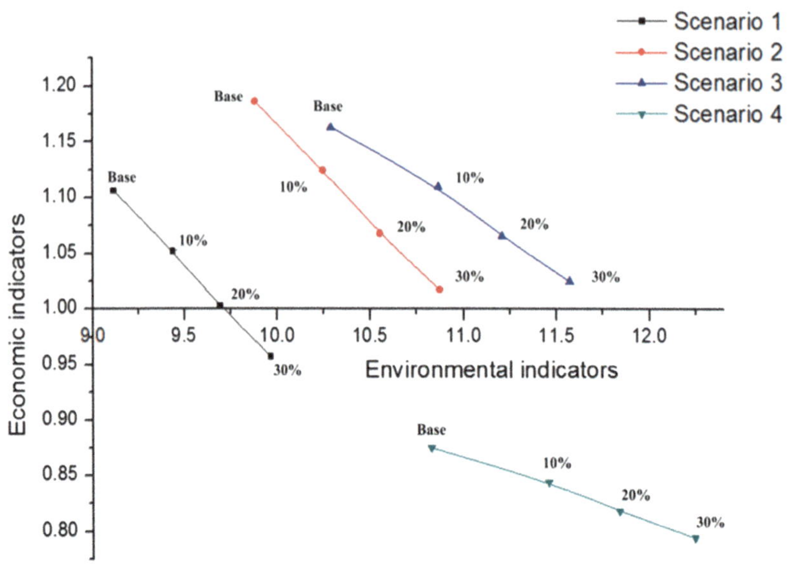

Fig. 8.10 Impact of pollutant treatment technology on sustainability assessment

technology to reduce pollutant emissions. In this manuscript, we assumed that the operational cost of the power plants would increase by 10% if they implemented pollutant reduction technology to bring pollution down by 10%. The variation trend of sustainability assessment results is shown in Fig. 8.10.

Figure 8.10 demonstrates that pollutant treatment technology has a great impact on the sustainability. Scenario 1 and scenario 2 are more sensitive to the treatment technology, and the economic indicators rapidly declined with the increase in environmental indicators. A comparison of scenario 2 and scenario 3 showed that when the pollutant reduction is greater than 10%, the economic and environmental indicators in scenario 3 exceed those in scenario 2, demonstrating that scenario 3 can promote the sustainability of the straw power generation system.

Impact of Straw Purchase Price

The purchase price of straw affects not only the income of enterprises but also the supply of straw. In order to determine the impact of straw price on straw power generation sustainability assessment results, the present study also conducted a price sensitivity analysis, as shown in Fig. 8.11. As it is clearly shown in Fig. 8.11, the economic indicators of straw power generation can be improved when the straw purchase price is low. The increase in straw purchase price will decrease the economic indicators, as the operational cost of the enterprise increases.

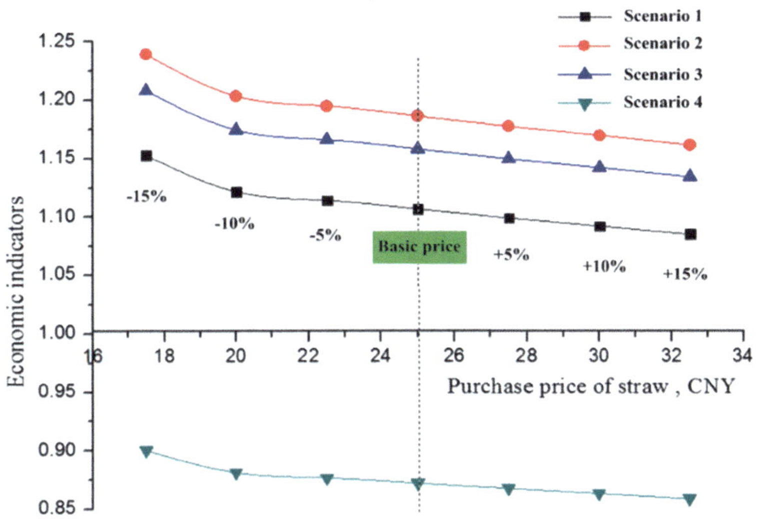

Fig. 8.11 Impact of straw purchase price on economic indicators

8.4.2.4 Emission Reduction Potential of Straw Power Generation

The emissions from coal-based power generation are major contributors to environmental pollution. Straw can replace coal for power generation and effectively alleviate environmental pollution. Our findings indicated that scenario 3 had the best sustainability; thus, the emission data of scenario 3 was selected to study the emission reduction potential of straw power generation. Taking CO_2, NO_X, SO_2, and PM_{10} as examples, it can be seen from Fig. 8.12 that straw power generation has great potential for emission reduction. The emission reduction effects of CO_2 and SO_2 are more obvious with the increase of the substituted coal.

8.4.3 Discussion

8.4.3.1 How to Select a Straw Power Generation Scheme

Our findings showed that straw power generation had great potential to alleviate environmental pollution, whereas sustainability of straw power generation strongly depends on economic indicators. By analyzing four straw power generation scenarios, it can be inferred that the sustainability was different for different straw power generation schemes, which supports decision-makers in selecting an appropriate power generation scheme. The straw power generation scenario 3 presents best sustainability when an environmentally friendly society with more economic benefits is required. When the goal of the enterprise is to achieve maximum benefits while reducing the impact on the environment, the straw power generation scheme

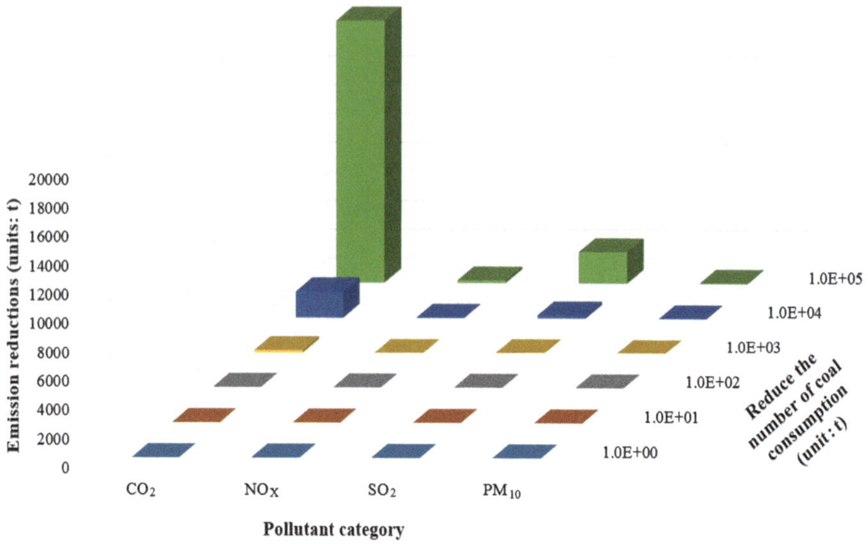

Fig. 8.12 Emission reduction potential of straw replacing coal-fired power generation

of scenario 2 has the best sustainability. The straw power generation scheme of scenario 4 can be a good choice when the goal of the enterprise is to maximize environmental benefits during a special period. Straw power plants can determine their own development goals and select appropriate power generation schemes based on these assessment results.

8.4.3.2 How to Improve the Sustainability of Straw Power Generation

We analyzed four main factors that affect the sustainability of straw power generation. We found that the scale of power plants, pollutant treatment technology, and straw purchase price were negatively related to sustainability; this was consistent with the research of Wang et al. (2018b). The selection of energy-saving transport vehicles can improve sustainability. In comparison to the current situation of straw power generation in China, it is not difficult to discern that the sustainability of straw power generation has not achieved a high level, mainly due to a lack of systematic planning and management and higher operating costs (Wang et al. 2017b). Through this study, we found that improving the environmental and economic performance of straw power generation can raise its sustainability. After analyzing the sensitivity of influencing factors, some methods can be used to improve the sustainability of straw power generation: establishing an efficient straw supply chain, using transport vehicles with low energy consumption, developing new technology for pollutant treatment, and reducing straw purchase price.

8.4.3.3 What Should Government Do

By reviewing the research results, we have found that straw power generation industry lacks economic vitality. The main reasons include higher operating costs, lower benefits, and complex technology of power generation. Through the analysis of emission reduction potential (see Fig. 8.12), straw power generation has great potential to reduce the emission of pollutants, which can help the government achieve the goal of emission reduction. It is worth popularizing straw power generation technology. While keeping environmental benefits in mind, the Chinese government should promote straw power generation projects along with financial support. Tax, fiscal subsidies, and electricity prices have an important impact on the economic performance of power plants, and the sustainability of straw power generation can be promoted through policy guidance. Our findings suggest that the government can guide farmers to plant on a large scale to reduce the cost of straw collection, make policies for automobile industry to develop low-carbon and efficient vehicles for transport, and lead straw power plants to develop innovative straw-processing and pollutant treatment technologies.

8.5 Conclusion

In this study, from the environmental and economic perspectives of straw substitution for coal, a sustainability assessment model was built by integrating life cycle assessment, life cycle cost, and Best-Worst Method. The environmental and economic indicators were calculated, and the sustainability of four different SPG scenarios was assessed. The findings showed that straw direct combustion power generation has great sustainability potential in China. In terms of environment, the results showed that straw power generation has great potential to reduce emissions and alleviate environmental pollution in comparison to coal-based power generation. However, from an economic aspect, sustainability has not reached an advanced level in the current situation. It shows weak economic sustainability potential, and there can be losses (e.g., scenario 4) when the scheme of straw power generation was not appropriate. Finally, sensitivity analyses of different factors affecting the sustainability of SPG were carried out. The sensitivity analysis results showed that straw supply chain model, pollutant treatment technology, transport vehicles, and straw purchase cost are the main factors leading to unfavorable economic performance. The research findings can provide support for the government in popularizing straw power generation technology and can also help to develop direct combustion SPG system with better environmental friendliness and economic sustainability.

Based on the research results, here are some suggestions which may help enterprises to improve the sustainability of straw direct combustion power generation system: (1) Power plants should design the appropriate straw power generation schemes. The scale of straw power plants affects the choice of power generation schemes, and different power generation schemes should be selected for different power plant scales (2). Power plants should choose energy-saving vehicles. The transportation cost of straw is about 60% of the total cost, and a lot of pollutants are emitted during the transportation stage. The use of energy-saving transportation is conducive to improving the sustainability of straw power generation projects (3). Enterprises should innovate in pollutant treatment technology. In the straw-burning power generation stage, the treatment technology of pollutants directly affects the discharge and operating costs of pollutants. The efficient treatment technology can save costs and reduce the discharge of pollutants (4). The government can provide subsidies to balance power generation enterprises and farmers. Through the results of this study, it can be found that the economic performance affects the sustainability of straw power generation. Government subsidies can reduce the production cost of enterprises and encourage farmers to supply more straw resources. This can improve economic indicators and promote the sustainability of straw power generation.

Acknowledgments This chapter has been reprinted from Z. Wang, Z. Wang, G. Xu, et al., *Sustainability Assessment of Straw Direct Combustion Power Generation in China: From the Environmental and Economic Perspectives of Straw Substitute to Coal*, Journal of Cleaner Production, 2020, 273: 122890. Copyright (2020), with permission from Elsevier.

References

Ahmad WNKW, Rezaei J, Sadaghiani S, Tavasszy LA (2017) Evaluation of the external forces affecting the sustainability of oil and gas supply chain using best worst method. J Clean Prod 153(1):242–252. https://doi.org/10.1016/j.jclepro.2017.03.166

Cao J, Pang B, Mo X, Xu F (2016) A new model that using transfer stations for straw collection and transportation in the rural areas of China: a case of Jinghai, Tianjin. Renew Energy 99:911–918. https://doi.org/10.1016/j.renene.2016.07.061

Ebadian M, Sowlati T, Sokhansanj S, Townley-Smith L, Stumborg M (2013) Modeling and analysing storage systems in agricultural biomass supply chain for cellulosic ethanol production. Appl Energy 102:840–849. https://doi.org/10.1016/j.apenergy.2012.08.049

El Hanandeh A (2015) Energy recovery alternatives for the sustainable management of olive oil industry waste in Australia: life cycle assessment. J Clean Prod 91:78–88. https://doi.org/10.1016/j.jclepro.2014.12.005

Giorio C, Pizzini S, Marchiori E, Piazza R, Grigolato S, Zanetti M et al (2019) Sustainability of using vineyard pruning residues as an energy source: combustion performances and environmental impact. Fuel 243:371–380. https://doi.org/10.1016/j.fuel.2019.01.128

Giuntoli J, Boulamanti AK, Corrado S, Motegh M, Agostini A, Baxter D (2013) Environmental impacts of future bioenergy pathways: the case of electricity from wheat straw bales and pellets. Glob Change Biol Bioenergy 5(5):497–512. https://doi.org/10.1111/gcbb.12012

Golecha R, Gan J (2016) Biomass transport cost from field to conversion facility when biomass yield density and road network vary with transport radius. Appl Energy 164:321–331. https://doi.org/10.1016/j.apenergy.2015.11.070

Hong J, Ren L, Hong J, Xu C (2016) Environmental impact assessment of corn straw utilization in China. J Clean Prod 112:1700–1708. https://doi.org/10.1016/j.jclepro.2015.02.081

Hu J, Lei T, Wang Z, Yan X, Shi X, Li Z et al (2014) Economic, environmental and social assessment of briquette fuel from agricultural residues in China—a study on flat die briquetting using corn stalk. Energy 64:557–566. https://doi.org/10.1016/j.energy.2013.10.028

Ji C, Cheng K, Nayak D, Pan G (2018) Environmental and economic assessment of crop residue competitive utilization for biochar, briquette fuel and combined heat and power generation. J Clean Prod 192:916–923. https://doi.org/10.1016/j.jclepro.2018.05.026

Karkania V, Fanara E, Zabaniotou A (2012) Review of sustainable biomass pellets production—a study for agricultural residues pellets' market in Greece. Renew Sustain Energy Rev 16(3):1426–1436. https://doi.org/10.1016/j.rser.2011.11.028

Kylili A, Christoforou E, Fokaides PA (2016) Environmental evaluation of biomass pelleting using life cycle assessment. Biomass Bioenergy 84:107–117. https://doi.org/10.1016/j.biombioe.2015.11.018

Liao H, Mi X, Yu Q, Luo L (2019) Hospital performance evaluation by a hesitant fuzzy linguistic best worst method with inconsistency repairing. J Clean Prod 232:657–671. https://doi.org/10.1016/j.jclepro.2019.05.308

Liu B, Wu Q, Wang F (2017) Regional optimization of new straw power plants with greenhouse gas emissions reduction goals: a comparison of different logistics modes. J Clean Prod 161:871–880. https://doi.org/10.1016/j.jclepro.2017.06.015

Ng RTL, Maravelias CT (2017) Design of biofuel supply chains with variable regional depot and biorefinery locations. Renew Energy 100:90–102. https://doi.org/10.1016/j.renene.2016.05.009

Nishiguchi S, Tabata T (2016) Assessment of social, economic, and environmental aspects of woody biomass energy utilization: direct burning and wood pellets. Renew Sustain Energy Rev 57:1279–1286. https://doi.org/10.1016/j.rser.2015.12.213

Nunes LJR, Matias JCO, Catalão JPS (2014) Mixed biomass pellets for thermal energy production: a review of combustion models. Appl Energy 127:135–140. https://doi.org/10.1016/j.apenergy.2014.04.042

Proskurina S, Alakangas E, Heinimö J, Mikkilä M, Vakkilainen E (2017) A survey analysis of the wood pellet industry in Finland: future perspectives. Energy 118:692–704. https://doi.org/10.1016/j.energy.2016.10.102

Ren J (2018) Technology selection for ballast water treatment by multi-stakeholders: a multi-attribute decision analysis approach based on the combined weights and extension theory. Chemosphere 191:747–760. https://doi.org/10.1016/j.chemosphere.2017.10.053

Ren J, Tan S, Dong L, Mazzi A, Scipioni A, Sovacool BK (2014) Determining the life cycle energy efficiency of six biofuel systems in China: a data envelopment analysis. Bioresour Technol 162:1–7. https://doi.org/10.1016/j.biortech.2014.03.105

Ren J, Dong L, Sun L, Goodsite ME, Tan S, Dong L (2015) Life cycle cost optimization of biofuel supply chains under uncertainties based on interval linear programming. Bioresour Technol 187:6–13. https://doi.org/10.1016/j.biortech.2015.03.083

Ren J, Ren X, Dong L, Manzardo A, He C, Pan M (2018) Multiactor multicriteria decision making for life cycle sustainability assessment under uncertainties. AICHE J 64(6):2103–2112. https://doi.org/10.1002/aic.16149

Rezaei J (2015) Best-worst multi-criteria decision-making method. Omega 53(jun.):49–57

Rezaei J, Nispeling T, Sarkis J, Tavasszy L (2016) A supplier selection life cycle approach integrating traditional and environmental criteria using the best worst method. J Clean Prod 135:577–588. https://doi.org/10.1016/j.jclepro.2016.06.125

Rountree V (2019) Nevada's experience with the renewable portfolio standard. Energy Policy 129:279–291. https://doi.org/10.1016/j.enpol.2019.02.010

Shafie SM, Mahlia TMI, Masjuki HH (2013) Life cycle assessment of rice straw co-firing with coal power generation in Malaysia. Energy 57:284–294. https://doi.org/10.1016/j.energy.2013.06.002

Song S, Liu P, Xu J, Chong C, Huang X, Ma L et al (2017) Life cycle assessment and economic evaluation of pellet fuel from corn straw in China: a case study in Jilin Province. Energy 130:373–381. https://doi.org/10.1016/j.energy.2017.04.068

Sun J, Peng H, Chen J, Wang X, Wei M, Li W et al (2016) An estimation of CO2 emission via agricultural crop residue open field burning in China from 1996 to 2013. J Clean Prod 112:2625–2631. https://doi.org/10.1016/j.jclepro.2015.09.112
Sun Y, Cai W, Chen B, Guo X, Hu J, Jiao Y (2017) Economic analysis of fuel collection, storage, and transportation in straw power generation in China. Energy 132:194–203. https://doi.org/10.1016/j.energy.2017.05.077
Tang, C. X. Z. (2011). Research on investment risk analysis and decision-making of biomass power generation project
Tauro R, Garcia CA, Skutsch M, Masera O (2018) The potential for sustainable biomass pellets in Mexico: an analysis of energy potential, logistic costs and market demand. Renew Sustain Energy Rev 82:380–389. https://doi.org/10.1016/j.rser.2017.09.036
Tsalidis GA, Joshi Y, Korevaar G, Jong WD (2014) Life cycle assessment of direct co-firing of torrefied and/or pelletised woody biomass with coal in The Netherlands. J Clean Prod 81(Oct.15):168–177
Valente C, Brekke A, Modahl IS (2018) Testing environmental and social indicators for biorefineries: bioethanol and biochemical production. Int J Life Cycle Assess 23(3):581–596. https://doi.org/10.1007/s11367-017-1331-x
van de Kaa G, Kamp L, Rezaei J (2017) Selection of biomass thermochemical conversion technology in The Netherlands: a best worst method approach. J Clean Prod 166:32–39. https://doi.org/10.1016/j.jclepro.2017.07.052
Wang C, Zhang L, Chang Y, Pang M (2015) Biomass direct-fired power generation system in China: an integrated energy, GHG emissions, and economic evaluation for Salix. Energy Policy 84:155–165. https://doi.org/10.1016/j.enpol.2015.04.025
Wang Z, Lei T, Yang M, Li Z, Qi T, Xin X et al (2017a) Life cycle environmental impacts of cornstalk briquette fuel in China. Appl Energy 192:83–94. https://doi.org/10.1016/j.apenergy.2017.01.071
Wang Z, Xu G, Ren J, Li Z, Zhang B, Ren X (2017b) Polygeneration system and sustainability: multi-attribute decision-support framework for comprehensive assessment under uncertainties. J Clean Prod 167:1122–1137. https://doi.org/10.1016/j.jclepro.2017.08.089
Wang J, Xi F, Liu Z, Bing L, Alsaedi A, Hayat T et al (2018a) The spatiotemporal features of greenhouse gases emissions from biomass burning in China from 2000 to 2012. J Clean Prod 181:801–808. https://doi.org/10.1016/j.jclepro.2018.01.206
Wang X e, Li K, Song J, Duan H, Wang S (2018b) Integrated assessment of straw utilization for energy production from views of regional energy, environmental and socioeconomic benefits. J Clean Prod 190:787–798. https://doi.org/10.1016/j.jclepro.2018.04.191
Wang Z, Xu G, Lin R, Wang H, Ren J (2019) Energy performance contracting, risk factors, and policy implications: identification and analysis of risks based on the best-worst network method. Energy 170:1–13. https://doi.org/10.1016/j.energy.2018.12.140
Wang ZW, Wang ZF, Xu GY, Ren JZ, Wang H, Li J (2020) Sustainability assessment of straw direct combustion power generation in China: from the environmental and economic perspectives of straw substitute to coal. J Clean Prod 273:122890. https://doi.org/10.1016/j.jclepro.2020.122890
Wen W, Zhang Q (2015) A design of straw acquisition mode for China's straw power plant based on supply chain coordination. Renew Energy 76:369–374. https://doi.org/10.1016/j.renene.2014.11.062
Zhao Z-y, Yan H (2012) Assessment of the biomass power generation industry in China. Renew Energy 37(1):53–60. https://doi.org/10.1016/j.renene.2011.05.017

Chapter 9
Policy Research on the Sustainable Development of Agricultural Waste Power Supply Chain

Zhanwu Wang, Guangyin Xu, Zhenfeng Wang, and Zhiping Zhang

Contents

Abstract The sustainable development of the agricultural waste power generation industry (AWPGI) can reduce environmental pollutant emissions and ameliorate energy shortages. However, the current development of AWPGI lacks sustainability. Hence, to address this issue, a Decision-Making Trial and Evaluation Laboratory (DEMATEL) model, considering relative index weight, is utilized, and the fuzzy composite DEMATEL model is employed to identify the interrelationship of the internal system criteria. According to the obtained results, the system's cause fac-

Z. Wang · G. Xu · Z. Wang (✉) · Z. Zhang
College of Mechanical and Electrical Engineering, Henan Agricultural University, Zhengzhou, Henan Province, China
e-mail: zfking@henau.edu.cn

Z. Wang et al. (eds.), *Agricultural Waste Power Generation Supply Chain*, Industrial Ecology and Environmental Management 7,
https://doi.org/10.1007/978-3-032-20087-7_9

tors include reducing greenhouse gas (GHG) emissions, investment costs, management costs, improving technology maturity, energy conversion efficiency, technical complexity, and safety and reliability. The effect factors include the non-renewable energy utilization, reduction of agricultural waste and production costs, competitiveness, social acceptability, regional development contribution, government support degree, water use, sound pollution, and job production. The internal relationships between the mentioned two factors are further verified, indicating that diverse cause factors can affect different effect factors. Moreover, systematic policies are designed to improve the power generation industry's sustainability based on the obtained results. The findings can help power generation companies and government decision-makers make effective decisions to promote the sustainability of the AWPGI.

Keywords Agricultural waste power generation · Policy design mechanism · Biomass energy · Decision-making trial · Evaluation laboratory

9.1 Introduction

The use of fossil fuels has caused global warming, energy shortages, and environmental issues, which also endanger human health and the living environment (Tsalidis et al. 2014). The Chinese government emphasized developing renewable energy in the "13th Five-Year Plan" (Guney 2019) to overcome the environmental problems. Agricultural waste (AW), as a vital renewable energy has a huge output of about 700 million tons annually in China, equivalent to 350 million tons of coal (Ji et al. 2018). However, only 100 million tons of AW were recycled for energy conversion (Fan et al. 2022). Substituting agricultural waste for coal-fired power generation could dispose of the waste resources, reduce pollution emissions, and ameliorate the energy shortage. However, most agricultural waste power generation enterprises in China cannot make profits due to the low operating efficiency of power plants, high power generation costs, and uncertain raw material supply. Therefore, it is meaningful to improve the sustainability of the agricultural waste power generation industry (AWPGI).

The AWPGI development has not yet reached a sustainable level in China. Some scholars have employed specific policies to improve sustainability. Xue et al. (2020) pointed out that the government's subsidy policies should be targeted and scientific. Spiegel et al. (2018) suggested that subsidies should be implemented to increase farmers' benefits. Wang et al. (2020a) studied the impact of government price subsidies on the collaborative development of the agricultural waste power supply chain. According to their studies, the optimal supply chain system could be obtained for a subsidy price of 53 yuan per ton. Salleh et al. (2020) indicated that biomass energy policy would lead to the sustainable development of renewable energy in Malaysia. A unilateral analysis of a specific policy was performed in these studies. The policies to promote AWPGI sustainability should be performed in different aspects and levels, including technological, environmental, economic, and social dimensions.

Previous studies on promoting AWGPI sustainability have mainly concentrated on economic, environmental, and social aspects. However, few of them have been devoted to identifying influencing factors and the policy design mechanism. This research studies the influencing factors of sustainable development of AWPGI from a systematic perspective to identify their relationship and provide a basis for policy design. The study of policy design mechanism revealed the policy design principle and explained the relationship between policy design and influencing factors. The obtained results can help policymakers employ more scientific policies for promoting the sustainable development of AWPGI.

9.2 Literature Review

9.2.1 Literature Review on Agricultural Waste Power Generation

Agricultural waste power generation substitutes coal to alleviate environmental pollution and improve renewable energy recycling. Some studies have been devoted to the sustainability of AWPGI. The sustainability assessment is one of the research fields. Sastre et al. (2015) considered the uncertainty of the Spanish wheat straw power generation system to construct a life cycle assessment (LCA) model. Their studies demonstrated that biomass straw power generation could generate less greenhouse gas and consume less fossil energy than natural gas power generation. Aberilla et al. (2019) evaluated the environmental sustainability of small-scale biomass power generation systems in developing countries. The results indicated that agricultural waste power generation could reduce the environmental impact by 85% compared to diesel power generation. The electricity generation from agricultural waste effectively improved waste management. Environmental sustainability is the primary indicator for the AWPGI assessment. Integrated assessments have been performed from the environmental, economic, and social aspects in the literature to better understand the sustainability of AWPGI. Bentsen et al. (2019) constructed a global biomass energy partnership framework and conducted a comprehensive sustainability assessment of the Danish straw heating and power system. The results demonstrated that straw could reduce greenhouse gas emissions and increase the farmers' income and the diversity of energy supplies. Dai et al. (2020) showed the significant potential of agricultural waste power generation technology for greenhouse gas emission reduction. Spagnolo et al. (2020) established an emergy analysis model to evaluate the sustainability of biomass power generation. Wang et al. (2020b) considered environmental and economic indicators to construct a comprehensive assessment model of agricultural waste direct-combustion power generation. According to their results, agricultural waste power generation had good environmental sustainability but lacked economic sustainability compared with coal-fired power generation.

Economic performance is the main factor hindering the sustainability of AWPGI. According to the relevant literature, investment, raw material, management, and maintenance costs are the main costs of agricultural waste power generation. The high raw material costs could compress the power plants' profits and restrict the AWPGI development. Sun et al. (2017) studied the collection, storage, and transportation systems of agricultural waste and analyzed the collection costs under different supply modes. Reducing raw material costs can promote the sustainability of AWPGI. Huang et al. (2019) evaluated the economic benefits of distributed AWPGI. They found that 5-MW gasification power generation technology was the best choice, with a power generation cost of 0.402 CNY/kWh, a net present value of 85.9 million CNY, and an investment profit rate of 49.7%. Lin and He (2017) studied the cost-benefit of government subsidies for biomass power generation. However, since biomass power generation has excellent emission reduction potential, they suggested developing different power generation technologies according to local conditions. Srivastava et al. (2020) showed that government subsidies can help reduce fuel prices and promote the utilization of biomass resources. Jung et al. (2020) indicated that agricultural waste power generation is an essential component of the circular bioeconomy. Tippayawong et al. (2020) established a cost model of biomass power generation and demonstrated that replacing eucalyptus barks with corncobs could reduce the production cost by more than 60%. Many factors affect the economic performance of agricultural waste power generation. The power generation cost reduction could improve the sustainability of AWPGI.

As an essential part of the power generation system, the supply chain of agricultural waste can ensure the sustainable development of AWPGI. Karkania et al. (2012) proposed that agricultural waste densification can increase the calorific value per unit volume to improve the transportation efficiency of the supply chain and reduce transportation and storage costs. Golecha and Gan (2016) developed a dynamic biomass transportation model to optimize the supply network's transportation costs with radius changing. The design and optimization of the supply mode of the agricultural waste supply chain can also promote the AWPGI development. Sun et al. (2017) investigated different transportation modes of straw brokers and developed methods to reduce the straw transportation costs. Tauro et al. (2018) studied the sustainability of the biomass supply chain from the perspective of logistics costs. The optimized agricultural waste supply chain structure can promote AWPGI, while the coordinated development of the supply chain can guarantee the power generation industry's sustainability. Xue and Wang (2017) established a dynamic incentive mechanism to analyze the participants' relationships and promote the supply chain's coordinated development. Zhang et al. (2017) used the game theory model to study the coordination mechanism of supply chain stakeholders. Logeswaran et al. (2020) conducted cost analysis on straw power generation to evaluate the potential of straw power generation. Accordingly, the transportation efficiency of raw materials and the economic performance of the AWPGI can be improved.

To sum up, the sustainability research on AWPGI mainly focuses on life cycle assessment, economic analysis, integrated evaluation, supply chain design and optimization, coordination, and promotion policies. We found the necessity of policy research, especially the systematic policy research and policy design mechanisms. This research aims to improve the sustainability of AWPGI by studying the interrelationship of internal factors, exploring the policy design basis, and determining the policy design mechanism.

9.2.2 *The DEMATEL Method*

Decision-Making Trial and Evaluation Laboratory (DEMATEL) is a systematic analysis method to study the causal relationships among many elements of decision-making problems in complex systems. This method employs matrix and graph theories as auxiliary expressions. Then, quantitative analysis can be performed to determine the critical criteria by calculating the central and cause degrees (Seyed-Hosseini et al. 2006). The DEMATEL method has been widely used in different studies, for example, obstacle factor identification (Shao et al. 2016), supply chain management (Ren et al. 2013a, b), and sustainable supply chain adoption (Li and Mathiyazhagan 2018). John et al. (2019) established the DEMATEL model to study the supply chain's independence and coordination and analyze the relevance of the influence factors in the relationship network. Since the traditional DEMATEL method cannot deal with fuzzy numbers and incomplete information states, the fuzzy DEMATEL method has been widely utilized to solve these problems. Liang et al. (2016) constructed a gray DEMATEL model to identify the main factors affecting the sustainability of the biomass fuel production industry. Lin et al. (2018) constructed a method based on D numbers and DEMATEL to identify the risk factors of China's new energy power system. Wei et al. (2019) studied the key obstacles affecting China's shale gas industry development using the fuzzy DEMATEL method.

Although the fuzzy DEMATEL method is a valuable tool to determine the relationships between system elements, it cannot consider the weight of the mutual influence of system factors. The relative weights of different indicators should be considered to make the results more reliable and realistic. Therefore, this study incorporates the weight of criteria into the fuzzy DEMATEL method for constructing a fuzzy composite DEMATEL model to study the sustainability of AWPGI.

9.3 Methods

Based on the previous research (Ortiz-Barrios et al. 2021), a fuzzy composite DEMATEL model is constructed, considering the weight of sustainability evaluation criteria. Compared with previous research methods, considering the weight of system criteria can reflect the influence of decision-makers on the system, making the research results more realistic. In this section, the criteria for assessing the sustainability of the AWPGI are chosen, and then the developed fuzzy composite-DEMATEL method is employed to study these criteria. First, stakeholders or decision-makers employ linguistic terms to determine the relationship between the criteria. Second, these linguistic terms are converted into triangular fuzzy numbers to obtain the fuzzy direct-influence matrix. Third, the fuzzy direct-influence matrix is standardized, and the total relation matrix is calculated. The Fuzzy Analytic Hierarchy Process (FAHP) is then utilized to determine the relative impact weights of the criteria, and the composite total relation matrix is obtained. Finally, the cause-effect relations of the criteria are identified through the calculation results. The methodological framework of this study is presented in Fig. 9.1.

9.3.1 Criteria System for Sustainability Assessment

Sustainable development or sustainability refers to the rational use of resources to meet humanity's current requirements without damaging the ability of future generations to meet their own needs (Ba et al. 2016). The sustainability assessment should be comprehensive, including environmental, economic, technical, and social aspects. Liang et al. (2016) studied the sustainability of China's biomass fuel industry from these four aspects. Wang et al. (2019) studied different energy conversion technologies from the mentioned aspects and employed a multi-attribute decision-making method to prioritize them. Since various factors can influence the sustainability of AWPGI, the criteria can be selected from the environmental, economic, technological, and social aspects. The selected criteria to assess the agricultural waste power generation industry are shown in Table 9.1.

9.3.2 Theory of Triangular Fuzzy Number

In multi-attribute decision-making theory, the decision-makers cannot describe the relationship between different factors with specific values. They are usually expressed in linguistic terms, such as "equally important," "slightly important," "obviously important," "extremely important," and "extremely unimportant." In natural science research, these linguistic terms cannot be calculated directly. The fuzzy theory converts these language terms into mathematical forms. Various studies have been performed on the application of fuzzy theory in sustainability assessment. Ren (2018) employed a fuzzy multi-attribute decision-making method to study the sustainability of multiple energy systems. Khishtandar et al. (2017) established a fuzzy multi-attribute decision-making framework to assess the sustainability of biomass energy production technology. Triangular fuzzy numbers can be defined as $\tilde{\alpha} = \left(\alpha^L, \alpha^M, \alpha^U\right)$. Two triangular fuzzy numbers, $\tilde{\alpha} = \left(\alpha^L, \alpha^M, \alpha^U\right)$ and

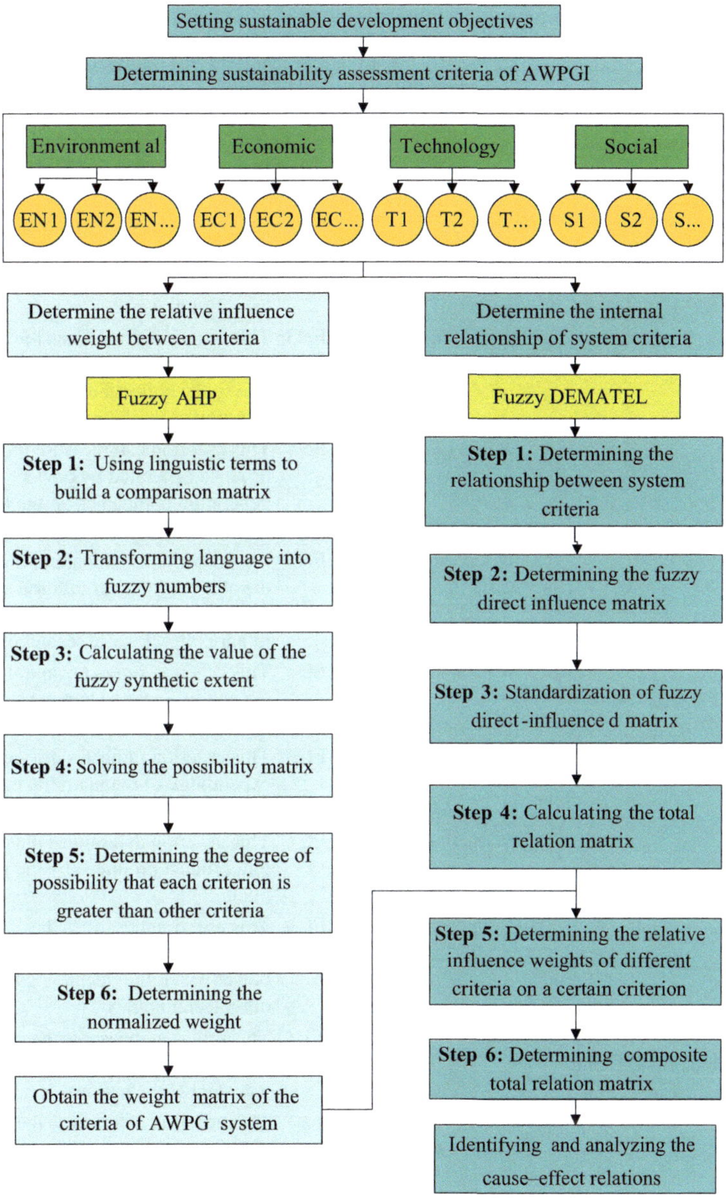

Fig. 9.1 Framework of the fuzzy composite DEMATEL method

Table 9.1 Selected criteria for sustainability assessment

Aspects	Criterion	Abb	Description
Environment (Giuntoli et al. 2016; Yang et al. 2014; Ren and Lutzen, 2015)	Reduction of GHG emissions	EN1	This criterion is employed to measure the reduction of greenhouse gases from the replacement of traditional fuel power generation by agricultural waste
	Water use	EN2	This criterion measures the consumption of water resources in the process of generating electricity from agricultural waste
	Nonrenewable energy use	EN3	This criterion determines the consumed nonrenewable energy during the collection and transportation of agricultural waste
	Noise pollution	EN4	This criterion can measure the impact of noise generated by power generation companies on the lives of residents
	Reduce agricultural waste	EN5	This criterion is adopted to measure the ability of the agricultural waste power generation industry to dispose of agricultural waste resources
Economic (Yang et al. 2014; Zhao et al. 2016)	Investment costs	EC1	This criterion refers to capital expenditure related to power plant construction
	Management costs	EC2	This criterion indicates the capital expenditure to maintain the power plant operation
	Production costs	EC3	This criterion determines the cost of unit product produced by the power plant
	Competitiveness	EC4	This criterion describes the advantages of agricultural waste power generation technology over other technologies
Technical (Yang et al. 2014; Liang et al. 2016; Wang et al. 2018)	Maturity	T1	This criterion measures the degree of maturity of the technology, indicating its extent at the national level
	Conversion efficiency	T2	This criterion determines the efficiency of agricultural waste power generation
	Complexity	T3	This criterion represents the number of required steps for a process of the agricultural waste power generation technology
	Safety and reliability	T4	This criterion indicates the risk and system reliability of the agricultural waste power generation process

(continued)

Table 9.1 (continued)

Aspects	Criterion	Abb	Description
Social (Wang et al. 2018; Liang et al. 2016; Yang et al. 2014)	Social acceptability	S1	This criterion measures the attitudes of local residents toward the AWPGI
	Job production	S2	This criterion determines the number of jobs provided by agricultural waste generation companies
	Government support degree	S3	This criterion represents political leaders' acceptance of agricultural waste power generation projects
	Regional development contribution	S4	This criterion measures the contribution of the introduction of power generation technology to the economic development of the local area

$\tilde{\beta} = \left(\beta^L, \beta^M, \beta^U\right)$, are considered. Now, the following mathematical operations can be defined for these numbers (Li and Mathiyazhagan 2018; Chen et al. 2011; Ren et al. 2015):

$$\tilde{\alpha} + \tilde{\beta} = \left(\alpha^L + \beta^L, \alpha^M + \beta^M, \alpha^U + \beta^U\right) \tag{9.1}$$

$$\tilde{\alpha} - \tilde{\beta} = \left(\alpha^L - \beta^U, \alpha^M - \beta^M, \alpha^U - \beta^L\right) \tag{9.2}$$

$$\tilde{\alpha} * \tilde{\beta} = \left(\alpha^L \beta^L, \alpha^M \beta^M, \alpha^U \beta^U\right) \tag{9.3}$$

$$\tilde{\alpha} / \tilde{\beta} = \left(\alpha^L / \beta^U, \alpha^M / \beta^M, \alpha^U / \beta^L\right) \tag{9.4}$$

$$\delta\tilde{\alpha} = \left(\delta\alpha^L, \delta\alpha^M, \delta\alpha^U\right) \tag{9.5}$$

Triangular fuzzy numbers can be de-fuzzy by the following equation into a crisp number (Chen et al. 2011).

$$\alpha = R\left(\tilde{\alpha}\right) = \frac{\alpha^L + 4\alpha^M + \alpha^U}{6} \tag{9.6}$$

9.3.3 *Fuzzy Analytic Hierarchy Process*

The Analytic Hierarchy Process (AHP), proposed by Professor A. L. Saaty, is a systematic analysis method that combines qualitative and quantitative analysis (Saaty 2004). In traditional AHP, it is not easy to master and adjust the consistency.

Table 9.2 Language terms and fuzzy numbers (Tseng et al. 2009)

Language terms	Abbreviations	Fuzzy scales
Equally important	E	(1)
Weakly important	W	(1/2,1,3/2)
Moderately important	M	(1,3/2,2)
Very strongly important	F	(3/2,2,5/2)
Fairly strongly important	V	(2,5/2,3)
Absolutely important	A	(5/2,3,7/2)
Reciprocal of all above	RE, RW, RM, RF, RV, RA	The reciprocals of the fuzzy numbers

The improved Fuzzy Analytic Hierarchy Process (FAHP) solves the judgment consistency problem. The specific steps of FAHP are given in Wang et al. (2008), Tseng et al. (2009), and Choudhary and Shankar (2012).

Step 1: Employing linguistic terms to construct a comparison matrix (X^1).

There exist n criteria in the system. The relationship matrix can be obtained using linguistic terms describing their interrelationships. Language terms and fuzzy scales are presented in Table 9.2.

Step 2: Obtaining the fuzzy matrix (Wang et al., 2020a, b; Tseng et al. 2009).

The elements in the comparison matrix X^1 can be converted into fuzzy numbers according to the semantic table (Table 9.2), while the matrix (X^2) is defined as:

$$X^2 = \begin{vmatrix} \tilde{x}_{11} & \tilde{x}_{12} & \cdots & \tilde{x}_{1n} \\ \tilde{x}_{21} & \tilde{x}_{21} & \cdots & \tilde{x}_{2n} \\ \vdots & \vdots & \ddots & \vdots \\ \tilde{x}_{n1} & \tilde{x}_{n2} & \cdots & \tilde{x}_{nn} \end{vmatrix} \tag{9.7}$$

where $\tilde{x}_{ij}$ represents the triangular fuzzy number of the relative importance of the i-th criterion relative to the j-th one, and $\tilde{x}_{ij} = \dfrac{1}{\tilde{x}_{ji}}, i, j = 1, 2, \ldots, n$.

Step 3: Calculating the value of the fuzzy synthetic extent of each criterion (Wang et al. 2020; Tseng et al. 2009).

$$K_i = \sum_{j=1}^{n} \tilde{x}_{ij} \times \left[\sum_{i=1}^{n} \sum_{j=1}^{n} \tilde{x}_{ij} \right]^{-1} \tag{9.8}$$

where

$$\sum_{j=1}^{n} \tilde{x}_{ij} = \left(\sum_{j=1}^{n} x_{ij}^{L}, \sum_{j=1}^{n} x_{ij}^{M}, \sum_{j=1}^{n} x_{ij}^{U} \right) \tag{9.9}$$

$$\left[\sum_{i=1}^{n}\sum_{j=1}^{n}\tilde{x}_{ij}\right]^{-1} = \left(\frac{1}{\sum_{i=1}^{n}\sum_{j=1}^{n}x_{ij}^{U}}, \frac{1}{\sum_{i=1}^{n}\sum_{j=1}^{n}x_{ij}^{M}}, \frac{1}{\sum_{i=1}^{n}\sum_{j=1}^{n}x_{ij}^{L}}\right) \tag{9.10}$$

where $K_i = \left(K_i^L, K_i^M, K_i^U\right)$ represents the value of the fuzzy synthetic extent of i-th criterion.

Step 4: Solving the possibility matrix. According to the fuzzy synthetic degree of the criteria, the following possibility matrix can describe the relative magnitude between each pair of criteria (Tseng et al. 2009).

$$V = \begin{vmatrix} / & \tilde{p}_{12} & \cdots & \tilde{p}_{1n} \\ \tilde{p}_{21} & / & \cdots & \tilde{p}_{2n} \\ \vdots & \vdots & \ddots & \vdots \\ \tilde{p}_{n1} & \tilde{p}_{n2} & \cdots & / \end{vmatrix} \tag{9.11}$$

where $\tilde{p}_{ij}$ represents the possibility of $K_i \geq K_j$, which can be determined as:

$$\tilde{p}_{ij} = V\left(K_i \geq K_j\right) = \begin{cases} 1 & K_i^M \geq K_j^M \\ 0 & K_j^L \geq K_i^U \\ \dfrac{K_j^L - K_i^U}{\left(K_i^M - K_i^U\right) - \left(K_j^M - K_j^L\right)} & others \end{cases} \tag{9.12}$$

Step 5: Determining the possibility degree that the value of the fuzzy synthetic extent of each criterion is more significant than other criteria. Taking the i-th criterion as an example, the method is described in the following equations (Tseng et al. 2009).

$$\begin{aligned} \mathrm{V}\left(K_i \geq K_1, K_2, \ldots, K_k, \ldots, K_n\right) &= V\left(K_i \geq K_1\right) \text{and } V\left(K_i \geq K_2\right) \text{and} \cdots \\ \text{and } V\left(K_i \geq K_n\right) &= \min V\left(K_i \geq K_k\right), \text{for } k = 1, 2, \ldots, n, \text{and } k \neq i \end{aligned} \tag{9.13}$$

$$\text{Assuming } d'\left(C_i\right) = \min V\left(K_i \geq K_k\right), \text{for } k = 1, 2, \ldots, n, \text{and } k \neq i \tag{9.14}$$

Now, an array of $d'(C_i)$ $i = 1, 2, \ldots, n$ can be formed to obtain the weight of system criteria as:

$$W' = \left(d'\left(C_1\right), d'\left(C_2\right), \ldots, d'\left(C_n\right)\right)^T \tag{9.15}$$

Step 6: Determining the normalized weight of the system criteria. The weight vector in Eq. (9.15) is normalized by Eq. (9.16), and the normalized weight vector W is shown in Eq. (9.17) (Tseng et al. 2009).

$$d(C_i) = \frac{d'(C_i)}{\sum_{i=1}^{n} d'(C_i)} \tag{9.16}$$

$$W = \left(d(C_1), d(C_2), \ldots, d(C_n)\right)^T = \left(w_1, w_2, \ldots, w_n\right)^T \tag{9.17}$$

9.3.4 *Fuzzy DEMATEL Method*

The DEMATEL method can identify the internal relationships between the system assessment criteria and rank them. The specific steps of the fuzzy DEMATEL method are summarized as follows (Seyed-Hosseini et al. 2006; Liang et al. 2016).

Step 1: Deriving the relationship between system criteria.

Stakeholders were invited to employ linguistic terms to determine the relationships between the system criteria. They were asked to judge how one criterion could influence the others. Suppose the system has n criteria, where x_{ij} indicates the influence of the i-th criterion on the j-th one. The language terms "no influence," "weak influence," "moderate influence," "strong influence," and "high influence" can be utilized to describe the degree of influence between the criteria.

Step 2: Determining the fuzzy direct-influence matrix.

Language terms can be transformed into triangular fuzzy numbers through a semantic table. The transformation relationship of language terms is shown in Table 9.3. Assuming that m stakeholders were invited, the average value method can be utilized to determine the mutual influence relationship of the system criteria (Ren et al. 2015), as shown in Eq. (9.19). Through the stakeholders' judgments on the relationships between the system criteria, the initialization fuzzy direct-influence matrix $\tilde{X}$ can be obtained as (Liang et al. 2016):

Table 9.3 Semantic table (Wu 2008)

Language terms	Triangular fuzzy number
No influence (NI)	(0, 0, 1/5)
Weak influence (WI)	(0, 1/5, 2/5)
Moderate influence(MI)	(1/5, 2/5, 3/5)
Strong influence (SI)	(2/5, 3/5, 4/5)
High influence (EI)	(3/5, 4/5, 1)

$$\tilde{X} = \begin{bmatrix} [0,0,1/5] & x_{12} & \cdots & x_{1n} \\ x_{21} & [0,0,1/5] & \cdots & x_{2n} \\ \vdots & \vdots & \ddots & \vdots \\ x_{n1} & x_{n2} & \cdots & [0,0,1/5] \end{bmatrix} \tag{9.18}$$

where

$$x_{ij} = \frac{\sum_{i=1}^{m} x_{ij}^{'}}{m} \tag{9.19}$$

Step 3: Standardization of fuzzy direct-influence matrix (Liang et al. 2016).

The following relation can determine the standardized fuzzy direct-influence matrix $\tilde{A}$.

$$\tilde{A} = \frac{\tilde{X}}{S} \tag{9.20}$$

where

$$S = \max_{1 \le i \le n} \sum_{j=1}^{n} X_{ij}^{U} \tag{9.21}$$

Step 4: Calculating the total relation matrix (Seyed-Hosseini et al. 2006; Liang et al. 2016).

It is necessary to calculate the system's total relation matrix to identify the criteria' relationship accurately. The total relation matrix $\tilde{\mathrm{T}}$ can be calculated as:

$$\tilde{\mathrm{T}} = \tilde{A} + \left(\tilde{A}\right)^2 + \left(\tilde{A}\right)^3 + \cdots + \left(\tilde{A}\right)^m = \sum_{m=1}^{\infty} \left(\tilde{A}\right)^m = \tilde{A} \times \left(\tilde{I} - \tilde{A}\right)^{-1} \tag{9.22}$$

where $\tilde{I}$ is the identity matrix.

Step 5: Determining the weight matrix of the criteria. Consider the i-th criterion as an example, and assume that it is affected by n criteria. The fuzzy AHP is utilized to determine the relative influence weights of different criteria on a specific criterion, where its process is described as follows:

$$W^1 = \left(A_1, A_2, \ldots, A_n\right)^T \tag{9.23}$$

$$W^2 = \left(B_1, B_2, \ldots, B_n\right)^T \tag{9.24}$$

$$W^3 = \left(A_1B_1, A_1B_2, \ldots, A_1B_n, A_2B_1, A_2B_2, \ldots, A_2B_n, \ldots, A_nB_1, A_nB_2, \ldots, A_nB_n\right)^T$$
$$= \left(w_1^3, w_2^3, \ldots, w_n^3\right)^T \tag{9.25}$$

$$W^4 = \left(d\left(w_1^4\right), d\left(w_2^4\right), \ldots, d\left(w_n^4\right)\right)^T = \left(w_1^4, w_2^4, \ldots, w_n^4\right)^T \tag{9.26}$$

$$d\left(w_i^4\right) = \frac{w_i^3}{\sum_{i=1}^{n} w_i^3} \tag{9.27}$$

where W^1 represents the weight of the criteria layer, W^2 and W^3 represent the local and global weights of the criterion, respectively, and W^4 describes the normalization weight.

After obtaining the relative weight of each criterion, a weight matrix W^5 can be obtained as:

$$W^5 = \begin{bmatrix} 0 & w_{12} & \cdots & w_{1n} \\ w_{21} & 0 & \cdots & w_{2n} \\ \vdots & \vdots & 0 & \vdots \\ w_{n1} & w_{n2} & \cdots & 0 \end{bmatrix} \tag{9.28}$$

Step 6: Determining the prior sequence of the criteria and identifying the cause-effect relations.

The composite total relation matrix $\tilde{T}_1$ can be obtained by multiplying the number in the total relation matrix with the number of the corresponding weight matrix. The elements of each column and row of the matrix $\tilde{T}_1$ should be added to derive the sum of each column and row, denoted by $\tilde{T}_{1c}$ and $\tilde{T}_{1r,}$ respectively. Consider that $\tilde{Z}$ indicates the central degree of the criteria, while $\tilde{Y}$ indicates the cause degree. These indicators can be calculated as follows (Liang et al. 2016):

$$\tilde{Z} = \tilde{T}_{1r} + \tilde{T}_{1c} \tag{9.29}$$

$$\tilde{Y} = \tilde{T}_{1r} - \tilde{T}_{1c} \tag{9.30}$$

Equation (9.6) can be employed to defuzzy the central degree and cause degree. Then, the central degree is prioritized. In the sequence, the former criterion is more important than the latter. The cause-effect relations of the criterion are then identified based on the value of $\tilde{Y}$. If the cause degree $\tilde{Y} > 0$, the criterion impacts other criteria, called the cause criterion, and categorized into cause groups. If $\tilde{Y} < 0$, this criterion is significantly influenced by other criteria and categorized into effect groups.

9.3.5 Data Collection

Since the sustainability of AWPGI requires the joint efforts of the stakeholders, their opinions should be considered when analyzing the interrelationships of the criteria. It makes the data more accurate and the calculation results more reliable. Three groups of stakeholders were selected, including agricultural waste brokers (DM#1), farmers (DM#2), and power plant decision-makers (DM#3). According to their own experience, these stakeholders were asked to judge the relationships between the system assessment criteria. The data were then processed using the average method. Finally, the mutual influence relationships of system criteria were obtained.

9.4 Results and Discussion

9.4.1 Calculation Results

The method of fuzzy composite DEMATEL was utilized to analyze the sustainability of AWPGI. First, the relationships between the system criteria were determined. Then, the language terms were converted into triangular fuzzy numbers. Accordingly, the fuzzy direct-influence matrix $\tilde{X}$ can be obtained. Equations (9.20) and (9.21) were employed to standardize the direct-influence matrix and obtain the standardized fuzzy direct-influence matrix $\tilde{A}$. Finally, Eq. (9.22) was adopted to calculate the total relation matrix $\tilde{\mathrm{T}}$. The relative influence weight of system criteria was calculated using FAHP. The fuzzy composite DEMATEL method was employed to obtain the composite total relation matrix $\tilde{\mathrm{T}}_1$. The triangular fuzzy numbers $\tilde{\mathrm{T}}_{1c}$ and $\tilde{\mathrm{T}}_{1r}$ could be obtained by summing each column and row of the matrix $\tilde{\mathrm{T}}_1$; simultaneously, the comprehensive influencing factors $\tilde{\mathrm{Z}}$ and $\tilde{\mathrm{Y}}$ could also be obtained. The results are presented in Table 9.4. Equation (9.6) was utilized to defuzzy the triangular fuzzy numbers $\tilde{\mathrm{T}}_{1c}$, $\tilde{\mathrm{T}}_{1r}$, $\tilde{\mathrm{Z}}$, and $\tilde{\mathrm{Y}}$. According to the ranking results, the primary factor affecting the sustainability of the AWPGI was technological maturity, followed by government support degree, safety and reliability, competitiveness of power generation enterprises, and investment costs. According to the cause degree, the system criteria can be divided into cause group and effect group. When $\tilde{\mathrm{Y}} > 0$, the criterion was called the cause factor; otherwise, it was called the effect factor. According to the defuzzy results of $\tilde{\mathrm{Y}}$, the reduction of GHG emissions, investment costs, management costs, technical maturity, energy conversion efficiency, technical complexity, and safety and reliability belong to the cause group. The nonrenewable energy use, water use, noise pollution, reduction of agricultural waste, production costs, competitiveness, social acceptability, regional development contribution, government support degree, and jobs provision belong to the effect group.

Table 9.4 Results of influencing factors

	$\tilde{Z}$	$\tilde{Y}$
EN1	(0.049, 0.088, 0.149)	(−0.037, 0.011, 0.062)
EN2	(0.004, 0.030, 0.063)	(−0.039, −0.009, 0.020)
EN3	(0.033, 0.063, 0.098)	(−0.048, −0.014, 0.017)
EN4	(0.003, 0.025, 0.050)	(−0.046, −0.021, 0.001)
EN5	(0.064, 0.105, 0.151)	(−0.069, −0.024, 0.019)
EC1	(0.076, 0.118, 0.175)	(−0.002, 0.050, 0.097)
EC2	(0.047, 0.081, 0.134)	(−0.019, 0.025, 0.067)
EC3	(0.053, 0.085, 0.126)	(−0.051, −0.010, 0.023)
EC4	(0.060, 0.096, 0.136)	(−0.078, −0.039, −0.003)
T1	(0.082, 0.186, 0.315)	(−0.057, 0.051, 0.176)
T2	(0.061, 0.129, 0.218)	(−0.019, 0.049, 0.137)
T3	(0.012, 0.036, 0.073)	(−0.027, 0.006, 0.034)
T4	(0.088, 0.146, 0.221)	(−0.011, 0.052, 0.123)
S1	(0.048, 0.074, 0.108)	(−0.089, −0.055, −0.028)
S2	(0.034, 0.057, 0.086)	(−0.065, −0.036, −0.012)
S3	(0.101, 0.151, 0.209)	(−0.061, −0.010, 0.046)
S4	(0.030, 0.057, 0.093)	(−0.061, −0.027, 0.001)

9.4.2 *Analysis of Cause Factors*

It is evident from Table 9.5 that there were seven "cause factors" affecting the sustainability of AWPGI. The ranking result was: T1 > T4 > T2 > EC1 > EN1 > EC2 > T3. The greater the cause degree was, the higher the influence of the criterion on other criteria was. The overall impacts of different factors on the power generation industry from the four levels of the environment, economy, technology, and society are analyzed to verify the impact of these factors on the sustainability of the AWPGI. The result is shown in Fig. 9.2.

As shown in Fig. 9.2, the technical dimension criterion has the most substantial influence of 0.164. It indicates that technical criteria are the main factors affecting the sustainable development of AWPGI. Zhao et al. (2016) pointed out that the lower production efficiency of agricultural waste power in China was due to the immature power generation technology. Asadullah (2014) found that boiler combustion slagging and tar production could confine the biomass power generation development. The total influence of criteria from the economic dimension was 0.074. It indicates that economic criteria also affect the sustainability of AWPGI. Zhang et al. (2013) showed that the higher investment costs were one of the main factors making many biomass power plants in China in deficit. Lv et al. (2014) studied the economics of biomass power generation and demonstrated that higher raw material purchase costs could hinder the sustainability of AWPGI. Wen and Zhang (2015) pointed out that higher production costs would hinder the development of the biomass power generation industry. Xu and Chen (2020) studied the

Table 9.5 The results of the fuzzy composite DEMATEL

	EN1	EN2	EN3	EN4	EN5	EC1	EC2	EC3	EC4	T1	T2	T3	T4	S1	S2	S3	S4
Rank	8	16	12	17	6	5	10	9	7	1	4	15	3	11	14	2	13
Cause	√					√	√			√	√	√	√				
Effect		√	√	√	√			√	√					√	√	√	√

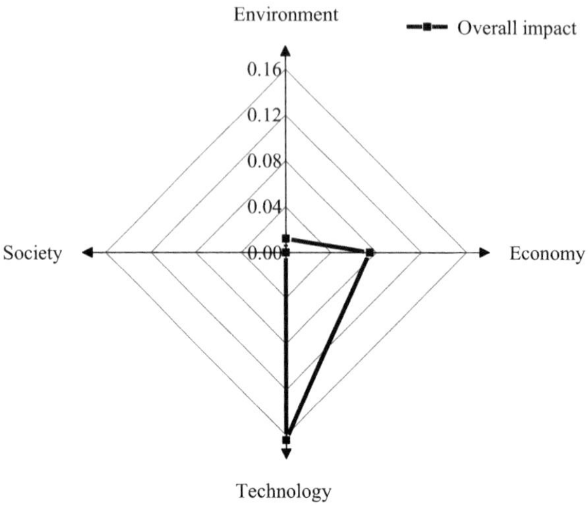

Fig. 9.2 Overall impact of different criteria dimensions

straw recycling logistics costs for biomass power generation. Their studies indicated that high production costs impeded the sustainable development of power generation companies. There was only one criterion in the environmental dimension as the "causal factor," with a total influence of 0.012, indicating that pollutant emissions should be considered to promote the AWPGI development. Feng et al. (2017) performed a life cycle assessment of agricultural waste power generation. The results demonstrated that the emission reduction potential of biomass power generation was superior to that of coal power generation. Yan et al. (2020) investigated carbon capture and CO_2 emissions as the influencing factors of biomass direct-fired power generation technology. According to their results, biomass power generation has good CO_2 emission reduction capabilities. It can be concluded that environmental benefits are the fundamental driving forces to develop the AWPGI. Through the analysis of these factors, the promotion aspect of the sustainability of the AWPGI can be determined.

9.4.3 Analysis of Effect Factors

It is known that there are ten "effect factors" (see Table 9.5), which can improve the sustainability of AWPGI. T_{1c} represents the total influence of system criteria on a specific criterion. The larger the value was, the greater the influence. According to the obtained values, the sequence of these factors is as follows: S3 > EC4 > S1 > EN5 > EC3 > S2 > S4 > EN3 > EN4 > EN2. The impact degree of different dimensions from four aspects of environment, economy, technology, and society was investigated to better understand the impact of different criteria. Figure 9.3 shows the results.

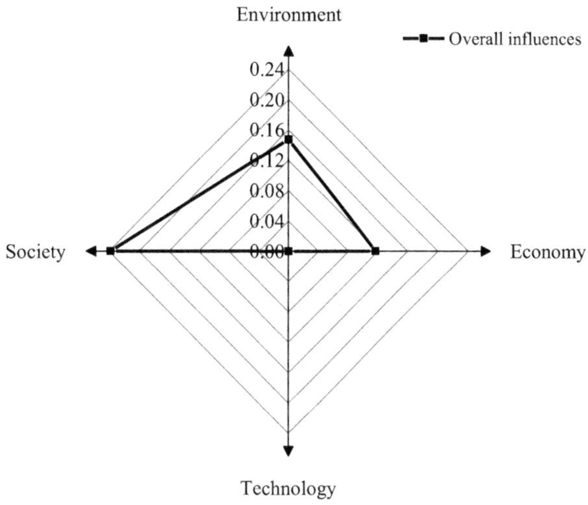

Fig. 9.3 Overall influences on the effect factors of different dimensions

It can be seen from Fig. 9.3 that the social dimension has the maximum influence of 0.238. It indicates that the social dimension criteria are the most critical "effect factors" affecting the sustainability of the AWPGI. Liang et al. (2016) demonstrated that government policy support was the primary factor in promoting the sustainability of the biofuel industry. Ren et al. (2013a) evaluated the sustainability of biomass production fuels and indicated that the social dimension indicators should be strengthened. Spiegel et al. (2018) and Moncada et al. (2018) found that policies could promote biomass energy utilization. Reddy (2018) pointed out that the biomass power generation development was related to the power generation and government investment costs. Wang et al. (2020a) evaluated the impact of policy incentives on the agricultural waste power generation supply chain and indicated that the AWPGI development required more support from the government at the present stage. The influence on the environmental dimension was 0.147, indicating that the AWPGI development should consider the impact of the environmental criteria. Qin et al. (2018) studied the sustainability of AWPGI and pointed out that agricultural waste power generation could reduce the use of fossil fuels and the emission of environmental pollutants. Hong et al. (2016) indicated that agricultural waste utilization has good environmental benefits. Due to the high greenhouse gas emission reduction potential of the agricultural waste power generation technology, improving the environmental benefits of the AWPGI can promote sustainable development (Zhao et al. 2016; Dai et al. 2020; Yan et al. 2020). The influence on the economic dimension was 0.117. Shafie et al. (2014) demonstrated that reducing power generation costs could promote agricultural waste power generation technology. Improving the economic efficiency of power generation can promote the sustainable development of AWPGI (Stich et al. 2017; Wang et al. 2020b). The above "effect factors" can be analyzed to identify the main factors clearly, which helps decision-makers formulate appropriate policies for different criteria to promote the utilization and development of agricultural waste power generation technology.

9.4.4 Sensitivity Analysis

The sustainability of AWGPI was analyzed from the aspects of the environment, economy, technology, and society. It is well known that human factors can affect the weight results of different criteria. Therefore, four scenarios were constructed to evaluate the influence of weight changes. Scenario 1 represents the current study; Scenario 2 represents the environmental preference expert group; Scenario 3 represents the economic preference expert group, and Scenario 4 represents the non-preference expert group. The sensitivity analysis results are shown in Figs. 9.4, 9.5, and 9.6. From these figures, it can be seen that the change in weight results has a weakly influence on the research results. Although there are some differences in the research results in different scenarios, they are generally consistent, demonstrating their robustness and relative consistency in the relationship evaluations.

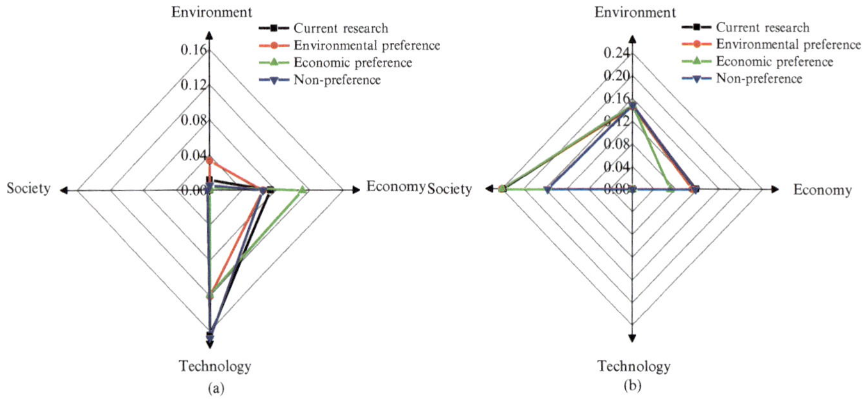

Fig. 9.4 Sensitivity analysis of (**a**) cause factors and (**b**) effect factors in different dimensions

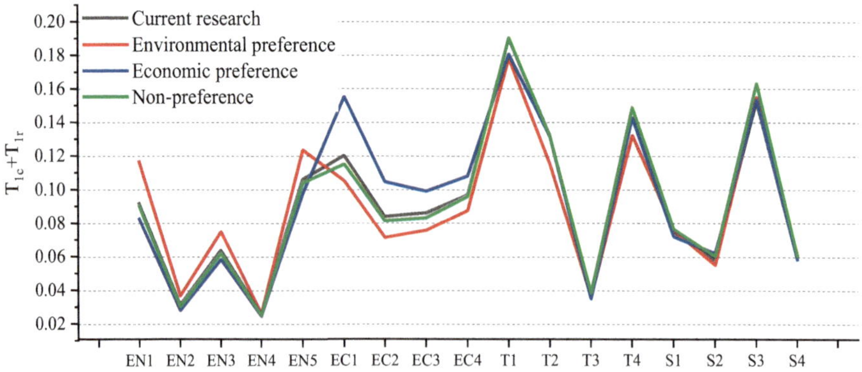

Fig. 9.5 Sensitivity analysis of the central degree of different criteria

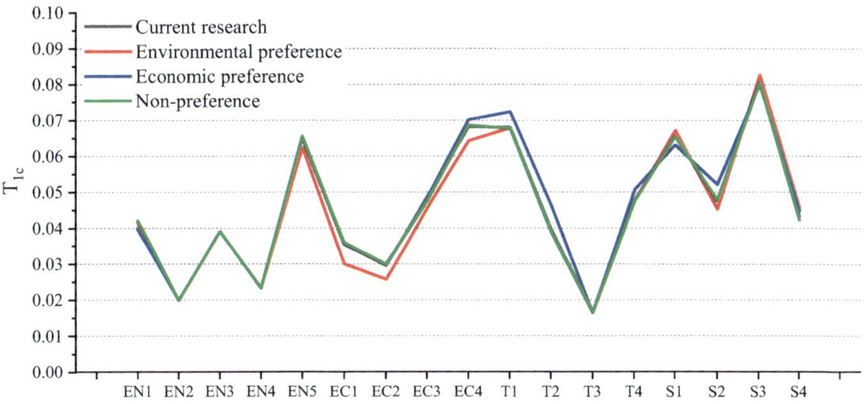

Fig. 9.6 Sensitivity analysis of the influence degree of different criteria

9.5 Policies to Promote the Sustainability of the AWPGI

Mechanism refers to the structural relationships and operation modes between various elements. The policy design mechanism refers to a set of rules for policymakers to identify the main influencing factors and formulate relevant policies to promote the system's sustainability. The research on policy design mechanisms can help the government clarify the work aspect and make efficient policy.

According to the results, S3, EC4, S1, EC3, EN5, and S2 were the most critical criteria affecting the sustainability of AWPGI. Therefore, decision-makers should consider these criteria when making decisions. The following suggestions could be proposed to promote the sustainability of AWPGI:

(1) Improving social acceptability: Power plants should employ technological updates to improve agricultural waste power generation's environmental and economic benefits and provide many more jobs to society. The government should increase publicity to raise farmers' environmental awareness and pay subsidies to farmers supplying agricultural waste.
(2) Increasing government support: Since agricultural waste is a renewable energy source, its utilization for power generation is in line with the national energy strategy and is an important project supported by the government. However, this support is still insufficient, mainly due to higher costs and lower energy conversion efficiency. Therefore, power plants should improve the maturity and energy conversion efficiency of agricultural waste power generation technology and reduce investment costs. More support should be attracted from national policies.
(3) Improving the competitiveness of the power generation industry: First of all, power plants improve their core capabilities from their perspective. Second, as the leader of renewable energy utilization, the government should provide preferential loans, tax relief, and other policies to enhance the competitiveness of

the power generation industry. Finally, the agricultural waste supply system dominated by power plants should improve management capabilities to promote the supply system's coordinated development.

(4) Increasing the social contribution of the power generation industry: Agricultural waste power plants employ benefit sharing to provide returns to society. It can provide participants with technical guidance and equipment investment, help stakeholders attain more benefits, and provide financial support for the government's public welfare undertakings. It leads to a positive interaction between the power generation industry and society, improving the sustainability of the AWPGI.

(5) Reducing power generation cost: Raw material purchase and operating costs increase the production costs. Therefore, the power plants should encourage farmers to engage in intensive farming, large-scale planting, and mechanized collection improvement while encouraging vehicles to consume bioethanol fuel to reduce raw material costs. Meanwhile, increasing the power generation efficiency and optimizing power plants' management structure can also reduce the production cost.

9.6 Conclusion

This chapter combined the triangular fuzzy number, DEMATEL, and FHAP methods to study the relationships between the sustainability assessment criteria of AWPGI. Accordingly, a policy design mechanism was presented to improve the sustainability of the AWPGI. According to the obtained results, reduction of GHG emissions, investment costs, management costs, technology maturity, energy conversion efficiency, technical complexity, and safety and reliability were chosen as the cause factors. The nonrenewable energy utilization, water use, noise pollution, agricultural waste, production cost, competitiveness, social acceptability, regional development contribution, government support degree, and job production were identified as the effect factors. The relationship between the factors was discussed to investigate how the cause factors can affect the effect factors from the internal system. Significant efforts made by the power plants and the support provided by the external stakeholders are necessary to improve the sustainability of the AWPGI. Accordingly, several policies were designed to promote the sustainability of the AWPGI in improving social acceptability, increasing government support, improving the competitiveness of power generation enterprises, increasing social development contributions, and reducing production costs. This research provided valuable information for promoting the sustainability of the AWPGI.

The main contribution of this research is to construct the fuzzy composite DEMATEL model to study the interrelationships of the internal influencing factors of the AWPGI. However, there are still some shortcomings. The influence of the relative relationship change between factors on the results was not considered.

Besides, the sustainable development of the AWPGI assumed a sufficient supply of raw materials without considering the impact of human factors. In future research, the participant's behavioral factors should be incorporated into the sustainability evaluation criteria. Moreover, the changes in the relative relations of the factors should be considered.

Acknowledgments This chapter has been reprinted from Z. Wang, G. Xu, Z. Wang, et al., *Sustainability of Agricultural Waste Power Generation Industry in China: Criteria Relationship Identification and Policy Design Mechanism*, Environment, Development and Sustainability, 2022, 24 (3): 3371–3395. Copyright (2022), with permission from Springer.

References

Aberilla JM, Gallego-Schmid A, Azapagic A (2019) Environmental sustainability of small-scale biomass power technologies for agricultural communities in developing countries. Renew Energy 141:493–506. https://doi.org/10.1016/j.renene.2019.04.036

Asadullah M (2014) Barriers of commercial power generation using biomass gasification gas: a review. Renew Sustain Energy Rev 29:201–215. https://doi.org/10.1016/j.rser.2013.08.074

Ba BH, Prins C, Prodhon C (2016) Models for optimization and performance evaluation of biomass supply chains: an operations research perspective. Renew Energy 87:977–989. https://doi.org/10.1016/j.renene.2015.07.045

Bentsen NS, Jorgensen JR, Stupak I, Jorgensen U, Taghizadeh-Toosi A (2019) Dynamic sustainability assessment of heat and electricity production based on agricultural crop residues in Denmark. J Clean Prod 213:491–507. https://doi.org/10.1016/j.jclepro.2018.12.194

Chen VYC, Lien H-P, Liu C-H, Liou JJH, Tzeng G-H, Yang L-S (2011) Fuzzy MCDM approach for selecting the best environment-watershed plan. Appl Soft Comput 11(1):265–275. https://doi.org/10.1016/j.asoc.2009.11.017

Choudhary D, Shankar R (2012) An STEEP-fuzzy AHP-TOPSIS framework for evaluation and selection of thermal power plant location: a case study from India. Energy 42(1):510–521. https://doi.org/10.1016/j.energy.2012.03.010

Dai Y, Zheng H, Jiang Z, Xing B (2020) Comparison of different crop residue-based technologies for their energy production and air pollutant emission. Sci Total Environ 707. https://doi.org/10.1016/j.scitotenv.2019.136122

Fan, Jianling, Guo, Dengwei, Han, Lu, Liu, Cuiying, Zhang, Chuanhong, Xie, Jianan, . . . Yin, Liwen. (2022). Spatiotemporal dynamics of carbon footprint of Main crop production in China. Int J Environ Res Public Health, 19(21). doi:https://doi.org/10.3390/ijerph192113896

Feng F, Song G, Shen L, Xiao J (2017) Environmental benefits analysis based on life cycle assessment of rice straw-based synthetic natural gas in China. Energy 139:341–349. https://doi.org/10.1016/j.energy.2017.07.167

Giuntoli J, Agostini A, Caserini S, Lugato E, Baxter D, Marelli L (2016) Climate change impacts of power generation from residual biomass. Biomass Bioenergy 89:146–158. https://doi.org/10.1016/j.biombioe.2016.02.024

Golecha R, Gan J (2016) Biomass transport cost from field to conversion facility when biomass yield density and road network vary with transport radius. Appl Energy 164:321–331. https://doi.org/10.1016/j.apenergy.2015.11.070

Guney T (2019) Renewable energy, non-renewable energy and sustainable development. Int J Sust Dev World 26(5):389–397. https://doi.org/10.1080/13504509.2019.1595214

Hong J, Ren L, Hong J, Xu C (2016) Environmental impact assessment of corn straw utilization in China. J Clean Prod 112:1700–1708. https://doi.org/10.1016/j.jclepro.2015.02.081

Huang, Yi, Zhao, Ying-jie, Hao, Yan-hong, Wei, Guo-qiang, Feng, Jie, Li, Wen-ying, . . . Nimmo, William. (2019). A feasibility analysis of distributed power plants from agricultural residues resources gasification in rural China. Biomass Bioenergy, 121, 1–12. doi:https://doi.org/10.1016/j.biombioe.2018.12.007

Ji C, Cheng K, Nayak D, Pan G (2018) Environmental and economic assessment of crop residue competitive utilization for biochar, briquette fuel and combined heat and power generation. J Clean Prod 192:916–923. https://doi.org/10.1016/j.jclepro.2018.05.026

John L, Gurumurthy A, Soni G, Jain V (2019) Modelling the inter-relationship between factors affecting coordination in a humanitarian supply chain: a case of Chennai flood relief. Ann Oper Res 283(1–2):1227–1258. https://doi.org/10.1007/s10479-018-2963-3

Jung S, Lee J, Park Y-K, Kwon EE (2020) Bioelectrochemical systems for a circular bioeconomy. Bioresour Technol 300. https://doi.org/10.1016/j.biortech.2020.122748

Karkania V, Fanara E, Zabaniotou A (2012) Review of sustainable biomass pellets production - a study for agricultural residues pellets' market in Greece. Renew Sustain Energy Rev 16(3):1426–1436. https://doi.org/10.1016/j.rser.2011.11.028

Khishtandar S, Zandieh M, Dorri B (2017) A multi criteria decision making framework for sustainability assessment of bioenergy production technologies with hesitant fuzzy linguistic term sets: the case of Iran. Renew Sustain Energy Rev 77:1130–1145. https://doi.org/10.1016/j.rser.2016.11.212

Li Y, Mathiyazhagan K (2018) Application of DEMATEL approach to identify the influential indicators towards sustainable supply chain adoption in the auto components manufacturing sector. J Clean Prod 172:2931–2941. https://doi.org/10.1016/j.jclepro.2017.11.120

Liang H, Ren J, Gao Z, Gao S, Luo X, Dong L, Scipioni A (2016) Identification of critical success factors for sustainable development of biofuel industry in China based on grey decision-making trial and evaluation laboratory (DEMATEL). J Clean Prod 131:500–508. https://doi.org/10.1016/j.jclepro.2016.04.151

Lin B, He J (2017) Is biomass power a good choice for governments in China? Renew Sustain Energy Rev 73:1218–1230. https://doi.org/10.1016/j.rser.2017.02.024

Lin S, Li C, Xu F, Liu D, Liu J (2018) Risk identification and analysis for new energy power system in China based on D numbers and decision-making trial and evaluation laboratory (DEMATEL). J Clean Prod 180:81–96. https://doi.org/10.1016/j.jclepro.2018.01.153

Logeswaran J, Shamsuddin AH, Silitonga AS, Mahlia TM, Indra. (2020) Prospect of using rice straw for power generation: a review. Environ Sci Pollut Res 27(21):25956–25969. https://doi.org/10.1007/s11356-020-09102-7

Lv H, Ding H, Zhou D, Zhou P (2014) A site selection model for a straw-based power generation plant with CO2 emissions. Sustainability 6(10):7466–7481. https://doi.org/10.3390/su6107466

Moncada JA, Verstegen JA, Posada JA, Junginger M, Lukszo Z, Faaij A, Weijnen M (2018) Exploring policy options to spur the expansion of ethanol production and consumption in Brazil: an agent-based modeling approach. Energy Policy 123:619–641. https://doi.org/10.1016/j.enpol.2018.09.015

Ortiz-Barrios M, Cabarcas-Reyes J, Ishizaka A, Barbati M, Jaramillo-Rueda N, de Jesus Carrascal-Zambrano G (2021) A hybrid fuzzy multi-criteria decision making model for selecting a sustainable supplier of forklift filters: a case study from the mining industry. Ann Oper Res 307(1–2):443–481. https://doi.org/10.1007/s10479-020-03737-y

Qin, Zhangcai, Zhuang, Qianlai, Cai, Ximing, He, Yujie, Huang, Yao, Jiang, Dong, . . . Wang, Michael Q. (2018). Biomass and biofuels in China: toward bioenergy resource potentials and their impacts on the environment. Renew Sustain Energy Rev, 82, 2387–2400. doi:https://doi.org/10.1016/j.rser.2017.08.073

Reddy BS (2018) Economic dynamics and technology diffusion in indian power sector. Energy Policy 120:425–435. https://doi.org/10.1016/j.enpol.2018.05.044

Ren J (2018) Multi-criteria decision making for the prioritization of energy systems under uncertainties after life cycle sustainability assessment. Sustainable Prod Consumption 16:45–57. https://doi.org/10.1016/j.spc.2018.06.005

Ren J, Lutzen M (2015) Fuzzy multi-criteria decision-making method for technology selection for emissions reduction from shipping under uncertainties. Transportation Research Part D-Transport and Environment 40:43–60. https://doi.org/10.1016/j.trd.2015.07.012

Ren J, Fedele A, Mason M, Manzardo A, Scipioni A (2013a) Fuzzy multi-actor multi-criteria decision making for sustainability assessment of biomass-based technologies for hydrogen production. Int J Hydrog Energy 38(22):9111–9120. https://doi.org/10.1016/j.ijhydene.2013.05.074

Ren J, Manzardo A, Toniolo S, Scipioni A (2013b) Ustainability of hydrogen supply chain. Part I: identification of critical criteria and cause-effect analysis for enhancing the sustainability using DEMATEL. Int J Hydrog Energy 38(33):14159–14171. https://doi.org/10.1016/j.ijhydene.2013.08.126

Ren J, Gao S, Tan S, Dong L, Scipioni A, Mazzi A (2015) Role prioritization of hydrogen production technologies for promoting hydrogen economy in the current state of China. Renew Sustain Energy Rev 41:1217–1229. https://doi.org/10.1016/j.rser.2014.09.028

Saaty TL (2004) Decision making — the analytic hierarchy and network processes (AHP/ANP). J Syst Sci Syst Eng 13(1):1–35. https://doi.org/10.1007/s11518-006-0151-5

Salleh SF, Mohd Roslan ME, Abd Rahman A, Shamsuddin AH, Tuan Abdullah TAR, Sovacool BK (2020) Transitioning to a sustainable development framework for bioenergy in Malaysia: policy suggestions to catalyse the utilisation of palm oil mill residues. Energy Sustain Soc 10(1). https://doi.org/10.1186/s13705-020-00269-y

Sastre CM, Gonzalez-Arechavala Y, Santos AM (2015) Global warming and energy yield evaluation of Spanish wheat straw electricity generation - a LCA that takes into account parameter uncertainty and variability. Appl Energy 154:900–911. https://doi.org/10.1016/j.apenergy.2015.05.108

Seyed-Hosseini SM, Safaei N, Asgharpour MJ (2006) Reprioritization of failures in a system failure mode and effects analysis by decision making trial and evaluation laboratory technique. Reliability Eng Syst Saf 91(8):872–881. https://doi.org/10.1016/j.ress.2005.09.005

Shafie SM, Masjuki HH, Mahlia TMI (2014) Rice straw supply chain for electricity generation in Malaysia: economical and environmental assessment. Appl Energy 135:299–308. https://doi.org/10.1016/j.apenergy.2014.08.101

Shao J, Taisch M, Ortega-Mier M (2016) A grey-DEcision-MAking trial and evaluation laboratory (DEMATEL) analysis on the barriers between environmentally friendly products and consumers: practitioners' viewpoints on the European automobile industry. J Clean Prod 112:3185–3194. https://doi.org/10.1016/j.jclepro.2015.10.113

Spagnolo S, Chinellato G, Cristiano S, Zucaro A, Gonella F (2020) Sustainability assessment of bioenergy at different scales: an emergy analysis of biogas power production. J Clean Prod 277. https://doi.org/10.1016/j.jclepro.2020.124038

Spiegel A, Britz W, Djanibekov U, Finger R (2018) Policy analysis of perennial energy crop cultivation at the farm level: short rotation coppice (SRC) in Germany. Biomass Bioenergy 110:41–56. https://doi.org/10.1016/j.biombioe.2018.01.003

Srivastava RK, Shetti NP, Reddy KR, Aminabhavi TM (2020) Biofuels, biodiesel and biohydrogen production using bioprocesses. A review. *Environ Chem Lett* 18(4):1049–1072. https://doi.org/10.1007/s10311-020-00999-7

Stich J, Ramachandran S, Hamacher T, Stimming U (2017) Techno-economic estimation of the power generation potential from biomass residues in Southeast Asia. Energy 135:930–942. https://doi.org/10.1016/j.energy.2017.06.162

Sun Y, Cai W, Chen B, Guo X, Hu J, Jiao Y (2017) Economic analysis of fuel collection, storage, and transportation in straw power generation in China. Energy 132:194–203. https://doi.org/10.1016/j.energy.2017.05.077

Tauro R, Garcia CA, Skutsch M, Masera O (2018) The potential for sustainable biomass pellets in Mexico: an analysis of energy potential, logistic costs and market demand. Renew Sustain Energy Rev 82:380–389. https://doi.org/10.1016/j.rser.2017.09.036

Tippayawong KY, Chaidi N, Ngamlertsappakit T, Tippayawong N (2020) Demand and cost analysis of agricultural residues utilized as biorenewable fuels for power generation. Energy Rep 6:1298–1302. https://doi.org/10.1016/j.egyr.2020.11.040

Tsalidis G-A, Joshi Y, Korevaar G, de Jong W (2014) Life cycle assessment of direct co-firing of torrefied and/or pelletised woody biomass with coal in The Netherlands. J Clean Prod 81:168–177. https://doi.org/10.1016/j.jclepro.2014.06.049

Tseng M-L, Lin Y-H, Chiu ASF (2009) Fuzzy AHP-based study of cleaner production implementation in Taiwan PWB manufacturer. J Clean Prod 17(14):1249–1256. https://doi.org/10.1016/j.jclepro.2009.03.022

Wang Y-M, Luo Y, Hua Z (2008) On the extent analysis method for fuzzy AHP and its applications. Eur J Oper Res 186(2):735–747. https://doi.org/10.1016/j.ejor.2007.01.050

Wang Z, Ren J, Goodsite ME, Xu G (2018) Waste-to-energy, municipal solid waste treatment, and best available technology: comprehensive evaluation by an interval-valued fuzzy multi-criteria decision making method. J Clean Prod 172:887–899. https://doi.org/10.1016/j.jclepro.2017.10.184

Wang B, Song J, Ren J, Li K, Duan H, Wang X'e (2019) Selecting sustainable energy conversion technologies for agricultural residues: a fuzzy AHP-VIKOR based prioritization from life cycle perspective. Resour Conserv Recycl 142:78–87. https://doi.org/10.1016/j.resconrec.2018.11.011

Wang Z, Wang Z, Tahir N, Wang H, Li J, Xu G (2020a) Study of synergetic development in straw power supply chain: straw price and government subsidy as incentive. Energy Policy 146. https://doi.org/10.1016/j.enpol.2020.111788

Wang Z, Wang Z, Xu G, Ren J, Wang H, Li J (2020b) Sustainability assessment of straw direct combustion power generation in China: from the environmental and economic perspectives of straw substitute to coal. J Clean Prod 273. https://doi.org/10.1016/j.jclepro.2020.122890

Wei D, Liu H, Shi K (2019) What are the key barriers for the further development of shale gas in China? A grey-DEMATEL approach. Energy Rep 5:298–304. https://doi.org/10.1016/j.egyr.2019.02.010

Wen W, Zhang Q (2015) A design of straw acquisition mode for China's straw power plant based on supply chain coordination. Renew Energy 76:369–374. https://doi.org/10.1016/j.renene.2014.11.062

Wu W-W (2008) Choosing knowledge management strategies by using a combined ANP and DEMATEL approach. Expert Syst Appl 35(3):828–835. https://doi.org/10.1016/j.eswa.2007.07.025

Xu, Xin Long, & Chen, Yu Jie. (2020). A comprehensive model to analyze straw recycling logistics costs for sustainable development: evidence from biomass power generation. Environ Prog Sustain Energ, 39(4). doi:https://doi.org/10.1002/ep.13394, Environ Progr Sustain Energy

Xue C, Wang X (2017) Study on government subsidy decision-making of straw power generation supply chain. Procedia Eng 174:211–218. https://doi.org/10.1016/j.proeng.2017.01.122

Xue, Shengrong, Song, Jinghui, Wang, Xiaojiao, Shang, Zezhou, Sheng, Chenjing, Li, Chongyuan, . . . Liu, Jingyu. (2020). A systematic comparison of biogas development and related policies between China and Europe and corresponding insights. Renew Sustain Energy Rev, 117. doi:https://doi.org/10.1016/j.rser.2019.109474

Yan L, Wang Z, Cao Y, He B (2020) Comparative evaluation of two biomass direct-fired power plants with carbon capture and sequestration. Renew Energy 147:1188–1198. https://doi.org/10.1016/j.renene.2019.09.047

Yang S, Xiao L, Yang S, Kraslawski A, Man Y, Qian Y (2014) Sustainability assessment of the coal/biomass to Fischer-Tropsch fuel processes. ACS Sustain Chem Eng 2(1):80–87. https://doi.org/10.1021/sc400336e

Zhang Q, Zhou D, Zhou P, Ding H (2013) Cost analysis of straw-based power generation in Jiangsu Province, China. Appl Energy 102:785–793. https://doi.org/10.1016/j.apenergy.2012.08.032
Zhang X, Luo K, Tan Q (2017) A game theory analysis of China's Agri-biomass-based power generation supply chain: a co-opetition strategy. Energy Procedia 105:168–173. https://doi.org/10.1016/j.egypro.2017.03.297
Zhao Z-Y, Zuo J, Wu P-H, Yan H, Zillante G (2016) Competitiveness assessment of the biomass power generation industry in China: a five forces model study. Renew Energy 89:144–153. https://doi.org/10.1016/j.renene.2015.12.035

Chapter 10
Research on the Sustainable Development of Agricultural Waste Power Generation Supply Chain Based on FCM

Zhanwu Wang

Contents

Abstract As a renewable energy technology, agricultural waste power generation holds considerable potential for substituting coal-based power and mitigating carbon emissions. Nevertheless, the stability and sustainable development of its supply chain encounter significant challenges. To systematically examine the key influencing factors and their interrelationships, this study employs the PESTEL framework to identify 21 critical factors across six dimensions—political, economic, social, technological, environmental, and legal—and constructs a fuzzy cognitive map (FCM) model to enable dynamic system reasoning and evolutionary analysis. The findings reveal that the cost of agricultural waste power generation, supply acces-

Z. Wang (✉)
College of Mechanical and Electrical Engineering, Henan Agricultural University, Zhengzhou, Henan Province, China
e-mail: wzw6997@henau.edu.cn

Z. Wang et al. (eds.), *Agricultural Waste Power Generation Supply Chain*, Industrial Ecology and Environmental Management 7,
https://doi.org/10.1007/978-3-032-20087-7_10

sibility, availability of financial market instruments, policy support, and societal awareness of corporate responsibility are the core drivers shaping the system's development. Notably, economic cost exhibits the highest sensitivity in the model, exerting the most pronounced influence on system stability, while political factors serve as a strong external catalyst. Furthermore, five major barriers to development are identified: cost pressures, logistical and supply chain inefficiencies, capital shortages, inadequate coordination mechanisms, and insufficient alignment of stakeholder interests. To address these challenges, this study recommends strengthening cost subsidies and establishing robust collection and storage infrastructure in the short term; in the medium to long term, it advocates enhancing industrial policy frameworks, developing digitalized coordination platforms, and instituting benefit-sharing mechanisms to foster multi-stakeholder collaboration—ultimately advancing the efficiency, resilience, and sustainability of the agricultural waste power generation supply chain.

Keywords Agricultural waste-to-energy power generation · Supply chain · Fuzzy cognitive maps · Sustainable development

10.1 Introduction

Agricultural waste-to-energy technology, as a form of renewable energy utilization, serves as a viable and sustainable alternative to coal-fired power generation by generating clean energy, significantly reducing greenhouse gas emissions, and contributing to climate change mitigation (Gao et al. 2024; Wang et al. 2020a, b). This technology has now reached a mature stage and has been implemented across multiple regions, gradually evolving into an integrated multi-product system. However, despite its technical readiness, agricultural waste-to-energy enterprises continue to encounter persistent operational and maintenance challenges, largely stemming from the lack of reliable and sustainable feedstock supply chains (Hou et al. 2025a). This critical bottleneck has prompted growing academic interest in the design and management of agricultural waste-to-energy supply chains in the recent years, resulting in substantial scholarly progress (Wang et al. 2023). Key research areas encompass supply chain modeling and optimization, economic feasibility assessment, and comprehensive lifecycle analysis, all aimed at enhancing system efficiency and long-term sustainability.

Among the various technological pathways for energy recovery from agricultural waste, life cycle assessment (LCA) has emerged as a critical tool for the systematic evaluation of their environmental impacts. Tomatis et al. (2019) investigated a biomass-derived dimethyl ether (DME) production process based on CO_2-enhanced gasification technology, demonstrating that this approach can achieve reductions in environmental impacts—including climate change, human toxicity, and ecotoxicity—by at least 20% compared to conventional pathways. The results highlight the significant potential of the gasification pathway for efficient carbon utilization and greenhouse gas emission mitigation. Prasad et al. (2020) conducted a systematic review of life cycle assessment (LCA) studies on various bioenergy conversion technologies from the perspective of comprehensive crop residue utilization, emphasizing that the careful selection of pretreatment and conversion pathways is essential for maximizing environmental benefits. By integrating experimental and simulation approaches, Wang and Cheng (2024) evaluated an integrated rice straw energy system based on pyrolysis-gasification cogeneration, demonstrating that its greenhouse gas emission intensity ranges from 2.92 to 3.51 g CO_2/MJ—significantly lower than that of conventional fossil fuels—thereby providing further substantiation of the carbon mitigation benefits of agricultural residue-based power generation. Furthermore, Chopra et al. (2024) emphasized the critical role of the pretreatment stage in the life cycle of biomass fuels, highlighting that efficient and environmentally sustainable pretreatment technologies are essential for the commercialization of biomass energy. Wang and Yang (2022) focused on domestic data from China to assess the full life cycle carbon emissions and economic performance of biomass gasification power generation technology. Their results show that carbon emissions are 14.7% lower than those of coal-fired power generation, demonstrating significant environmental and climate change mitigation benefits. Overall, although multiple technological pathways exist for the energy utilization of agricultural waste, power generation technologies—such as gasification and pyrolysis—that offer large-scale processing capacity, high energy recovery efficiency, and substantial carbon reduction potential have become key focal points in both LCA research and practical applications.

As an important component of renewable energy, the large-scale utilization of agricultural waste for power generation holds significant importance for optimizing energy structures and mitigating environmental pollution. However, the economic viability and environmental performance of biomass power generation projects largely depend on supply chain efficiency, with power plant location serving as a critical determinant of both logistical costs and carbon emissions. Consequently, the scientific and precise selection of plant locations to optimize resource collection, transportation, and conversion processes has become a focal point of current research. Teng and Wang (2022) developed a calculation model for estimating the availability of straw resources, enabling an accurate assessment of their spatial distribution across 31 provinces in China and providing essential data support for power plant site selection. Mao et al. (2024) further accounted for supply uncertainty and facility disruption risks by developing a stochastic programming model to optimize the straw supply chain, demonstrating that this approach effectively reduces total costs and enhances system resilience. Driven by low-carbon objectives, Mao et al. (2021) incorporated carbon emission reduction into the facility location model and achieved Pareto optimality between cost and carbon emissions through bi-objective programming, thereby providing a valuable reference for decision-making in both government and industry. Furthermore, Zhao et al. (2017) using Hubei Province as a case study, innovatively integrated Geographic Information Systems (GIS) with spatially explicit optimization models for power plant siting, significantly improving the accuracy and feasibility of assessing the spatial distribution of straw resources and developing optimal siting plans. These studies collectively indicate that a comprehensive siting strategy—integrating resource assessment, supply chain optimization, and spatial information technology—is essential for the successful implementation and sustainable development of agricultural waste-to-energy projects.

Agricultural waste, as an important renewable energy source, has emerged as a key strategy for achieving energy transition and resource utilization through power generation. Currently, technological pathways for agricultural waste-to-energy conversion are evolving in a diversified manner, primarily including gasification, fuel cells, pyrolysis, and anaerobic digestion. These technologies exhibit distinct characteristics in terms of energy conversion efficiency, output products, and application scenarios. For instance, Shahavi et al. (2022) conducted a study on seven types of agricultural waste within a gasification-based combined cycle system and found that crushed straw and rice bran achieved the highest performance in terms of cold gas efficiency (88.41%) and power generation output (8328 kW), respectively. Hou et al. (2025b) focused on direct carbon solid oxide fuel cells (DC-SOFCs) and increased the maximum power output of biochar fuel from 279.6 MW cm^{-2} to 371.5 MW cm^{-2} through a strategy involving co-modification of the anode and carbon fuel. Potnuri et al. (2023) conducted a systematic review of the production of biochar from agricultural waste via microwave-assisted pyrolysis, highlighting this process as a highly promising conversion technology. The studies by Chandra et al. (2025) and Yang et al. (2024) demonstrated the pathway of generating biogas through anaerobic digestion and subsequently converting it into electricity. The latter study specifically highlighted that the biogas-based energy system imposes a lower environmental burden and offers higher sustainability compared to straw return to fields and direct combustion for power generation. Kamusoko and Mukumba (2025) also highlighted the significant potential of converting corn stalks into biogas for combined heat and power generation in South Africa. These studies collectively indicate that optimizing the technological pathway is central to enhancing the efficiency and feasibility of power generation from agricultural waste.

Microbial fuel cell (MFC) technology offers an innovative pathway for the energy utilization of agricultural waste through "synergistic waste treatment and energy recovery," directly converting the chemical energy in organic waste into electrical energy. This technology has demonstrated strong adaptability and significant environmental benefits in treating various types of organic waste. Rojas-Flores et al. (2024) utilized Andean tuber waste as a substrate in MFCs, achieving a voltage of 0.99 V and a power density of 373.8 MW/cm^2, and successfully powered an light-emitting diode (LED). The study by Rojas-Villacorta et al. (2023) further confirmed that lettuce waste can serve as an effective substrate in MFCs, with the system achieving a voltage of 0.959 V. To enhance the performance of MFCs, Liu et al. (2022) developed a cathode catalyst composed of peanut shell biochar and conductive carbon black, which significantly increased the isopropanol removal rate and power output. Zhang et al. (2024) systematically investigated Pb(II) removal and electricity generation performance using different pretreated agricultural wastes as carbon sources in MFCs. They found that alkali- and heat-pretreated straw achieved the highest efficiency, with a Pb (II) removal rate of 87.66%. The innovative research by Magotra et al. (2021) incorporated rice husk ash into white cement bricks to create "brick fuel cells" capable of generating electricity at a power density of 2008.93 MW/m^2, thereby expanding the application scope of MFCs. These studies provide robust technical support for the development of distributed and low-environmental-impact rural energy supply systems.

The sustainable development of the agricultural waste-to-energy industry depends not only on technological advancement but also on economic feasibility, environmental benefits, and social acceptance, among other comprehensive factors. Therefore, numerous studies have conducted systematic evaluations from the perspectives of technology and economics, environmental impact, and social dimensions. From a technical and an economic perspective, the study conducted by Mana et al. (2021) on the Fez-Meknes region in Morocco demonstrates that utilizing olive residue for electricity generation could yield an annual power output of up to 2828.11 GWh, with a levelized cost of electricity as low as 15.03 cents/kWh. Ali et al. (2025) evaluated the potential for biomass-based power generation in Pakistan and concluded that the annual electricity output could reach 609,964 GWh, with generation costs substantially lower than those of conventional fossil fuel technologies. The analysis of rice waste in Sri Lanka by Illankoon et al. (2022) and the model optimization study in the states of São Paulo and Paraná, Brazil, conducted by Crotman et al. (2024) both demonstrate the economic feasibility of electricity generation from agricultural waste. From the perspective of environmental sustainability, Havrysh et al. (2021) evaluated the energy potential of agricultural waste in Poland and concluded that it could meet 15% of the country's electricity demand while significantly reducing CO_2 emissions. Wang et al. (2022) identified the key causal factors influencing the sustainable development of China's agricultural waste power generation industry by applying the Decision-Making Trial And Evaluation Laboratory (DEMATEL) model, thereby providing a foundation for policy design. In the social dimension, Zhu et al. (2022) used social network analysis to reveal weak stakeholder connections—such as those between enterprises and farmers—in the biomass energy industry chain in northern Jiangsu, China. Finally, Chen et al. (2025), under the "dual carbon" policy vision, conducted a comprehensive Political, Economic, Social, Technological (PEST) and Strengths, Weaknesses, Opportunities, Threats (SWOT) analysis to explore systematic solutions for the resource utilization of agricultural waste. These integrated assessment studies offer comprehensive decision-making support for the scientific planning and policy formulation of agricultural waste-to-energy projects.

Economic benefits serve as a key enabler for the sustainable development of biomass power generation technology. The research by Dassanayake and Kumar (2012) indicates that although the economic competitiveness of straw-based power generation falls short of that of coal-fired power generation, biomass power generation still holds certain economic viability. Abdelhady et al. (2018) assessed the technical and economic performance of biomass power generation and found that straw-based power generation exhibits promising development potential within a specific load range. Wang et al. (2019) demonstrated that a 25-MW-biomass power plant can achieve an annual net present value of 2.746 million yuan, yielding an enterprise investment return rate of 7.72%. Huang et al. (2019) analyzed the economic performance of straw gasification power generation systems at different scales and identified a 5 MW capacity as the most economically viable option. Zhao and Li (2016) proposed a bi-objective 0–1 integer programming model to optimize the siting of biomass power plants. Despite these advancements, the current economic performance of agricultural waste-to-energy power generation remains relatively low, indicating the need for further research to enhance its economic competitiveness.

In summary, the current research on the supply chain of agricultural waste power generation primarily focuses on life cycle assessment, facility siting, economic benefit analysis, sustainability evaluation, and comprehensive performance assessment. However, studies on carbon migration within the supply chain and on the coordinated development of its various components remain relatively limited.

10.2 Analysis of the External Environment in the Agricultural Waste-to-Energy Supply Chain

The utilization of agricultural waste for power generation offers environmental and social benefits; however, economic constraints have hindered its sustainable development. During the research process, it becomes evident that the external production environment significantly influences the agricultural waste-to-energy supply chain system. The external environment refers to the impact of macro-level external factors on the operation of this supply chain. It encompasses key dimensions such as the political, economic, sociocultural, technological, environmental, and legal environments. The PESTEL framework serves as the primary tool for conducting macro-environmental analysis. Also known as the macro-environmental analysis model, the PESTEL is an effective approach for identifying and evaluating external factors affecting organizational operations, particularly through six core aspects: political, economic, sociocultural, technological, environmental, and legal. Given the multitude of factors influencing the agricultural waste-to-energy supply chain, the PESTEL model enables their systematic classification into these six dimensions, thereby facilitating comprehensive environmental assessment.

1. Analysis of political dimension factors: The resource utilization of agricultural waste is closely linked to societal development and the advancement of ecological civilization. During the research process, political factors were categorized into four dimensions: political stability (P1), restrictive policies (P2), supportive policies (P3), and policy opportunities (P4). Political stability refers to the consistency and predictability of the political environment, which can create a favorable external context for the stable operation and long-term development of the agricultural waste-to-energy supply chain. Restrictive policies include regulations such as bans on open-field burning and emission limits on pollutants. The

enforcement of these policies ensures a reliable supply of agricultural waste feedstock while contributing to environmental protection. Supportive policies encompass government-issued incentives and measures that actively promote the resource utilization of agricultural waste, such as subsidies, tax exemptions, and preferential tariffs. Policy opportunities refer to strategic development prospects arising from national-level initiatives, including the rural revitalization strategy, the green and sustainable development agenda, and the "carbon peak and carbon neutrality" goals, all of which provide significant momentum for the expansion of agricultural waste-based energy systems.

2. Analysis of economic dimension factors: The establishment and operation of agricultural waste-to-energy enterprises require substantial capital investment, making the economic environment a critical determinant of the sustainable development of the agricultural waste power generation supply chain. In this study, three key factors are identified to examine the influence of economic conditions on the resource utilization of agricultural waste: social and economic development status (EC1), financial market instruments (EC2), and the cost of agricultural waste power generation (EC3). Social and economic development status reflects the overall level of economic advancement and societal progress. A higher level of development promotes industrial growth across sectors, accelerates the modernization and intensification of agriculture, and thereby creates favorable conditions for the efficient utilization of agricultural waste. Financial market instruments refer to the mechanisms and channels through which enterprises engaged in agricultural waste valorization access financing. An improved and accessible financial market enables these enterprises to secure funding more efficiently, supporting their operational stability and long-term management. The cost of agricultural waste power generation encompasses all expenses incurred throughout the production process; elevated costs can constrain profitability and scalability, thus limiting the widespread adoption of waste-to-energy technologies. Data collection and analysis reveal that economic dimension factors exert a significant influence on the feasibility and development potential of agricultural waste resource utilization.
3. Analysis of social dimension factors: Open burning of agricultural waste leads to environmental pollution, and utilizing it for power generation can effectively reduce the occurrence of such practices. Social group attitudes (S1), the convenience of power generation from agricultural waste (S2), social responsibility awareness (S3), and other factors influence the supply and utilization of agricultural waste. Therefore, these factors are selected to analyze the social environment within the agricultural waste power generation supply chain. Social group attitude refers to the public's perceptions and willingness toward the resource utilization of agricultural waste. A higher level of social group attitude facilitates the adoption and promotion of agricultural waste utilization. The convenience of agricultural waste utilization encompasses the ease of collection, transportation, and storage, as well as the accessibility of resource utilization technologies and funding. Greater convenience in resource utilization increases farmers' willingness to participate, thereby enhancing the effective treatment of agricultural

waste. Social responsibility awareness refers to the commitment of stakeholders—such as farmers and agricultural waste power generation enterprises—to advancing social sustainable development and ecological civilization construction. A strong sense of social responsibility contributes to the sustainable development and coordination of the agricultural waste power generation supply chain.

4. Analysis of technical dimension factors: Technical maturity (T1), technical equipment stability (T2), technical complexity (T3), and technical safety and reliability (T4) are commonly selected as key indicators for assessing the technological environment in agricultural waste resource utilization. Technical maturity refers to the extent to which a technology has reached industrial-scale production capability, reflecting its readiness for widespread application. Technical equipment stability denotes the consistent and reliable performance of machinery and equipment during continuous operation. Technical complexity pertains to the intricacy of agricultural waste-to-energy conversion technologies; higher complexity often leads to increased maintenance costs and a greater likelihood of operational failures. Technical safety and reliability refer to the safety performance of the system and the probability of failure during the production process. Technologies characterized by high maturity, stable equipment performance, low complexity, and strong safety and reliability are more conducive to promoting the efficient and sustainable power generation utilization of agricultural waste.
5. Analysis of environmental dimension factors: The resource utilization of agricultural waste is influenced by natural environmental factors (EN1), such as the timing of collection and storage temperature of agricultural waste feedstock. Meanwhile, the development status of the agricultural waste power generation industry (EN2), the level of media attention (EN3), and sustainable development potential (EN4) also significantly affect the sustainability of agricultural waste-to-energy systems and their supply chains. These factors are therefore selected as key components of the environmental dimension for analyzing the agricultural waste power generation supply chain system. The development status of the industry refers to the overall maturity, scale, and geographical distribution of current agricultural waste power generation activities. The level of media attention reflects the extent to which external media report on agricultural waste treatment methods and pollutant emissions. Media supervision plays a critical role in reducing open-field burning and unauthorized stacking of agricultural waste. Sustainable development potential refers to the long-term viability and future growth prospects of agricultural waste-based power generation. As a renewable resource, agricultural waste can substitute fossil fuels in energy production. Moreover, the ash and slag generated after combustion can be recycled as raw materials for organic fertilizer production, thereby enhancing the environmental and economic value of power generation utilization.
6. Analysis of legal dimension factors: The indiscriminate disposal of agricultural waste has led to environmental pollution and disrupted ecological balance. To address this issue, China has enacted relevant legislation prohibiting the open burning of agricultural waste. The three key factors, legal completeness (L1),

legal enforceability (L2), and legal supervision (L3) are selected to analyze the legal environment for agricultural waste-to-energy utilization. Legal completeness refers to the comprehensiveness of national laws governing the management, utilization methods, and pollutant emissions associated with agricultural waste. Higher levels of legal completeness assist enterprises engaged in agricultural waste resource utilization in establishing industry standards and reducing environmental emissions. Legal enforceability denotes the effectiveness and consistency of law enforcement during the handling and conversion of agricultural waste. Legal supervision refers to the monitoring capacity of the public, media, and government authorities over the implementation and enforcement of relevant regulations. The factors within the legal dimension collectively contribute to the sustainable and orderly development of the agricultural waste power generation industry by ensuring that enterprises operate within a clear regulatory framework. Furthermore, robust legal mechanisms prevent the unauthorized disposal of agricultural waste, thereby securing a stable supply of raw materials for energy production.

10.3 Construction of a Fuzzy Cognitive Map (FCM) Model for the Agricultural Waste-to-Energy Supply Chain System

10.3.1 Fuzzy Cognitive Map Method

Fuzzy cognitive map (FCM) was first proposed by Kosko, who integrated fuzzy mathematics theory into cognitive map theory to construct a directed graph that represents causal relationships among concepts. The FCM is a soft computing method derived from the integration of fuzzy logic and neural networks. The FCM model consists of nodes and directed arcs. Nodes, also known as concept nodes, represent key variables or entities within a system; directed arcs denote causal influences and propagation paths between these nodes, with the arrow direction indicating the flow from cause to effect. The FCM enables the construction of a dynamic reasoning model capable of simulating temporal changes. The reasoning process incorporates both positive and negative causal interactions simultaneously, allowing for the analysis of complex, nonlinear system behaviors.

The construction of the fuzzy cognitive map method primarily involves five key steps.

Step 1: Determine the concept nodes. The factors, indicators, and other relevant elements selected within the research system are defined as concept nodes.

Step 2: Determine the correlations and causal weights between concept nodes. Apply fuzzy theory to assess relational patterns and compute the causal weights among the concept nodes.

Step 3: Construct a fuzzy cognitive map. Utilize a professional software to create a fuzzy cognitive map based on the causal relationships among the concept nodes.
Step 4: Fuzzy cognitive map inference: Evolutionary inference is a core analytical feature of fuzzy cognitive maps. Concept nodes in a fuzzy cognitive map undergo evolutionary inference through directed arcs, causal weights, and predefined inference functions, enabling the dynamic simulation of system behavior over time.
Step 5: Judgment of fuzzy cognitive map system stability.

10.3.2 Construction of Fuzzy Cognitive Maps

Step 1: Determine the concept nodes of the agricultural waste power generation supply chain system.

The improvement of agricultural waste utilization for power generation requires comprehensive consideration of both internal and external influencing factors. Through investigation and analysis, it is evident that multiple factors affect the development of the agricultural waste power generation supply chain. In the context of external environment analysis, these factors can be defined as the concept nodes of the system. Let C denote the set of concept nodes, where $C = \{C_1, C_2, \ldots, C_n\}$, with each C_i representing a specific concept node in the fuzzy cognitive map model and n indicating the total number of nodes.

Step 2: Determine the relevance and causal weights of the concept nodes.

The correlations and causal weights among the influencing factors of the agricultural waste power generation supply chain system can be determined through a questionnaire survey. The relationships among these factors are assessed by leveraging the professional knowledge and experience of an expert panel, while the degree of mutual influence is evaluated using the fuzzy linguistic terms presented in Table 10.1. The membership function of the fuzzy numbers is illustrated in Fig. 10.1.

After determining the judgment results of the expert groups, the centroid method can be used to integrate the discrimination results of different expert groups. The calculation of the centroid method is shown in Eq. (10.1).

$$x = \frac{\int_a^b xU(x)dx}{\int_a^b U(x)dx} \tag{10.1}$$

In the fuzzy cognitive map method, concept nodes are interconnected through directed arcs, representing their interrelationships. Let $V = \{(C_i, C_j) | C_i, C_j \in C\}$, where V denotes the set of directed arcs between the concept nodes.

Let W denote the causal weight matrix of the fuzzy cognitive map; then, W can be expressed as:

Table 10.1 Fuzzy language and semantic table

Fuzzy language	Abb	Membership function
Strong negative	NVS	$U_{NVS}(x) = -4x - 3 \quad -1.00 \le x \le -0.75$
Negative strength	NS	$U_{NS}(x) = \begin{cases} 4x+4 & -1.00 \le x \le -0.75 \\ -4x-2 & -0.75 \le x \le -0.50 \end{cases}$
Negative to medium strong	NM	$U_{NM}(x) = \begin{cases} 4x+3 & -0.75 \le x \le -0.50 \\ -4x-1 & -0.50 \le x \le -0.25 \end{cases}$
Weakness	NW	$U_{NW}(x) = \begin{cases} 4x+2 & -0.50 \le x \le -0.25 \\ -4x & -0.25 \le x \le -0.00 \end{cases}$
None	Z	$U_{Z}(x) = \begin{cases} 4x+1 & -0.25 \le x \le 0 \\ -4x+1 & 0 \le x \le 0.25 \end{cases}$
Weak and declining	PW	$U_{PW}(x) = \begin{cases} 4x & 0 \le x \le 0.25 \\ -4x+2 & 0.25 \le x \le 0.5 \end{cases}$
Central strong	PM	$U_{PM}(x) = \begin{cases} 4x-1 & 0.25 \le x \le 0.5 \\ -4x+3 & 0.5 \le x \le 0.75 \end{cases}$
Strong	PS	$U_{PS}(x) = \begin{cases} 4x-2 & 0.5 \le x \le 0.75 \\ -4x+4 & 0.75 \le x \le 1 \end{cases}$
Very strong	PVS	$U_{PVS}(x) = 4x - 3 \quad 0.75 \le x \le 1$
Other	OT	$U_{OT}(x) = 0 \quad x < -1 \,\&\, x > 1$

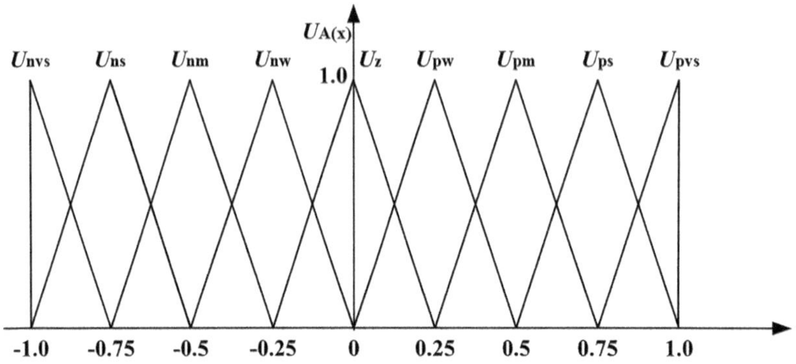

Fig. 10.1 Membership function of fuzzy numbers

$$W = \begin{bmatrix} w_{11} & w_{12} & \cdots & w_{1n} \\ w_{21} & w_{22} & \cdots & w_{2n} \\ \cdots & \cdots & \ddots & \vdots \\ w_{n1} & w_{n2} & \cdots & w_{nn} \end{bmatrix} \tag{10.2}$$

In the formula, w_{ij} represents the causal weight between the concept node i and node j. If $w_{ij} > 0$, it indicates a positive causal influence from node i to node j, such that an increase in node i leads to an increase in node j; If $w_{ij} = 0$, there is no causal influence between the two nodes, and changes in node i do not affect node j; If $w_{ij} < 0$, it signifies a negative causal influence from node i to node j, where an increase in node i results in a decrease in node j. The causal weight satisfies $|w_{ij}| \leq 1$; the closer $|w_{ij}|$ is to 1, the stronger the influence of node i on node j; conversely, the closer $|w_{ij}|$ is to 0, the weaker the influence.

Step 3: Construct a fuzzy cognitive map.

FCM Expert is a software tool developed by Nápoles et al. for modeling fuzzy cognitive maps. It offers robust capabilities in graphical representation and reasoning computation, enabling the direct import of predetermined system causal weight matrices for subsequent reasoning analysis. Given these functionalities, FCM Expert was selected in this study to construct the fuzzy cognitive map.

Step 4: Reasoning in fuzzy cognitive map.

Fuzzy cognitive maps are capable of simulating the dynamic behavior of a system. By leveraging their reasoning function, the temporal evolution of concepts within the system can be effectively modeled. This reasoning mechanism is formalized in Eq. (10.3).

$$K_i^{(t+1)} = f\left(K_i^{(t)} + \sum_{i=1, i \neq j}^{n} K_j^{(t)} w_{ij}\right) \tag{10.3}$$

In the equation, $K_i^{(t+1)}$ denotes the state value of concept node C_i at time $t + 1$, $K_i^{(t)}$ denotes the state value of C_i at time t, and w_{ij} represents the causal weight from concept node C_i to concept node C_j. The function f is a threshold function designed to normalize the outputs and map the state values of the fuzzy cognitive map onto the interval $[0, 1]$. The Sigmoid function is a commonly used threshold function in fuzzy cognitive map modeling due to its smooth and differentiable characteristics. The Sigmoid function is defined in Eq. (10.4).

$$f(x) = \frac{1}{1 + e^{-\gamma x}} \tag{10.4}$$

Step 5: Stability analysis of fuzzy cognitive map system.

Through the reasoning formula, it is evident that the state value of each node in the fuzzy cognitive map at a given time influences its connected nodes via directed arcs. Consequently, the system state is updated iteratively. After multiple iterations and successive updates, the system ultimately converges to one of the three possible stable states:

1. The system reaches an equilibrium point. This means that, after several iterations, the mutual causal interactions among the nodes in the system reach a balanced state, that is $K_i^{(t+1)} = K_i^{(t)}$.
2. The system enters a cyclic state, meaning that, from a certain time onward, the system's state evolves into a repeating cycle, exhibiting a periodic behavior.
3. The system is in a chaotic state, characterized by the absence of a discernible pattern and the inability to converge to an equilibrium point or enter a cyclic state.

10.4 Case Study

10.4.1 Determining the Causal Weights of External Environmental Influencing Factors

The stakeholders involved in agricultural waste for power generation include farmers, agricultural waste brokers, agricultural waste power generation enterprises, and the government. These stakeholders hold differing assessments of the causal relationships among the influencing factors. Collecting and integrating opinions from multiple groups enhance the objectivity and fairness of the results. On-site investigations and questionnaire surveys are commonly employed data collection methods. First, an expert group with practical experience in production and management is identified; then, this group is invited to express the interrelationships among the influencing factors using fuzzy linguistic terms. Subsequently, the collected information is processed through statistical analysis, and the fuzzy linguistic inputs are de-fuzzified using the center of gravity method. Finally, the weight matrix of the influencing factors is derived.

Take the calculation of the causal weight w_{14} between P1 and P4 as an example. Based on the statistical analysis of the collected data, it is determined that two expert groups identify a moderately strong positive relationship between P1 and P4, while another two expert groups indicate a strongly positive relationship. The aggregation of causal weights is performed using the centroid method, with the detailed computation presented in Eq. (10.5).

$$Z = \frac{\begin{array}{c} 2\int_{0.25}^{0.5}(4x-1)xdx + 2\int_{0.5}^{0.625}(-4x+3)xdx \\ +2\int_{0.625}^{0.75}(4x-2)xdx + 2\int_{0.75}^{0.1}(-4x+4)xdx \end{array}}{\begin{array}{c} 2\int_{0.25}^{0.5}(4x-1)dx + 2\int_{0.5}^{0.625}(-4x+3)dx \\ +2\int_{0.625}^{0.75}(4x-2)dx + 2\int_{0.75}^{0.1}(-4x+4)dx \end{array}} \tag{10.5}$$

The result Z = 0.67 is obtained, indicating that $w_{14} = 0.67$. Similarly, the mutual causal weights among the other influencing factors can be calculated. Given that this study includes a total of 21 influencing factors, a total of 441 causal weights must be determined. The results are presented in Table 10.2.

10.4.2 Results of the Fuzzy Cognitive Map

After determining the concept nodes and their mutual causal weights in the agricultural waste power generation supply chain system, a fuzzy cognitive map was constructed using FCM Expert software, as shown in Fig. 10.2.

Figure 10.2 presents a fuzzy cognitive map of the influencing factors in the agricultural waste power generation supply chain system. Each circle represents an influencing factor, and the lines denote directed arcs between these factors, where the node at the start of a directed arc represents the causal factor and the node at the end represents the effect factor. The directed arcs illustrate the transmission of causal relationships among the factors.

FCM Expert software was used to perform reasoning and evolutionary calculations based on Eq. (10.3) and (10.4). The system's reasoning and evolutionary process, along with the corresponding results, are shown in Fig. 10.3. As can be observed, the system reaches a stable state after seven iterations. The detailed reasoning process and the steady-state results are presented in Table 10.3.

Table 10.2 Causal weights matrix of the influencing factors

	P1	P2	P3	P4	EC1	EC2	EC3	S1	S2	S3
P1	0.00	0.13	0.25	0.67	0.74	0.63	0.00	0.05	0.00	0.00
P2	0.00	0.00	0.05	0.00	0.00	0.00	0.00	0.38	0.00	0.00
P3	0.00	0.00	0.00	0.45	0.00	0.75	−0.50	0.00	0.75	0.00
P4	0.00	0.00	0.25	0.00	0.00	0.30	−0.13	0.00	0.30	0.13
EC1	0.00	0.00	0.00	0.00	0.00	0.67	0.00	0.00	0.00	0.05
EC2	0.00	0.00	0.00	0.00	0.00	0.00	0.00	0.00	0.00	0.00
EC3	0.00	0.00	0.00	0.00	0.00	0.00	−0.05	0.00	0.32	0.00
S1	0.00	0.00	0.00	0.00	0.00	0.00	0.00	0.00	−0.79	0.00
S2	0.00	0.00	0.00	0.00	0.00	0.00	0.00	0.00	0.00	0.25
S3	0.00	0.00	0.00	0.00	0.00	0.05	−0.75	0.00	0.00	0.00
T1	0.00	0.00	0.00	0.00	0.00	0.00	0.00	0.00	0.00	0.75
T2	0.00	0.00	0.00	0.00	0.00	0.00	0.00	0.00	0.00	0.00
T3	0.00	0.00	0.00	0.00	0.00	0.00	−0.79	0.00	0.79	0.00
T4	0.00	0.00	0.00	0.00	0.00	0.00	−0.75	0.00	0.50	0.00
EN1	0.00	0.00	0.00	0.00	0.00	0.00	0.25	0.00	−0.25	0.00
EN2	0.00	0.00	0.00	0.00	0.00	0.00	−0.38	0.00	0.50	0.00
EN3	0.00	0.00	0.00	0.00	0.00	0.25	0.00	0.00	0.13	0.30
EN4	0.00	0.00	0.00	0.00	0.00	0.00	0.00	0.00	0.00	0.00
L1	0.00	0.00	0.00	0.00	0.00	0.00	0.00	0.00	0.00	0.00
L2	0.00	0.00	0.00	0.00	0.00	0.00	−0.92	0.00	0.00	0.00
L3	0.00	0.00	0.00	0.00	0.00	0.00	0.00	0.25	0.00	0.00

	T1	T2	T3	T4	EN1	EN2	EN3	EN4	EN5	EN6	L1	L2	L3
P1	0.00	0.00	0.00	0.00	0.25	0.00	0.19	0.00	0.05	0.25	0.25	0.00	0.00
P2	0.00	0.00	0.00	0.00	−0.25	0.00	0.00	0.00	0.25	0.00	0.00	0.00	0.00
P3	0.00	0.00	0.00	0.00	0.55	0.00	0.00	0.00	0.00	0.00	0.00	0.00	0.00
P4	0.00	0.00	0.00	0.00	0.13	0.00	0.75	0.00	0.00	0.00	0.00	0.00	0.00
EC1	0.00	0.00	0.00	0.00	0.42	0.00	0.00	0.00	0.00	0.00	0.00	0.00	0.00
EC2	0.00	0.00	0.00	0.00	0.25	0.00	0.00	0.00	0.00	0.00	0.00	0.00	0.00
EC3	0.00	0.00	0.00	0.00	0.00	0.00	0.00	0.00	0.00	0.00	0.00	0.00	0.00
S1	0.00	0.00	0.00	0.00	−0.20	0.00	0.00	0.00	0.00	0.00	0.00	0.00	0.00
S2	0.00	0.00	0.00	0.00	0.00	0.00	0.00	0.00	0.00	0.00	0.00	0.00	0.00
S3	0.00	0.00	0.00	0.00	0.05	0.00	0.00	0.00	0.00	0.00	0.00	0.00	0.00
T1	0.00	0.00	0.00	0.00	0.50	0.00	0.00	0.00	0.00	0.00	0.00	0.00	0.00
T2	0.00	0.00	0.00	0.00	0.00	0.00	0.00	0.00	0.00	0.00	0.00	0.00	0.00
T3	0.00	0.38	0.00	0.13	0.00	0.00	0.00	0.00	0.00	0.00	0.00	0.00	0.38
T4	0.25	0.00	0.00	0.25	0.00	0.00	0.00	0.00	0.00	0.00	0.00	0.25	0.00
EN1	0.00	−0.25	0.00	−0.25	0.00	0.00	0.00	0.00	0.00	0.00	0.00	0.00	−0.25
EN2	0.50	0.00	0.00	0.00	0.00	0.00	0.00	0.00	0.00	0.00	0.00	0.50	0.00
EN3	0.00	0.00	0.00	0.00	0.00	0.00	0.00	0.00	0.00	0.00	0.00	0.00	0.00
EN4	0.00	0.00	0.00	0.00	0.00	0.00	0.00	0.00	0.00	0.00	0.00	0.00	0.00
L1	0.00	0.00	0.00	0.00	0.00	0.00	0.00	0.00	0.00	0.00	0.00	0.00	0.00
L2	0.00	0.00	0.00	0.00	0.00	0.00	0.00	0.00	0.00	0.00	0.00	0.00	0.00
L3	0.00	0.00	0.00	0.00	0.00	0.00	0.00	0.00	0.00	0.00	0.00	0.00	0.0.

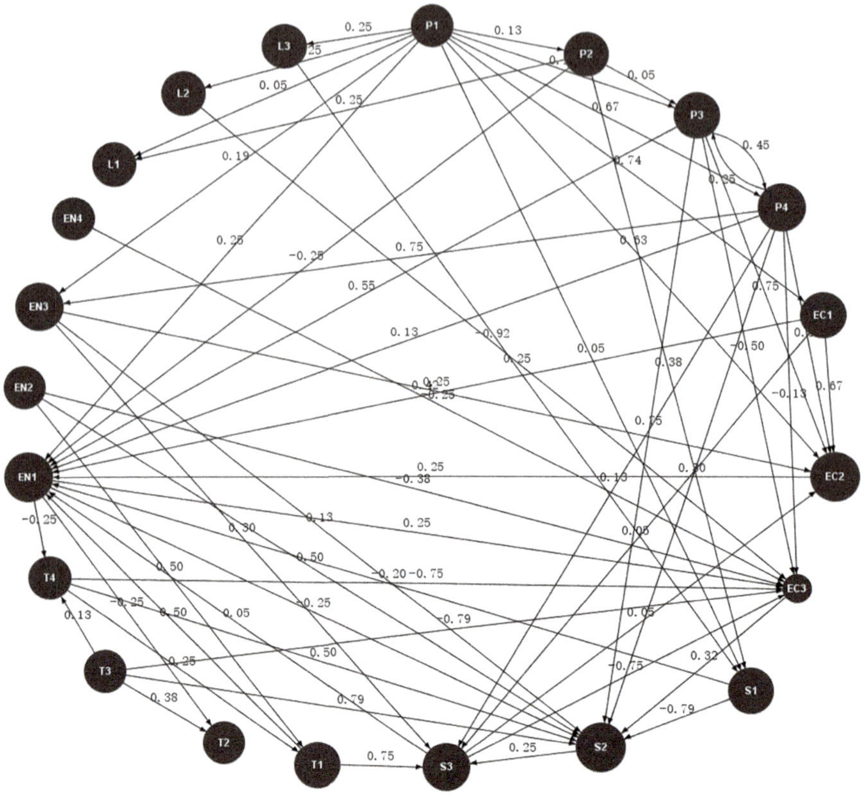

Fig. 10.2 Fuzzy cognition map of factors of agricultural waste power utilization

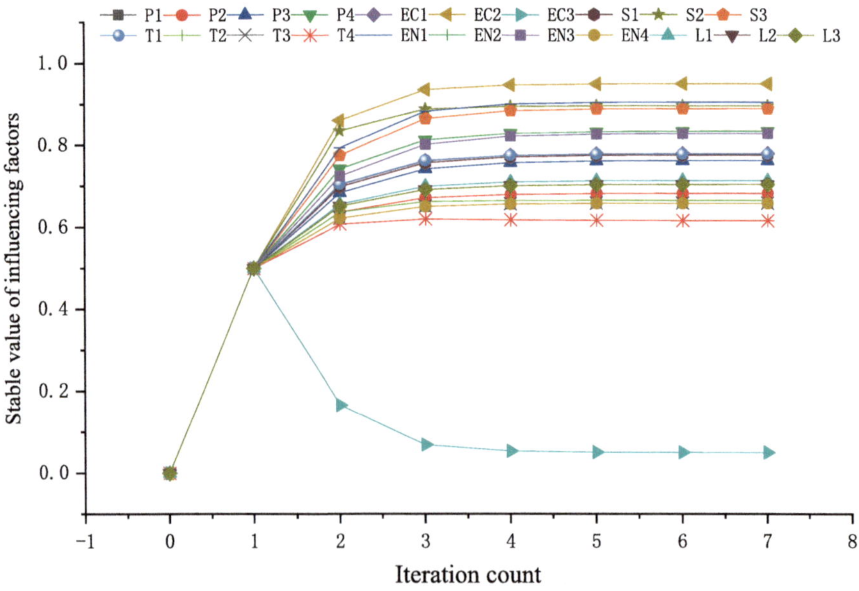

Fig. 10.3 System evolution results

Table 10.3 System evolution process and stable value

Step	P1	P2	P3	P4	EC1	EC2	EC3	S1	S2	S3
0	0.00	0.00	0.00	0.00	0.00	0.00	0.00	0.00	0.00	0.00
1	0.50	0.50	0.50	0.50	0.50	0.50	0.50	0.50	0.50	0.50
2	0.62	0.64	0.68	0.74	0.70	0.86	0.17	0.70	0.84	0.78
3	0.65	0.67	0.74	0.81	0.76	0.94	0.07	0.76	0.89	0.87
4	0.66	0.68	0.76	0.83	0.78	0.95	0.05	0.77	0.90	0.89
5	0.66	0.68	0.76	0.83	0.78	0.95	0.05	0.78	0.90	0.89
6	0.66	0.68	0.76	0.83	0.78	0.95	0.05	0.78	0.90	0.89
7	0.66	0.68	0.76	0.83	0.78	0.95	0.05	0.78	0.90	0.89

Step	T1	T2	T3	T4	EN1	EN2	EN3	EN4	EN5	EN6	L1	L2	L3
0	0.00	0.00	0.00	0.00	0.00	0.00	0.00	0.00	0.00	0.00	0.00	0.00	0.00
1	0.50	0.50	0.50	0.50	0.50	0.50	0.50	0.50	0.50	0.50	0.50	0.50	0.50
2	0.71	0.64	0.62	0.61	0.79	0.62	0.73	0.62	0.66	0.65	0.65	0.71	0.64
3	0.76	0.66	0.65	0.62	0.88	0.65	0.80	0.65	0.70	0.69	0.69	0.76	0.66
4	0.78	0.67	0.66	0.62	0.90	0.66	0.82	0.66	0.71	0.70	0.70	0.78	0.67
5	0.78	0.67	0.66	0.62	0.91	0.66	0.83	0.66	0.71	0.70	0.70	0.78	0.67
6	0.78	0.67	0.66	0.62	0.91	0.66	0.83	0.66	0.71	0.70	0.70	0.78	0.67
7	0.78	0.67	0.66	0.62	0.91	0.66	0.83	0.66	0.71	0.70	0.70	0.78	0.67

10.4.3 Analysis and Discussion of the Results

1. Calculating the performance indicators for each concept node

The concept nodes in the fuzzy cognitive map are interrelated. To clarify the influence of each concept node, it is necessary to calculate the performance indicators of the map (see Table 10.4). These performance indicators clearly reflect the out-degree, in-degree, centrality, and the type of each concept node. In the agricultural waste power generation utilization system, the analysis of the performance results enables the identification of the main influencing factors and their categories.

Mental-Modeler software, developed by Dr. Steven Gray, is a modeling and simulation tool specifically designed for fuzzy cognitive maps. It enables the performance calculation of fuzzy cognitive maps by importing the causal weight matrix data of concept nodes, with the resulting performance indicators presented in Table 10.5. As shown in Table 10.5, the centrality values of agricultural waste power generation cost, farmers' behavioral characteristics, the convenience of agricultural waste power generation, market financial tools, and policy support exceed 4. These factors are therefore identified as the key influences on the utilization of agricultural waste for power generation.

Figure 10.4 is constructed based on the in-degree (X-axis) and out-degree (Y-axis) values of the influencing factors. Using the average values of in-degree and out-degree as thresholds, these factors are categorized into four quadrants: Region 1 includes factors with both out-degree and in-degree above the average; Region 2 contains factors with out-degree above but in-degree below the average; Region 3

Table 10.4 FCM performance indicators

Indicator	Indicator description
OUT-degree (OUT)	Out-degree refers to the total influence of a concept node on its adjacent nodes, calculated as the sum of the absolute values of all elements in the corresponding row of the association matrix
IN-degree (IN)	In-degree refers to the total influence received by a concept node from its adjacent nodes, calculated as the sum of the absolute values of all elements in the corresponding column of the association matrix
Centrality (CEN)	Centrality is an indicator that measures the importance of a concept node, defined as the sum of its in-degree and out-degree
Concept node type	There are three main types of concept nodes. A concept node with IN = 0 and OUT $\neq$ 0 is defined as a driving node; one with IN $\neq$ 0 and OUT = 0 is defined as a receiving node; and one with IN $\neq$ 0 and OUT $\neq$ 0 is classified as a common concept node

Table 10.5 Performance indicators of the concept nodes

Order	Concept node	In-degree	Out-degree	Centrality	Type
1	EC3	4.77	0.37	5.14	Ordinary
2	S2	4.33	0.25	4.58	Ordinary
3	EN1	2.6	1.00	3.60	Ordinary
4	P3	0.55	3. 00	3.55	Ordinary
5	P1	0.00	3.46	3.46	Driver
6	P4	1.12	1.99	3.11	Ordinary
7	EC2	2.65	0.25	2.90	Ordinary
8	T4	0.63	1.75	2.38	Ordinary
9	S3	1.48	0.85	2.33	Ordinary
10	T3	0.00	2.09	2.09	Driver
11	T1	0.75	1.25	2.00	Ordinary
12	EC1	0.74	1.14	1.88	Ordinary
13	S1	0.68	0.99	1.67	Ordinary
14	EN3	0.94	0.68	1.62	Ordinary
15	EN2	0.00	1.38	1.38	Driver
16	L2	0.25	0.92	1.17	Driver
17	P2	0.13	0.93	1.06	Ordinary
18	T2	0.63	0.00	0.63	Receiver
19	L3	0.25	0.25	0.50	Ordinary
20	L1	0.30	0.00	0.30	Receiver
21	EN4	0.00	0.25	0.25	Driver

consists of factors where both measures are below the average; and Region 4 includes factors with out-degree below but in-degree above the average. As shown in Fig. 10.4, factors exhibiting higher in-degree—such as EC3, S2, EC2, and EN1—are more susceptible to influence from other factors, whereas those with higher out-degree—including P1, P3, T3, P4, and T4—exert stronger influence on others. The effective management of these high-impact factors enables control over the

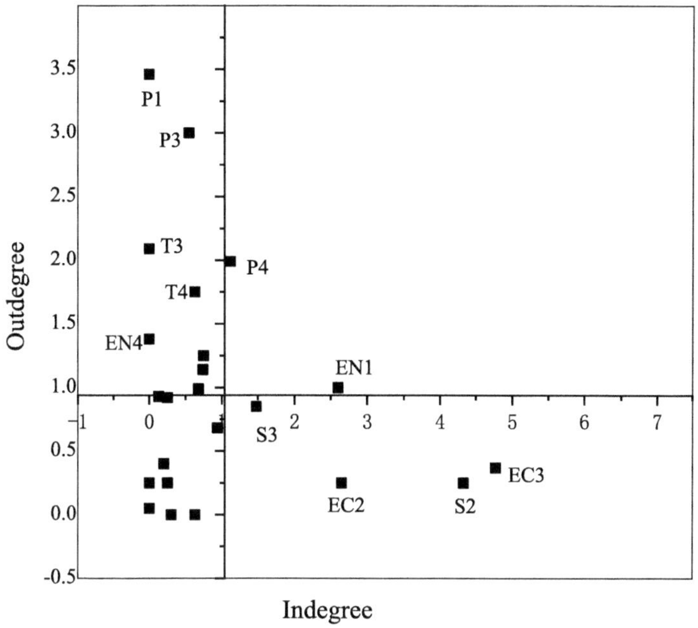

Fig. 10.4 Classification of influencing factors

behavioral trends of other elements within the system. The stable concept nodes are legal completeness, legal enforceability, technical complexity, disposable income level, and media attention. These factors demonstrate strong independence and are less influenced by external changes.

2. The rate of change of the influencing factors

Figure 10.5 shows the average rate of change of the influencing factors during the reasoning process of the agricultural waste power generation utilization system. As can be observed, the cost of agricultural waste power generation exhibits the highest rate of change at 67.68%, indicating that this factor has the greatest influence on system dynamics. This is followed by the convenience of agricultural waste power generation utilization (67.1%), financial market tools (62.25%), development of the agricultural waste power generation industry (58.82%), social responsibility awareness (55.12%), and policy opportunities (48.54%). These factors are therefore identified as the key drivers affecting the utilization of agricultural waste for power generation. The implication for enterprises involved in agricultural waste power generation is that the effective management of these high-sensitivity factors can significantly enhance the level of utilization.

3. Analysis of the values of in-degree and out-degree across different dimensions

The utilization of agricultural waste for power generation is influenced by multiple factors. To clarify the impact of influencing factors across different dimensions

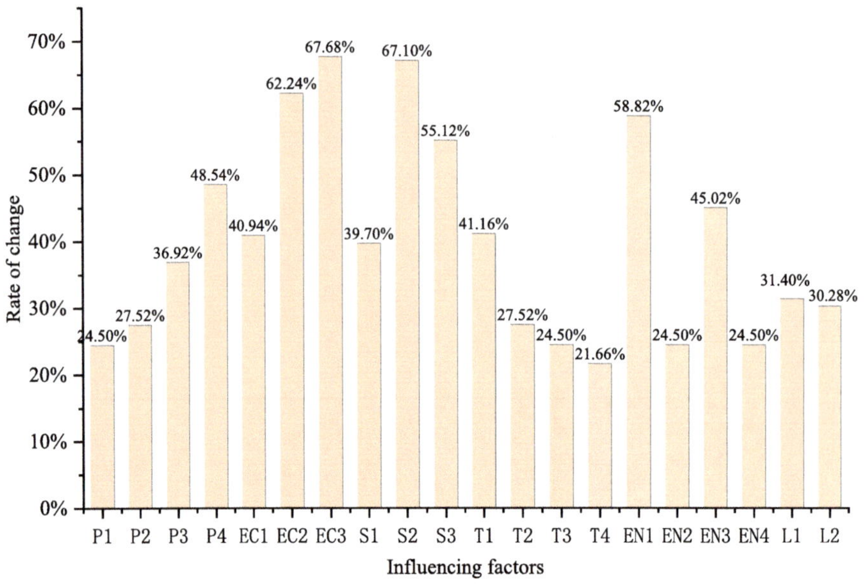

Fig. 10.5 The rate of change of the influencing factors

Fig. 10.6 Out-degree and in-degree analysis of the influence factors

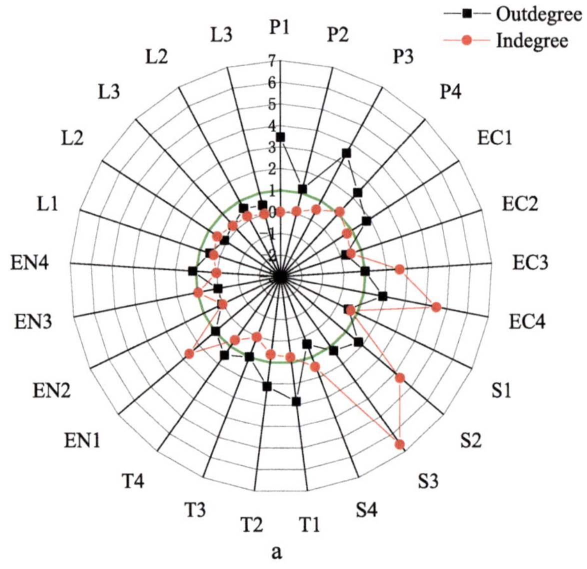

on the agricultural waste power generation utilization system, Fig. 10.6 visually presents the out-degree and in-degree values of each factor. As shown in Fig. 10.6, political factors exhibit a higher out-degree, whereas economic and social factors show a higher in-degree.

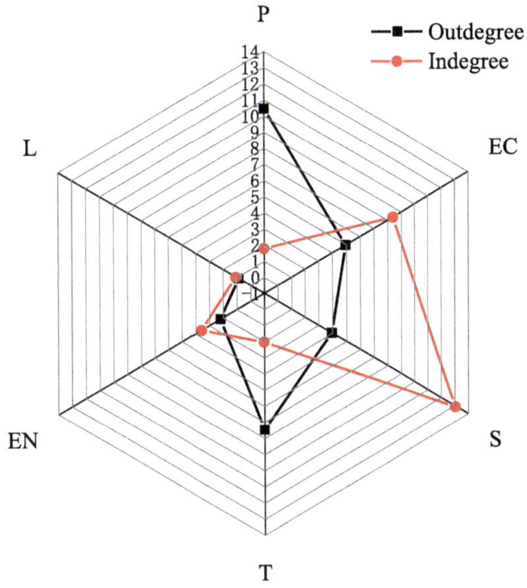

Fig. 10.7 Out-degree and in-degree analysis of the different dimensions

Figure 10.7 presents the out-degree and in-degree values across different dimensions. The political dimension exhibits the highest out-degree, followed by the technical dimension; the social dimension has the highest in-degree, with the economic dimension ranking second. This indicates that, within the agricultural waste power generation supply chain system, political factors exert a stronger influence on other factors, whereas social and economic factors are more susceptible to external influences. To enhance the development level of the agricultural waste power generation supply chain, the targeted management of these key dimensional factors is required.

10.5 The Main Obstacles to the Sustainable Development of the Agricultural Waste Power Generation Supply Chain

Agricultural waste power generation, as a key enabler of the circular economy, not only facilitates the disposal of agricultural solid waste—such as crop residues and livestock manure—to reduce environmental pollution, but also contributes to clean energy supply. This aligns closely with national "dual carbon" goals and the rural revitalization strategy. However, through the systematic analysis of factors including the external policy environment, market competition dynamics, technological adoption levels, and stakeholder interests using the fuzzy cognitive map method, it is revealed that its sustainable development still faces multiple structural and deep-seated barriers. These challenges have resulted in slow industrial scaling and suboptimal efficiency realization, leading to a persistent gap between abundant resource potential and low utilization efficiency.

1. The significant economic cost pressures have become the primary constraint on industry development.

Economic feasibility is a fundamental prerequisite for the sustainable operation of the supply chain. However, excessively high costs across the entire value chain have become a critical bottleneck constraining enterprise survival and development. In the power generation stage, the calorific value of agricultural waste ranges from one-half to two-thirds that of standard coal, and its composition is complex with a high impurity content, leading to low combustion efficiency. Specialized combustion equipment and pretreatment processes are therefore required, significantly increasing capital and operational expenditures—particularly in equipment acquisition, maintenance, and energy consumption—compared to conventional thermal power generation. Furthermore, to ensure stable power output, additional investments in pretreatment facilities for crushing, drying, and compressing agricultural waste further elevate fixed asset outlays and operating costs. At the collection stage, agricultural waste is widely dispersed and sourced from numerous decentralized locations. Small-scale farming practices and fragmented land plots result in collection radii extending up to tens or even hundreds of kilometers. Collection relies heavily on manual labor and small machinery, while rising labor costs exacerbate financial pressure. Moreover, decentralized on-farm storage leads to mold-related losses of 15–30%, compelling enterprises to bear higher procurement premiums and loss compensation, thereby inflating collection expenses. The combined costs of collection, storage, and transportation account for 30–50% of total system costs—far exceeding those in traditional energy supply chains. The "bulky but lightweight" nature of agricultural waste results in low transport value density, high vehicle empty return rates, and poor loading efficiency. In certain rural areas, narrow roads and poor infrastructure restrict access for large vehicles, necessitating transshipment via smaller or medium-sized trucks, which increases both transportation costs and delivery time. Under the cumulative impact of these cost drivers, even with national feed-in tariff support, most enterprises struggle to cover full-chain expenses and commonly face profitability challenges or sustained financial losses. This severely undermines their operational motivation and long-term sustainability.

2. Insufficient supply and transportation accessibility constrain the stable operation of the supply chain.

The physical properties and supply characteristics of agricultural waste create inherent weaknesses in transportation and supply, significantly undermining the stability and efficiency of the supply chain. During transportation, the "bulky, lightweight, and low-density" nature of agricultural waste often leads to overloading and oversized load issues. For example, agricultural waste weighs only 100–150 kg/m^3. A 10-ton truck can carry approximately 70 m^3 when fully loaded by volume, whereas the same vehicle could transport 10 tons of standard coal by weight—highlighting a substantial disparity in transportation efficiency. If loading volumes are reduced to comply with size regulations, unit transportation costs rise further, resulting in a trade-off between economic loss and regulatory violation: "No oversize means loss, oversize means violation." Moreover, agricultural waste transportation faces seasonal bottlenecks. Most agricultural waste is generated within 1–2 months following summer and autumn harvests, necessitating intensive collection and transport over a short timeframe. However, rural transportation capacity is limited, creating a dual challenge: raw material accumulation and spoilage at the source, alongside raw material shortages and forced plant shutdowns at power generation facilities. In the supply phase, the "dispersed, seasonal, and volatile" nature of agricultural waste supply is fundamentally mismatched with the "continuous and stable production" requirements of power plants. Farmers' willingness to supply is influenced by cropping patterns, farming schedules, and alternative uses such as returning waste to fields or using it as animal feed. In some regions, farmers prefer on-site burning or discarding, leading to inconsistent feedstock collection. Furthermore, both the quantity and quality of available waste fluctuate markedly across seasons—declining sharply in winter and exhibiting high moisture content during rainy periods. Enterprises must therefore invest in additional storage infrastructure for inventory buffering, while mold-related losses during storage further exacerbate supply uncertainty. As a result, some power plants operate only 3000–4000 h annually, significantly below the utilization rates of conventional energy power plants.

3. The industry's development foundation is weak, and funding shortages constrain long-term growth.

The capital-intensive nature of the agricultural waste power generation industry stands in sharp contrast to its weak industrial foundation, with funding shortages representing the core constraint that hinders sectoral scale expansion and sustainable development. In terms of profitability, high costs and low operational efficiency have resulted in most domestic projects operating at a loss or achieving only marginal profits. From 2020 to 2023, the industry's loss rate remained between 40% and 60%, forcing some small- and medium-sized enterprises (SMEs) to suspend operations or pivot to alternative business models. This persistent financial deficit has diminished the sector's attractiveness to capital markets, leading to low investment willingness from social capital. Compared with mature clean energy sectors such as wind and photovoltaic power, agricultural waste power generation projects face longer payback periods (8–12 years), lower returns (with internal rates of return ranging from 6% to 8%), and weaker profit stability—factors that deter venture capital, industrial funds, and other market-driven investors. At the same time, the industry suffers from severely inadequate financing capacity: Most enterprises are SMEs with limited asset bases, insufficient collateral, and low credit ratings, making it difficult to secure adequate funding through traditional channels such as bank loans or bond issuance. Furthermore, policy-based financial support covers only a narrow segment of the industry, and subsidy policies are frequently revised—such as through reductions or delayed disbursements—further exacerbating cash flow pressures. The lack of capital not only constrains technological upgrading, equipment modernization, and capacity expansion among existing firms but also stalls new project investments. As a result, the industry fails to achieve economies of scale, remaining trapped in a vicious cycle of "funding shortage → insufficient scale → difficulty reducing costs → profit challenges → even more severe funding shortage."

4. The absence of a supply chain collaboration mechanism hinders the realization of overall efficiency.

The agricultural waste power generation supply chain involves multiple stakeholders, including farmers, collectors, transportation enterprises, pretreatment centers, power plants, and grid companies. Coordinated operation across all stages is essential to achieving cost reduction, efficiency improvement, and stable supply. In theory, a well-integrated collaborative system can enable large-scale raw material collection, intensive transportation, and efficient power generation through resource integration, process optimization, and information sharing—such as coordinated planning of collection routes to reduce empty return rates, real-time exchange of inventory data to minimize storage losses, and joint development of pretreatment technologies to enhance combustion efficiency. However, the current supply chain lacks an effective collaboration mechanism, resulting in a fragmented structure characterized by "disconnected stages and isolated operations." On the one hand, the supply chain system remains incomplete, with no unified coordination platform or standardized operational procedures. The stages of collection, transportation, pretreatment, and power generation are poorly integrated. Collectors and power plants often lack long-term contractual agreements, leading to mismatches between raw material supply and power generation demand in terms of quantity, quality, and timing. In some regions, professional pretreatment centers have not been established, forcing raw materials to be fed directly into power plants, which increases energy consumption and operational losses during power generation. On the other hand, competition among participants outweighs cooperation: Farmers and collectors may raise prices or compromise quality for short-term gains; transportation providers compete aggressively on price, undermining service reliability; and power plants compete for feedstock procurement and subsidy allocations rather than engaging in industry-wide resource coordination. This prevailing "competition-driven, cooperation-deficient" model prevents the realization of economies of scale and synergistic effects, thereby increasing overall chain-wide costs and diminishing operational efficiency.

5. Weak coordination among stakeholders and intensified interest conflicts heighten systemic risks.

The interests of various supply chain participants differ significantly. Farmers prioritize revenue from raw material sales, collectors and transportation enterprises focus on operational profits, power plants emphasize power generation performance and policy subsidies, while grid companies are concerned with power supply stability and grid connection costs. However, the absence of an effective mechanism for interest coordination and distribution results in pronounced interparty conflicts and weak collaborative capacity, severely undermining the orderly operation and risk resilience of the supply chain. Interest conflicts manifest in three main dimensions:

First, frequent price negotiations—A market-based pricing mechanism for raw materials is lacking, making farmers susceptible to short-term market fluctuations or alternative disposal options when setting selling prices. Meanwhile, power plants, under cost pressure, seek to minimize procurement costs. These opposing incentives lead to unstable raw material supply. Second, ambiguous responsibility allocation—Losses, spoilage, and noncompliance issues during transportation, as well as power supply instability caused by substandard feedstock quality during generation and grid integration, often trigger disputes and mutual blame among involved parties. Third, unequal risk sharing—The sector faces multiple risks, including raw material price volatility, subsidy reductions, and electricity price adjustments. Currently, power plants bear the majority of these risks, while other stakeholders lack both awareness and institutional mechanisms for shared risk management. This imbalance imposes excessive operational burdens on power plants; once exposed to external shocks, they may suspend operations, thereby disrupting the stability of the entire supply chain.

10.6 Conclusion

This study systematically explores the sustainable development challenges of the agricultural waste power generation supply chain, covering its technological pathways, external environment, internal influencing factors, and dynamic system behaviors. Through a comprehensive literature review, the PESTEL macro-environmental analysis, and fuzzy cognitive map (FCM) modeling and simulation, the research identifies the key barriers and their underlying mechanisms affecting industry development and provides a solid theoretical foundation and practical guidance for policy formulation and business decision-making.

First, at the technical pathway level, research indicates that agricultural waste power generation has developed multiple technologies, including gasification, direct combustion, microbial fuel cells, and anaerobic digestion for biogas production. Literature analysis reveals significant differences among these technological routes in terms of energy efficiency, environmental benefits, and economic viability. For example, gasification technology exhibits strong carbon reduction potential, whereas microbial fuel cells offer distinct advantages in decentralized, small-scale treatment of specific organic wastes. However, no single technology is universally applicable, as its feasibility heavily depends on feedstock characteristics, geographical conditions, and end-use energy demands. Therefore, future technology selection should be guided by comprehensive local assessments rather than the pursuit of a single "optimal" solution. Second, regarding supply chain management and the external environment, this study identified 21 key influencing factors across six dimensions—political (P), economic (E), social (S), technological (T), environmental (EN), and legal (L)—using the PESTEL analysis framework. More importantly, by constructing a fuzzy cognitive map model and conducting dynamic simulations, this study extends beyond static factor identification to uncover causal transmission pathways and the system's dynamic evolution patterns. Simulation results show that the system stabilizes after seven iterations, with economic cost (EC3), supply convenience (S2), and financial market tools (EC2) exhibiting the highest centrality, thus serving as the most sensitive and influential core nodes. Meanwhile, political factors (e.g., P1, P3, and P4) demonstrate the highest out-degree, indicating that policy-driven forces act as an "engine" within the system, effectively driving changes across technological, economic, and social dimensions.

To promote the sustainable development of the agricultural waste power generation industry, a systematic governance strategy is required. In the short term, efforts should focus on reducing core costs and enhancing supply stability. Governments could implement targeted transportation subsidies, provide financial support for the purchase of pretreatment equipment, and encourage the establishment of regional collection, storage, and transportation centers to enable large-scale and standardized operations. In the medium and long term, it is essential to build a robust industrial ecosystem. On the one hand, policy continuity and stability—such as feed-in tariff subsidies—must be maintained, and mechanisms should be explored to incorporate carbon emission reduction benefits into project revenue models to strengthen market confidence. On the other hand, digital supply chain platforms should be promoted to improve information transparency and operational coordination, and contractual arrangements should be designed to establish "benefit-sharing, risk-sharing" partnerships. However, the causal weights in the FCM model developed in this study are primarily based on expert judgment. While scientifically justified and representative, this approach may have limited generalizability, as the importance ranking and interrelationships of influencing factors can vary across regions and across projects with different technological configurations. Agricultural waste power generation is not only a critical pathway for energy transition but also a key driver for green agricultural development, rural revitalization, and the advancement of the "dual carbon" goals. Only through coordinated efforts among governments, enterprises, farmers, and research institutions, along with systematic progress across technological, economic, institutional, and social dimensions, can current development bottlenecks be overcome and the substantial environmental and economic potential of this sector be fully realized, paving the way toward sustainability.

References

Abdelhady S, Borello D, Shaban A (2018) Techno-economic assessment of biomass power plant fed with rice straw: Sensitivity and parametric analysis of the performance and the LCOE. Renew Energy 115:1026–1034. https://doi.org/10.1016/j.renene.2017.09.040

Ali SU, Bin Nadeem T, Naqvi SMAA, Kamran M (2025) Comprehensive techno-economic assessment of biomass power generation for sustainable solution of Pakistan's energy crisis. Energy Sustain Dev 89. https://doi.org/10.1016/j.esd.2025.101856

Chandra S, Pandit S, Roy A, Rab SO, Roy AK, Saeed M et al (2025) Bioenergy recovery from jackfruit waste via biohydrogen production through dark fermentation and power generation through stacked microbial fuel cells. Int J Hydrog Energy 102:845–855. https://doi.org/10.1016/j.ijhydene.2025.01.044

Chen SCI, Dang X, Xu QQ, Own CM (2025) Transforming waste into value: sustainable recycling of agricultural resources under the 'carbon peak and carbon neutrality' vision. Sustainability 17(1). https://doi.org/10.3390/su17010055

Chopra J, Rangarajan V, Rathnasamy S, Dey P (2024) Life cycle assessment as a key decision tool for emerging pretreatment technologies of biomass-to-biofuel: unveiling challenges, advances, and future potential. Bioenergy Res 17(2):857–876. https://doi.org/10.1007/s12155-024-10741-8

Crotman B, Barros MO, Oliveira LB (2024) Assessing the feasibility of electric energy generation from agricultural waste in Brazil. Sustain Energy Technol Assess 68. https://doi.org/10.1016/j.seta.2024.103853

Dassanayake GDM, Kumar A (2012) Techno-economic assessment of triticale straw for power generation. Appl Energy 98:236–245. https://doi.org/10.1016/j.apenergy.2012.03.030

Gao J, Wang C, Wang Z, Lin J, Zhang R, Wu X et al (2024) Site selection decision for biomass cogeneration projects from a sustainable perspective: a case study of China. Energy 286. https://doi.org/10.1016/j.energy.2023.129518

Havrysh V, Kalinichenko A, Brzozowska A, Stebila J (2021) Agricultural residue management for sustainable power generation: the Poland case study. Appl Sci 11(13). https://doi.org/10.3390/app11135907

Hou T, Wang Z, Wang H, Xu G, Wang Z (2025a) Sustainability ranking of straw resource utilization technologies: life cycle assessment model based on IDEMATEL-IVIKOR. Energy 323. https://doi.org/10.1016/j.energy.2025.135683

Hou JW, Yang FJ, He JZ, Hong JH, Yang YJ, Li YZ et al (2025b) Synergistic modification engineering for optimizing reaction kinetics of direct carbon solid oxide fuel cells with agricultural solid waste as fuel. Fuel 398. https://doi.org/10.1016/j.fuel.2025.135579

Huang Y, Zhao YJ, Hao YH, Wei GQ et al (2019) A feasibility analysis of distributed power plants from agricultural residues resources gasification in rural China. Biomass & Bioenergy 121:1–12. https://doi.org/10.1016/j.biombioe.2018.12.007

Illankoon W, Milanese C, Girella A, Rathnasiri PG, Sudesh KHM, Llamas MM et al (2022) Agricultural biomass-based power generation potential in Sri Lanka: a techno-economic analysis. Energies 15(23). https://doi.org/10.3390/en15238984

Kamusoko R, Mukumba P (2025) Valorization of maize stover into biogas for heat and power generation: a south African perspective. Fermentation 11(6). https://doi.org/10.3390/fermentation11060338

Liu SH, You SS, Lin CW, Cheng YS (2022) Optimizing biochar and conductive carbon black composites as cathode catalysts for microbial fuel cells to improve isopropanol removal and power generation. Renew Energy 199:1318–1328. https://doi.org/10.1016/j.renene.2022.09.069

Magotra VK, Lee SJ, Inamdar AI, Kang TW, Walke PD, Hogan SC et al (2021) Development of white brick fuel cell using rice husk ash agricultural waste for sustainable power generation: a novel approach. Renew Energy 179:1875–1883. https://doi.org/10.1016/j.renene.2021.08.003

Mana AA, Allouhi A, Ouazzani K, Jamil A (2021) Feasibility of agriculture biomass power generation in Morocco: techno-economic analysis. J Clean Prod 295. https://doi.org/10.1016/j.jclepro.2021.126293

Mao J, Sun Q, Ma CH, Tang M (2021) Site selection of straw collection and storage facilities considering carbon emission reduction. Environ Sci Pollut Res. https://doi.org/10.1007/s11356-021-15581-z

Mao J, Zhou YZ, Shan LJ, Cheng JY (2024) Optimization of straw supply chain considering carbon emissions, supply uncertainty and facility disruption risk. Environ Dev Sustain. https://doi.org/10.1007/s10668-024-05165-5

Potnuri R, Surya DV, Rao CS, Yadav A, Sridevi V, Remya N (2023) A review on analysis of biochar produced from microwave-assisted pyrolysis of agricultural waste biomass. J Anal Appl Pyrolysis 173. https://doi.org/10.1016/j.jaap.2023.106094

Prasad S, Singh A, Korres NE, Rathore D, Sevda S, Pant D (2020) Sustainable utilization of crop residues for energy generation: a life cycle assessment (LCA) perspective. Bioresour Technol 303. https://doi.org/10.1016/j.biortech.2020.122964

Rojas-Flores S, De La Cruz-Noriega M, Cabanillas-Chirinos L, Otiniano NM, Soto-Deza N, Terrones-Rodriguez N, De La Cruz-Cerquin M (2024) Potential use of Andean tuber waste for the generation of environmentally sustainable bioelectricity. Molecules 29(9). https://doi.org/10.3390/molecules29091978

Rojas-Villacorta W, Rojas-Flores S, Benites SM, Nazario-Naveda R, Romero CV, Gallozzo-Cardenas M et al (2023) Preliminary study of bioelectricity generation using lettuce waste as substrate by microbial fuel cells. Sustainability 15(13). https://doi.org/10.3390/su151310339

Shahavi MH, Esfilar R, Golestani B, Sadeghabad MS, Biglaryan M (2022) Comparative study of seven agricultural wastes for renewable heat and power generation using integrated gasification combined cycle based on energy and exergy analyses. Fuel 317. https://doi.org/10.1016/j.fuel.2022.123430

Teng JY, Wang PY (2022) Study on the calculating methodology of the collectible straw resources for biomass power plants based on the indirect measurement. Sustain Energy Technol Assess 52. https://doi.org/10.1016/j.seta.2022.102259

Tomatis M, Parvez AM, Afzal MT, Mareta S, Wu T, He J, He TB (2019) Utilization of CO2 in renewable DME fuel production: a life cycle analysis (LCA)-based case study in China. Fuel 254. https://doi.org/10.1016/j.fuel.2019.115627

Wang B, Song J, Ren J, Li K et al (2019) Selecting sustainable energy conversion technologies for agricultural residues: A fuzzy AHP-VIKOR based prioritization from life cycle perspective. Resour Conserv Recycl 142:78–87. https://doi.org/10.1016/j.resconrec.2018.11.011

Wang SN, Cheng JK (2024) Co-generation and GHG emission from agricultural waste based on pyrolysis/gasification: experimental and LCA approaches. Energy Sci Eng 12(3):860–872. https://doi.org/10.1002/ese3.1658

Wang Y, Yang YZ (2022) Research on greenhouse gas emissions and economic assessment of biomass gasification power generation technology in China based on LCA method. Sustainability 14(24). https://doi.org/10.3390/su142416729

Wang Z, Wang Z, Tahir N, Wang H, Li J, Xu G (2020a) Study of synergetic development in straw power supply chain: straw price and government subsidy as incentive. Energy Policy 146. https://doi.org/10.1016/j.enpol.2020.111788

Wang ZW, Wang ZF, Xu GY, Ren JZ, Wang H, Li J (2020b) Sustainability assessment of straw direct combustion power generation in China: from the environmental and economic perspectives of straw substitute to coal. J Clean Prod 273:122890. https://doi.org/10.1016/j.jclepro.2020.122890

Wang ZW, Xu GY, Wang ZF, Zhang ZP (2022) Sustainability of agricultural waste power generation industry in China: criteria relationship identification and policy design mechanism. Environ Dev Sustain 24(3):3371–3395. https://doi.org/10.1007/s10668-021-01570-2

Wang Z, Hou T, Wang Z (2023) Revenue coordination of straw power generation supply chain under carbon trading environment. Waste Biomass Valoriz 14(5):1769–1785. https://doi.org/10.1007/s12649-022-01960-1

Yang B, Jia WG, Yu Y, Zhang H (2024) Sustainability assessment of agricultural waste biogas production system in China based on energy and carbon evaluation methods. Agriculture (Basel) 14(11). https://doi.org/10.3390/agriculture14111912

Zhang QY, Wang L, Xu DY, Tao ZK, Li JJ, Chen Y, ... Wang SY (2024) Optimization of biomass carbon sources for simultaneous Pb(II) removal and bioelectricity generation in microbial fuel cell: based on agricultural waste types and pretreatment approaches. Biomass Convers Biorefin 14(13):13889–13900. https://doi.org/10.1007/s13399-022-03451-8

Zhao X, Li A (2016) A multi-objective sustainable location model for biomass power plants: Case of China. Energy 112:1184–1193. https://doi.org/10.1016/j.energy.2016.07.011

Zhao X, Ma XY, Wang K, Long YQ, Zhang DJ, Xiao ZC (2017) A spatially explicit optimization model for agricultural straw-based power plant site selection: a case study in Hubei Province, China. Sustainability 9(5). https://doi.org/10.3390/su9050832

Zhu JA, Lin HX, Yang XY, Yang XH, Jiang P, Del Valle TM (2022) Social network relationships between biomass industry stakeholders in the agricultural waste power generation industry-a case of northern Jiangsu, China. Sustainability 14(1). https://doi.org/10.3390/su14010571